Environmental Politics and Policy in Industrialized Countries

D0974430

8/1/04

American and Comparative Environmental Policy
Sheldon Kamieniecki and Michael E. Kraft, editors

Critical Masses: Citizens, Nuclear Weapons Production, and Environmental Destruction in the United States and Russia
Russell J. Dalton, Paula Garb, Nicholas P. Lovrich, John C. Pierce, and John M. Whiteley

Toward Sustainable Communities: Transition and Transformations in Environmental Policy
Daniel A. Mazmanian and Michael E. Kraft, editors

Domestic Sources of International Environmental Policy: Industry, Environmentalists, and U.S. Power
Elizabeth R. DeSombre

Waste Trading among Rich Nations: Building a New Theory of Environmental Regulation
Kate O'Neill

Reflections on Water: New Approaches to Transboundary Conflicts and Cooperation
Joachim Blatter and Helen Ingram, editors

Environmental Leadership in Developing Countries: Transnational Relations and Biodiversity Policy in Costa Rica and Bolivia
Paul F. Steinberg

Environmental Politics and Policy in Industrialized Countries
Uday Desai, editor

Environmental Politics and Policy in Industrialized Countries

edited by Uday Desai

The MIT Press
Cambridge, Massachusetts
London, England

© 2002 Massachusetts Institute of Technology

All rights reserved. No part of this book may be reproduced in any form by any electronic or mechanical means (including photocopying, recording, or information storage and retrieval) without permission in writing from the publisher.

This book was set in Sabon by SNP Best-set Typesetter Ltd., Hong Kong and was printed and bound in the United States of America.

Library of Congress Cataloging-in-Publication Data

Environmental politics and policy in industrialized countries / edited by Uday Desai.
p. cm.—(American and comparative environmental policy)
Includes index.
ISBN 0-262-04210-X (alk. paper)—ISBN 0-262-54137-8 (pbk. : alk. paper)
1. Environmental policy—Case studies. I. Desai, Uday. II. Series.

GE170 .E5773 2002
363.7′056—dc21 2002021531

For my brothers and sisters-in-law: Kirit and Panna, Sharad and Minaxi, Kamal and Vibha—my lifelong friends

Contents

Foreword

Although the United States is the biggest polluter and energy consumer in the world, policy analysts generally believe that it has the most comprehensive environmental programs of any nation. The United States, for example, has made significant progress in controlling air and water pollution and managing the disposal of hazardous waste. It has successfully protected endangered species and has preserved millions of acres of land containing valuable natural resources and possessing spectacular scenic beauty. Energy efficiency has increased substantially since the 1970s and more communities are recycling today than ever before.

It is wrong, however, to think that the United States is the only nation to have significantly reduced pollution and conserved natural resources. Other advanced industrialized countries in Europe and other parts of the world also have adopted and implemented effective policies that promise to improve environmental quality. This book, edited by Uday Desai, assesses the progress made by major industrialized nations, including the United States, and attempts to explain the successes and failures in each nation through an institutional perspective. As readers will see, varying political and economic institutions, the use of diverse policy approaches, and domestic politics help to explain environmental policymaking in wealthy nations.

The importance of this book is reflected in the need for all nations of the world to begin to take serious action to balance economic growth with environmental protection. It is disingenuous for consumption-oriented Western nations to demand that less developed countries improve environmental quality and conserve natural resources when they themselves fail to adopt appropriate and effective environmental policies.

As this book shows, while most advanced industrialized nations have taken critical steps in this policy area, they have not done all they can to solve today's most serious environmental problems.

Students of comparative politics and public policy who read this book will learn a great deal about how wealthy nations have addressed environmental problems over time. Clearly, no nation, including the United States, can boast a perfect record in handling important environmental and natural resource issues. Yet much can be learned from what has worked and not worked in different nations. Policy analysts and policymakers in advanced industrialized nations can benefit from this knowledge, as can analysts and policy actors in developing countries.

Clearly, researchers must take into account divergent governmental, economic, and political conditions in individual nations as part of any effort to analyze policy history, policy processes, and policy performance. Such analysis is crucial for building knowledge of how institutions, policy processes, and policy outcomes are interrelated. It is important as well for designing more effective environmental policies for the future. Through their explorations of these relationships, the contributors to this volume help to identify promising areas for future research on comparative environmental politics and policy. They also add considerably to our knowledge of how environmental policies have evolved in selected developed nations and the effects these policies have had on environmental quality.

Sheldon Kamieniecki and Michael E. Kraft, editors
American and Comparative Environmental Policy series

Acknowledgments

It has taken several years to bring this book to fruition. I want to thank the seven chapter authors for their patience and cooperation in completing this book. Several graduate students have helped me with the project, and I especially want to thank Karen Schwander, Lynn Callaly, and Felix Anebo for their assistance. Rhonda Musgrave provided valuable assistance with word processing and in various other ways to keep the project going. Lisa Lindhorst was also helpful with the preparation of the manuscript. I would also like to thank Clay Morgan, acquisition editor for environmental studies at The MIT Press, and Sheldon Kamieniecki, series editor, for their patience and support. My wife, Christina, helped as she has done for over thirty years, in ways too numerous to mention. I am much indebted to them all.

1

Institutions and Environmental Policy in Developed Countries

Uday Desai

The material wealth of the developed countries at the beginning of the twenty-first century is unprecedented in history. The creation and consumption of this wealth have required the use of massive quantities of the earth's fossil fuels, forests, animals, minerals, and materials and have resulted in widespread fouling of the air, water, and land and destruction of plant and animal species, forests, wildlife, and habitats. Modern science and technology, the material wealth of the developed countries, the disparities between rich and poor, and the massive exploitation of nature and resulting pollution are all part of one seamless whole (Athanasiou 1996; Hampson and Reppy 1996; Cohen 1998). Globally, industrialized countries have been the major consumers of the earth's resources and polluters of the earth's environment (Grubb et al. 1993, 31). They have a voracious appetite for resources. On a per capita basis, in 1987 they consumed 13 times more paper products and iron, 16 times more aluminum, 17.5 times more copper, and 18 times more chemicals and had over 23 times more cars than did developing countries (Porter and Brown 1996, 112–113). Such high levels of consumption and production have seriously degraded the global, as well as their national, environment. They have faced serious national problems of air pollution, water pollution, toxic and hazardous waste disposal, loss of wildlife and wildlife habitats, urban sprawl, and loss of open space, among others.

Industrialized countries have contributed most to the global environmental pollution. Between 1950 and 1970, "world sulphur dioxide emissions rose by 50% and world consumption of fossil fuels more than doubled, and emissions of toxic heavy metals rose proportionately" (Brenton 1994, 20). Industrialized countries accounted for most of these increases. They are largely responsible for the depletion of the ozone layer and climate warming (due to greenhouse gases) because, historically, they have been dominant producers and consumers of chlorofluorocarbons (CFCs) and voracious consumers of fossil fuels (Porter and Brown 1996, 112). They have created a vast legacy of ecological degradation. At least partly as a result of this vast ecological imprint, the exploitation of natural resources and environmental pollution have become increasingly important issues in the industrialized countries over the last three decades (Doyle and McEachern 1998; Doyle 1997; Dowie 1995; Bramwell 1994). At the beginning of the twenty-first century, the environment remains a salient issue in the politics of developed nations. While its importance in national political agendas varies considerably from country to country, and over time (Dowie 1995; Bramwell 1994; Gottlieb 1993), concern about the environment among a wide section of the populace and its place on the political agendas of the developed nations are now firmly established.

There is a large and growing body of literature on environmental problems and policies in the developed world. Much is written about the air, water, land, and other forms of pollution and about the evolution, effectiveness, costs, and benefits of specific environmental policies. This literature focuses on environmental policies and their causes, consequences, and effectiveness. Not surprisingly, much of this literature concerns environmental policies in individual industrialized countries. The literature on environmental policies in the United States, for instance, is vast. While not as extensive, there is substantial and growing literature comparing environmental policies in two or more countries. These studies describe specific environmental policies—for example, policies to protect the ozone layer, to reduce water pollution, or to protect forests. They often also consider causes of variations in these policies and in their effectiveness among two or more countries. A number of perspectives or lenses are employed to explain these differences (see Sabatier 1999).

Given the large national and global ecological impacts that industrialized countries have had, and continue to have, study of their environmental policy and politics is important for protecting the earth's environment. Environmental policy in seven major industrialized nations—Australia, Canada, Germany, Italy, Japan, the United Kingdom, and the United States—is discussed in this book. Environmental policy process is the principal focus of each country chapter; an institutional perspective provides the common framework for the book. The book has three main purposes. The first purpose is to provide an overview of major environmental problems and policies in each of the seven countries. The second purpose is to discuss actors, values, interests, and, especially, institutions that have shaped and continue to shape environmental policies in these countries. The third purpose is to provide some sense of the effectiveness of the environmental policies and the policy processes in each country.

This introductory chapter presents the conceptual framework that informs the seven country chapters. The next section briefly discusses the three main elements of the conceptual framework. The third and main section of the chapter focuses on the central element of the conceptual framework—that is, the institutional perspective on policy process. It discusses the three institutions that provide the organizing structure for the country chapters. Each of the country chapters that follow provides a richly textured, historically grounded discussion of the environmental policy process. As noted, these chapters employ the institutional perspective and emphasize three major institutions. The concluding chapter provides summary comparisons of the role of institutional arrangements, as well as other forces in environmental policy processes in the seven countries. Based on these comparisons, several hypotheses concerning institutions and environmental policy are presented for further investigation.

Central Questions

Policy studies encompass three major concerns. First, they are concerned with describing the policies themselves. They include descriptions of substantive policies in the specific policy areas—for example, environmental policy, trade policy, or foreign policy. They may also include historical accounts of changes between current and previous policy. Both the detailed descriptions of current policies and the historical perspective are essential for fuller understanding and further study of the policies.

Second, the study of public policy is concerned with understanding how policies come about and how they change. These studies are concerned with explaining the variations in policies in different countries or different states in the same country over time. They are concerned with determinants or causes of public policies. They include large-N comparative policy studies. These are studies "involving a large number of cases [20 or more] and employing sophisticated data analysis techniques" (Blomquist 1999, 201). The focus of these studies often is on understanding and modeling variations in policy process. They are interested in developing theories of policy process.

Third, policy studies are concerned with evaluating the effectiveness of policies and identifying the factors that explain this effectiveness. They focus on comparing the actual effects of policies against their intended effects; in other words, they evaluate consequences of policies against the intentions of policymakers. They then study the factors that explain the differences between policy intentions and policy accomplishments. Comparative policy studies are often concerned with identifying the causes for the differences in the effectiveness of policies in different countries or states.

The organizing framework for the chapters in this book has three principal elements, which are discussed below. They are derived from the three major concerns of policy studies. While providing a common structure for each chapter, this framework provides flexibility to the authors, allowing them to emphasize the specific context, important circumstances, principal interests, and major institutions of environmental policy and politics in each country. The framework covers three theoretically and practically important dimensions of environmental policy studies: policy history, policy process, and policy performance.

The first element concerns the description of environmental policies and problems. What are the major environmental policies and problems in the country? What important changes have occurred in the policies over the years? How have the environmental issues and policies developed over time? Policy history is important in contextualizing policy. It helps us understand "the deep roots of institutional development" (Hays 1987, xi). Policy history, often seen as "history-as-continuing" (Rose

1993, 78), can also provide "a sensible account of how one period of time differs from another" (Hays 1987, xi). Understanding the history of environmental policy in a country is to recognize its continuity, as well as its discontinuity. It is to understand both the constraints it places on, and the opportunities it creates for, potential policy directions. The importance of policy history is often underestimated in policy theories and policy studies (Desai, Holden, and Shelley 1998).

While industrialized countries face many common environmental problems—for instance, air and water pollution or toxic and hazardous waste disposal—the magnitude, geographic distribution, and severity of these problems vary considerably among them. The development of these problems over time and space also varies from country to country. The evolution of major environmental issues in different countries varies considerably as well. The chapters in this book discuss the historical context of the major environmental problems and provide a "narrative account of policy change over time" (Blomquist 1999, 223).

The second and central element for this book concerns the environmental policy process. It includes the politics of policy formulation, enactment, and implementation. The vast policy literature provides a variety of conceptual schemes and theoretical frameworks for the policy process. Sabatier (1999, 7) identifies 11 theoretical frameworks, 7 of which he considers "more promising." There is little agreement yet about the best framework for comparative policy process studies (Blomquist 1999, 222). A combination of approaches over a single theoretical scheme is sometimes recommended (see Kamieniecki 2000).

The chapters in this book employ an institutional perspective for describing and understanding the environmental policy process. This institutional perspective encompasses both political and economic institutions, in a historical context. Institutions and history are central in shaping environmental policies: "The historical process of learning from particular national experiences, the strategic choices made by human agents, the shaping power of institutional contexts understood as a whole—these are important features of political reality" (Heclo 1990, 480). A "historical perspective is required," Hays (1987, xi) argues, to understand "the deep roots of institutional development" and their ramifications.

What roles do the institutions play in environmental policy formation and implementation? How have these institutional arrangements structured environmental politics and shaped policies? How have their roles and influence evolved over time? An institutional perspective for understanding the policy process forms the core of this book. I elaborate on this perspective by identifying and discussing three major types of institutions that have been most influential in shaping and changing the environmental policy process in industrialized nations. Because the institutional perspective is so central to an understanding of the policy process, it is discussed at greater length in the next section.

The third element of the framework for this book addresses the effectiveness of environmental policies and regulations in dealing with environmental problems. How effective have the policies been? This is an important, but very difficult, question to answer. It is, as Kraft and Vig (1997, 19) point out, "difficult, both conceptually and empirically, to measure the success or failure of environmental policies." As a result, "estimating environmental performance has become a highly contested field" (Jahn 1997, 3). However, it is also central to the political debate on environmental policies in every country. As environmental values have become more firmly established in the political agenda of the developed countries, the dispute about the need to protect the environment has lost momentum. Most of the arguments now center on the effectiveness of specific environmental policies. Much of the political conflict between environmentalists and their opponents is based, often implicitly, on their differing assessment of the effectiveness of various environmental policies and approaches.

To summarize, the organizing framework for this book consists of three elements, each addressing a major concern in the study of environmental policy: policy history, policy process, and policy performance. Each chapter in the book is structured to address all three elements. Generally, each chapter begins with a history of environmental problems and policy in one country, then discusses the environmental policy process, and concludes with a discussion of policy performance. The discussion of the environmental policy process emphasizes the role and influence of one or more of the three major institutions discussed below. I elaborate on the institutional perspective below because the environmental policy process is the central concern of the book. While the other two elements of the framework—policy history and policy performance—are discussed in each chapter, this book was designed primarily to examine the policy process. This is its disciplinary bias. It is grounded specifically in political science and more generally in the social sciences.

The next section surveys the three kinds of institutions included in the perspective employed for understanding the environmental policy process.

Institutions and the Environment

The institutional perspective asserts that "institutions matter" (March and Olsen 1996, 251). In a study of variations in environmental performance in eighteen Organization for Economic Cooperation and Development (OECD) countries in the last two decades, Jahn (1997, 3) finds that "institutional arrangements have profound impact on the environmental performance of a country." The institutional perspective does not ignore individual- or group-interest-driven accounts of policy choices. Rather, it emphasizes the institutional context within which individual and group interests and identities are constructed and pursued (March and Olsen 1996). It focuses on "the ways in which institutions shape the definition of alternatives and influence the perception and construction of reality within which action takes place" (March and Olsen 1995, 29). The basic institutional arrangements, "the organization of political life" (March and Olsen 1989, 1), are important in determining policy choices, as well as the effectiveness of these choices. Institutional arrangements "shape the processes through which [policy] decisions are made and implemented" (Weaver and Rockman 1993, 7). In addition, many noninstitutional factors, such as policy histories, political culture, and dominant-elite interests and beliefs, as well as socioeconomic and demographic conditions, often play an important role in policy choices and in the way institutions and institutional arrangements function.

Many institutions influence environmental policy choices and their implementation. Business and industry,[1] federal and provincial government agencies, political parties, legislatures, judiciary, media, and international organizations all play important roles. The variations in their roles and powers and in the rules governing their interactions have significant influence on environmental policy in each nation. The chapters in this book focus on three kinds of important institutions: (1) business and industry, (2) federal and provincial governments, and (3) international organizations. The important role of business and industry in environmental policy and politics is well established. The existing literature (among many other sources, see Ophuls and Boyan 1992; Schnaiberg and Gould 1994; Schnaiberg, Watts, and Zimmerman 1986) clearly shows them to be central actors in the environmental policy arena in every industrialized country. Moreover, the centrality of the roles and relationships of national and provincial/state governments in environmental policy has become increasingly clear and well established in the literature (Rabe 1999, 1997; Harrison 1996; Holland, Morton, and Galligan 1996; Ringquist 1993; Lowry 1992). The increasing importance and influence of international institutions on environmental policy and politics in industrialized countries is also now well documented in the literature (Haas 2000, 1990; Caldwell 1996; Brenton 1994; Hurrell and Kingsbury 1992). Thus, based on the vast environmental studies literature, these three institutions and their relationships appear central to understanding environmental policy in industrialized countries. Legislatures, judiciary, media, political parties, and movements are also important. However, it is not feasible to provide in one book a comprehensive account of the influence of all important institutions. All rich industrialized countries, including the seven covered in this book, are political democracies and capitalist market economies. Economic organizations—that is, business and industry—have a powerful influence on the politics and policies of all these countries. They are also all democratic states with a major government role, especially of executive agencies, in formulating and implementing public policies. Finally, over the last three decades, international organizations have had increasing influence on environmental policy.

These three institutions have the most general and overarching influences in the policy area. Business, industry, and their trade associations and lobbyists are perhaps the principal institutions that "shape the process through which" environmental policy "decisions are made and implemented" (Weaver and Rockman 1993, 7) in industrialized nations. They have a decisive influence on the ways environmental problems are defined, as well as on the alternative solutions that are given serious consideration. Business, industry, and their associated organizations embody a strong belief in economic growth and the value of material wealth. They are important actors in defining the values and norms that guide and constrain environmental policy decisions of public officials. The changing, and yet enduring, relationships between economic institutions and the environment represent the most important influence on environmental policies and their implementation. There is an extensive discussion in policy and political science literature on the influence of business and industry on public policies, including environmental policy.

Institutional arrangements for relations between national and subnational governments have been particularly important determinants of environmental politics in many industrialized countries. In recent years, there has been a general movement toward a transfer of environmental policy powers from the national governments to state/provincial and local governments in the industrialized countries. Paradoxically, the pressure to delegate authority to subnational levels of government has been especially strong in the countries with federal systems, especially the United States, Canada, and Australia. This is paradoxical since in federal systems generally, and especially, in these three countries, states/provinces have always had substantial authority over natural resources and environmental policies. However, in all industrialized nations the major responsibility for implementing environmental regulations is now decentralized to the state or local levels. The changing distribution of power and authority between levels of governments and among different agencies of government has had significant influence on environmental policies and their implementation.

International institutions, defined as "the array of persistent and connected sets of rules (formal and informal) that prescribe behavioral roles, constrain activity, and shape expectations" (Keohane 1989, 3), represent the third major type of institution in the environmental policy arena. In practice, these institutions take the form of formal international organizations (Haas 2000, 558) and conventions. Environmental organizations and activities of the United Nations—for example, the United Nations Environmental Programme (UNEP) and the United Nations Conference on the Human Environment (UNCHE)—have played an important role in shaping environmental policy in industrialized countries over the last three decades. International environmental conventions have increasingly shaped national environmental policies and priorities in the industrialized countries. Important conventions have included the 1972 Convention on the Prevention of Marine Pollution by the Dumping of Waste and Other Matter, the 1973 Convention on International Trade in Endangered Species (CITES), the 1979 Convention on Long-Range Transboundary Air Pollution (CLRTAP), the 1982 Convention on the Law of the Sea, the 1985 Vienna Convention for the Protection of the Ozone Layer, the 1987 Montreal Protocol on Substances that Deplete the Ozone Layer, and the 1992 United Nations Convention on Climate Change and on Biological Diversity.

In addition to the United Nations agencies and protocols, multinational regional integration organizations, especially the European Union, but also the North American Free Trade Agreement (NAFTA), have been important factors in the environmental policy of the industrialized countries (McCormick 1998). In addition to governmental organizations, both national and international, environmental nongovernmental organizations (ENGOs), acting as interest groups representing general "public" environmental interests, have also become increasingly significant actors in shaping the environmental policies of industrialized countries. While still loosely knit, the ENGOs in these countries have developed networks of relationships with their counterparts in both developed and developing countries, providing them with the capacity for coordinated activities. This has enhanced their influence on the environmental policies of their countries, as well as on international environmental policies and institutions.

These three institutional arrangements and the emerging changes in them are discussed further below. I then briefly highlight the historical and institutional context of environmental policy provided in the following chapters on individual countries.

Economic Organizations and the Environment

Business and industry are the principal economic organizations in industrialized countries. They are major political actors. They gain political access and influence through campaign contributions and other economic rewards. They have well-funded and well-staffed trade associations, lobbying firms, and, increasingly, think tanks that work continually to influence elected officials, senior bureaucrats, policy professionals, the media, and public opinion. Their influence on public policies, including environmental policy, is pervasive and fundamental. They influence public policy discourse and define boundaries of acceptable policy alternatives. The extent of influence of business and industry varies considerably among industrialized countries. In the United States, for example, the relationship between business and industry and government is much more adversarial than in Japan, where government and business collaborate closely to ensure economic growth, often at the expense of the environment and citizen participation in policymaking. Such variations in government-business relationships are partly the result of history. The United States has a much longer and stronger history of laissez-faire economics than Japan and European nations do. More recently, industry in Japan and Europe suffered much destruction in World War II. Rebuilding the industrial capacity in Japan and Europe after the Second World War required strong government interventions. Their governments worked closely with business and industry, both supporting and regulating them, to rebuild their economies. They have continued to do so. This has not, of course, been the case in the United States. Industry influence is also partly a result of the differences in the extent to which industries, such as utilities, transportation, communication, and defense, are nationalized or regulated among industrialized countries. These variations reflect differences in the historical development of these institutions and in the overarching political and legal culture. In some countries, there is close coordination between government and industry in the development and implementation of environmental policies, as well as trade, economic, and other policies affecting the environment. The relationship between government and industry in the development and enforcement of environmental policy is cooperative in some countries but adversarial in others (Vogel 1986).

The conflict between the economic institutions and the environment is rooted in the habit of applying a narrow, strictly monetary valuation to nature (Evernden 1993; Wright 1994). A cursory scan of daily newspapers or news reports shows a continuous stream of stories about the conflicts between economy and ecology. For instance, U.S. Congresswoman JoAnn Emerson (R-Missouri) called the 1997 Kyoto Protocol a form of "economic warfare" in a speech to a group of business and labor organization leaders opposed to the protocol ("Emerson Says U.S. Should Beware Agreement to Fight Global Warming," 1998). However, this notion of a negative relationship between economic growth and environmental protection is increasingly being challenged. Some have argued the exact opposite: that there is a positive relationship between environmental protection and jobs and economic prosperity (Abramovitz 1997; Repetto, Rothman, and Faeth 1997; Templet 1995; "Jobs for the Future," 1995; Organization for Economic Cooperation and Development, 1993). Hawken, Lovins, and Lovins (1999) describe dozens of examples (actual and potential) for combining sustained material prosperity with increased environmental protection. While the number of voices questioning the assumption of a negative relationship between the economy and environmental protection is increasing, that assumption does remain widespread and is a potent weapon against environmental policies (Ophuls and Boyan 1992).

While high levels of consumption and material wealth remain important popular goals, public opinion surveys and other studies show that the environment has become a serious public concern (Luke 1993; Milbrath 1984; Mann 1981). The remarkable rise of environmentalism in the developed countries in the short span of 30 years has been attributed to a fundamental shift in social values in these countries in the postwar years (Inglehart 1977, 1981, 1982, 1990; Kempton, Boster, and Hartley 1995). This cultural shift in values, sometimes labeled *postmaterialism*, is closely linked to advances in educational levels, economic prosperity, and security (Abramson and Inglehart 1995; Paehlke 1995; Laquatra and Potter 2000; Mertig and Dunlap 2001; Skogen 2001). In the United States over the last three decades, support for environmental protection has remained consistently high (Mitchell 1978; Gallup and Newport 1990; Inglehart 1990; Dunlap 1991; Dunlap, Gallup, and Gallup 1992; Hunter 2000; Weintraub 2000).

Environmental values have undoubtedly acquired a permanent place in advanced industrial societies. However, personal environmental values have not translated broadly into societal environmental action. Paradoxically, postmaterialist personal values and materialist behavior coexist in developed nations. Gundersen argues that this is due to a failure to connect our institutions with our values (Gundersen 1995, 3). The materialist values underpinning all the dominant societal institutions of government, business, and labor continue to hinder effective environmental action in spite of widespread support for environmental protection. However, the problem may not simply be the result of failure to connect institutions with values. Institutions may, indeed, reflect public preferences. After a critical review of over two decades of polling studies of citizen concerns for the environment in the United States, Ladd and Bowman (1995, 51) concluded that "impressive evidence" exists that "for most Americans, the urgency [to protect the environment] has been removed, and the battle to protect the environment is being waged satisfactorily."

In addition to the prevalence of postmaterialist personal values, the collective willingness to embrace the "precautionary principle" (for extensive discussion of this principle, see Freestone and Hey 1996) and the principle of "intergernational equity" (see De-Shalit 1995; World Commission on Environment and Development, 1987) has significant influence on environmental policy. The concept of "sustainable development" unites the desire for continued material prosperity with environmental protection (Moffatt 1996). It also encompasses precautionary and international equity principles (Dernbach 1998). Its advocates urge that nature and its resources be used so as to be sustainable indefinitely. While there is little agreement on what sustainable development means in practice (Fischer and Black 1995), proponents acknowledge a significant relationship between the economy and the environment and encourage the exploration of that relationship to find a balance between

economic growth and environmental sustainability. This approach does strike a chord with ordinary people concerned with the environment (Toner and Doren 1994).

In 1992, the United Nations Conference on Environment and Development, more commonly known as the Earth Summit, put sustainable development at the center of its deliberations (for a detailed account of international agreements concerning sustainable development and their implications, see Dernbach 1998). In many developed countries, including the United States, Australia, Canada, the United Kingdom, and the Netherlands, the concept of sustainable development is now being promoted by public officials, environmentalists, and policy experts (President's Council on Sustainable Development, 1996; *Sustainable Development: The U.K. Strategy*, 1994; Moffatt 1992; Keating 1989). For many years, corporations adamantly opposed policies to encourage environmentally sustainable practices (Hawken, Lovins, and Lovins 1999; Shabecoff 1996). However, things have begun to change. In recent years, some businesses have accepted a need for sustainable development policies. There is reason to believe that a growing number of corporations around the world are beginning to understand that doing the right environmental thing is in their own interest, at least in the long run (Shabecoff 1996). In the United States, businesses have taken leading roles in organizations and groups such as the President's Council on Sustainable Development (Schmidheiny and Zorraquin 1996).

Intergovernmental Relations and Environmental Protection

The way political and policymaking powers and responsibilities are divided between national and subnational governments significantly affects environmental politics, shapes environmental policy, and conditions the implementation of this policy. Among the industrialized countries, there are substantial differences in the responsibility and authority of national and state or local governments in formulating and implementing environmental policy and regulations. Federal political systems structure these relations differently than nonfederal systems. Even federal systems vary considerably in their allocation of authority and responsibility for environmental policy and its implementation (Rabe and Lowry

1999). In addition, the role federalism plays in structuring politics and influencing policies varies considerably in different countries. Both the variations in the role federalism itself plays and the variations in the allocation of authority and responsibility in different federal systems influence environmental politics and policy.

The predominant approach since the beginning of the current environmental movement in the late 1960s has been to protect the environment through policies and regulations formulated and implemented largely by national government. This regulatory approach, often referred to as "command-and-control," usually consists of the national government setting standards for allowable environmental pollution and enforcing those standards. Over the last two decades, this approach has come under increasing criticism. It is often claimed that regulatory programs are "poorly conceived, cumbersome, time-consuming, arbitrary and vulnerable to political interference" (Kraft 1996, 172–173). Serious difficulties in enforcing regulations are commonplace. Noncompliance is widespread (DiMento 1989; "Punishing Polluters," 1999). Government regulators in Britain have been reluctant to prosecute industry for environmental violations. In Canada, as Glen Toner notes in chapter 3 of this book, there is reluctance to strictly enforce environmental laws. As regulations to control a larger and larger number of different types of environmental pollution were promulgated, they not only created a complex system of regulations, but also created contradictory regulations and regulatory agencies that often worked at cross-purposes (Cairncross 1991; Ackerman and Hassler 1981; Swift 2000; Inhofe et al. 2000).

These factors contributed to what B. G. Rabe (1997, 31) has described as the "decentralization mantra" calling for "the extended transfer of environmental policy resources and regulatory authority from Washington, D.C. to states and localities" in the United States. In the United States, business and industry's dissatisfaction with federal government regulations because of their high costs as well as dissatisfaction on the part of the public (including environmentalists) because of the failure of these regulations to stop environmental degradation have contributed to a growing movement to grant greater authority to state and local governments (Rabe 1997; Conlan, Riggle, and Schwartz 1995). Given the historical distrust of national government in the United States, it is not surprising that there have been strong pressures to decentralize and to delegate responsibility for protecting the environment back to the states (Lowry 1992; Ringquist 1993). However, the intergovernmental relations play a significant role in environmental policy in all industrialized countries. In Australia and Canada, the provinces have always had primary legislative and regulatory authority for environmental protection. In both countries, "federal legislation on the environment is still the exception rather than the rule" (Holland 1996, 3). In Europe, also, the role of national governments in enforcing environmental policies has been limited. In Germany, local governments have a high degree of flexibility in environmental protection. The quality of the environment in Germany is thus greatly dependent on local administrators and citizens (see Weidner, chapter 5, this book). In Italy, municipalities have often been the leaders in environmental protection. The national government has enacted environmental policies and regulations only after strong pressures from municipal and regional governments (see Lewanski and Liberatore, chapter 6, this book). In Britain, local governments have traditionally been responsible for much environmental regulation (see McCormick, chatper 4, this book). Institutional arrangements for intergovernmental relations play a major role in shaping environmental politics and policy in industrialized countries.

International Organizations and Global Environmental Protection

One of the most important changes in environmental policy in developed countries over the last decade or so has been increasing focus on global environmental issues. The concern with the global environment is, of course, not entirely new. By the late 1960s, there was growing recognition that many environmental problems were transnational in scope and required a multinational response (Porter and Brown 1996; Weintraub 2000).

There are several reasons for this focus on global environmental concerns. The 1972 United Nations Conference on the Human Environment (UNCHE), in Stockholm, helped to legitimize the global environment as an important national concern. It helped move the environment from a protest movement to an established international policy agenda item

(Caldwell 1992). A series of environmental disasters during the 1980s also helped focus popular attention on the global dimension of environmental problems. Solutions to many environmental problems—climate change, stratospheric ozone, acid rain, pollution of the oceans, protection of biodiversity, endangered species, and the Antarctic environment—require international efforts and cooperation. Science has also played a significant role in this shift of focus. International organizations, such as the United Nations Environmmental Programme (UNEP) and the United Nations Economic Commission for Europe (ECE), have played an important role in shaping environmental policy in industrialized countries. In addition, multinational regional organizations, especially the European Union (EU) and the North American Free Trade Agreement (NAFTA), as well as global trade organizations, such as the General Agreements on Tariffs and Trade (GATT) and the World Trade Organization (WTO), have become important actors in the environmental policy arena (Brenton 1994; Dernbach 1998, 94–97; Von Moltke 2000; McKinney 2000). Armed with greater scientific knowledge, tighter international networks, and more resources, nongovernmental organizations around the world have demanded that their national governments cooperate with each other to deal with environmental problems (Brenton 1994; Caldwell 1992; Haas 1990; Alm 2000; Mongillo and Booth 2001; "Water Management in the Next 25 Years," 2000). Environmental nongovernmental organizations (ENGOs) and epistemic communities (for definition, see Haas 1990, xviii and chap. 2), especially those in the developed countries, have become major actors in global environmental politics.

The EU represents perhaps the best example of the influence of international organizations on the environmental policy of its member states. The EU has become deeply involved in creating a harmonized environmental policy and regulatory system among the member states. The cooperation among member states in addressing common environmental problems is the clearest example yet of the remarkable influence of international organizations on environmental issues (Brenton 1994). WTO and NAFTA also have significant influence on environmental policies in the member countries (Hafbauer et al. 2000; Kourous and Carter 2000a, 2000b, 2001). Widespread protests against NAFTA in the United

States and against WTO around the world are, in part, based on the belief that such influence would lead to serious negative consequences for the global environment (Mittelman 2000; Guruswamy 2000). This is not to suggest, however, that the UNEP, EU, or other international institutions have an overarching or integrated environmental approach. There are great differences among member states, both in their ability to deal with environmental problems and in their attitudes toward those problems.

While international organizations have played an important role in shaping international environmental policies and protocols, they have little authority over implementation, monitoring, or enforcement of these policies and protocols. Such authority resides largely with nation-states. The integration of international environmental agreements into national objectives and actions remains one of the major problems facing international environmental politics. There is also a general shortage of international implementation or monitoring institutions (Caldwell 1992; Hurrell and Kingsbury 1992). In fact, "there is no global authority responsible for proposing and enforcing environmental regulations affecting multiple states" (McCormick 1998, 1). At the national level, international agreements are often resolved according to domestic considerations. In most countries, while international environmental agreements are the domain of the national government, implementation of those policies is the domain of regional, local, or provincial governments. Thus, if the lower-level governments do not support the environmental agreement, implementation may be halfhearted. In addition, there is still no effective formal sanction mechanism for noncompliance with agreements (Haas 1990). As a result, the effectiveness of international treaties and agreements varies widely from country to country (Caldwell 1996). While most countries recognize this as a serious limitation, they are reluctant to agree to effective international monitoring and enforcement mechanisms since this raises the larger issue of national sovereignty (Brenton 1994; Birnie 1992). On balance, "it is widely acknowledged that implementation and enforcement has been the weakest part of international environmental law and related regimes" (Hurrell and Kingsbury 1992, 28).

In spite of these obstacles, the movement toward internationalization continues. Most countries in the world are now party to at least one international environmental agreement (Haas 1990). There is increasing, if modest, evidence of effective implementation and enforcement of international treaties by signatory sovereign countries. In the case of the ozone layer, for example, implementation has occurred widely; in fact, most countries in the world have taken steps to reduce CFC emissions (Caldwell 1996). In the EU, member states have been more and more willing to cooperate with the EU institutions and agreements. While Bush administration decisions in 2001 not to participate in Kyoto and Bonn agreements represent a setback for international cooperation in protecting the global environment, international institutions are certain to be major actors in the environmental policy arena in the twenty-first century.

This Book

All the large industrialized countries, with the exception of France, are incuded in the book. The seven countries profiled cast a wide geographic net. The United Kingdom, Germany, and Italy, three major industrialized nations, are included to cover Western Europe. Both rich industrialized nations in the Western Hemisphere, the United States and Canada, and the only rich industrialized country in Asia, Japan, and one of the two rich industrialized countries in the Southern Hemisphere, Australia, are also included. The book covers six of the seven countries that make up G7, a group of the world's seven wealthiest industrial economies. It includes the most populous and powerful, as well as the most profligate, industrialized countries in the world. Together, these seven chapters provide a comprehensive view of environmental policy and politics in the rich industrial world. These seven countries, collectively, have a vast impact on global ecology. Their environmental policies have grave consequences for all the nations of the world. Understanding their environmental politics is thus of great importance.

These seven countries vary significantly in their institutional arrangements. Australia, Canada, Germany, and the United States are federal states, while Italy, Japan, and the United Kingdom are unitary states. While all seven have capitalist market economies, the role of government and degree of market regulation vary considerably among them. In

Japan, for example, business and government both emphasize economic growth at the expense of the environment and work closely to ensure it with little concern for citizen participation in policymaking. In the other six industrialized countries, the policy process seems more open to citizen influence, to varying extents. These seven countries also vary considerably in the extent to which their environmental policy process is influenced by and is responsive to international organizations, treaties, and conventions. Environmental policies in Italy and the United Kingdom appear more responsive to multinational and international organizations and conventions than is the case with the United States. These seven countries are major industrial and economic powers and have a huge global environmental footprint. They also represent significant variations in the institutional arrangements that are a central element of the conceptual framework of the book. Thus, they allow comparative insights into the effects of institutional variations on the environmental policy process.

Collectively, the seven chapters provide an overall picture of environmental policies and the politics of environmental policymaking and implementation in the industrialized world. While many of these countries face common environmental problems, such as air and water pollution, these problems vary considerably. The historical development of environmental policies in each country is distinctive, as are the politics of environmental policy. While no explicit comparisons between the countries are made, discussions of environmental policies and politics, guided by the institutional and historical perspectives, should allow lesson drawing and "policy pinching" (deLeon and Resnick-Terry 1998). Each chapter provides a highly contextualized discussion of environmental policy and politics in the individual country. This allows deeper understanding of the range of potential policy responses to similar problems. It also highlights the importance of the historical and institutional context in developing effective environmental policy.

Note

1. Business and industry are considered an "interest group" by some scholars, while others exclude them from their definition of interest groups (see Baumgartner and Leech 1998). Sociologists generally define membership associations as interest groups. They "distinguish associations and interest groups from

such primary groups as the family, the corporation, and the bureaucracy" (Baumgartner and Leech 1998, 25). Scholars represented in the pressure-group literature in political science consider business and industrial corporations and their lobbying arms as interest groups. However, other political scientists in the last three decades have excluded business and industry, as well as government bureaucracies, from their definition of interest group (Baumgartner and Leech 1998, 25–30). Following Max Weber, Thorstein Veblen, John Commons, W. W. Willoughby, and more recent neoinstitutionalists (March and Olsen 1989), here I have included business and industry as an institution instead of an interest group. Institutions represent and channel, often even embody, interests. Every institution, however, generally represents more than one interest. Specific interests an institution represents are not assumed a priori, but instead are a matter of investigation. Interests are also not assumed to be constant over time. Changes in societal values, which affect "interests," are pursued by various institutions. I consider business and industry as an institution that channels the pursuit of primarily economic interests in the policy process. Thinking of business and industry simply as an interest group would perhaps understate the socially established role patterns and relationships they represent. In the context of the chapters in the book, this distinction is not critical, since chapter authors have treated them as both institutions and interests.

References

Abramovitz, J. N. (1997). Report turns "economy-versus-environment" argument on its head. *World Watch*, *10*(5), 9–10.

Abramson, P. R., and Inglehart, R. (1995). *Value change in global perspective.* Ann Arbor: University of Michigan Press.

Ackerman, B. A., and Hassler, W. T. (1981). *Clean coal/dirty air*. New Haven: Yale University Press.

Alm, L. R. (2000). *Crossing borders, crossing boundaries: The role of scientists in the acid rain debate.* Westport, CT: Praeger.

Athanasiou, T. (1996). *Divided planet: The ecology of rich and poor*. Boston: Little, Brown.

Baumgartner, F. R., and Leech, B. L. (1998). *Basic interests: The importance of groups in politics and in political science*. Princeton: Princeton University Press.

Birnie, P. (1992). International environmental law: Its adequacy for present and future needs. In A. Hurrell, and B. Kingsbury, eds., *The international politics of the environment: Actors, interests, and institutions* (pp. 51–84). Oxford: Clarendon Press.

Blomquist, W. (1999). The policy process and large-N comparative studies. In P. A. Sabatier, ed., *Theories of the policy process* (pp. 201–230). Boulder: Westview Press.

Bramwell, A. (1994). *The fading of the greens: The decline of environmental politics in the West*. New Haven: Yale University Press.

Brenton, T. (1994). *The greening of Machiavelli: The evolution of international environmental politics*. London: Earthscan.

Cairncross, F. (1991). *Costing the earth*. London: Economist Books.

Caldwell, L. (1992). Globalizing environmentalism: Threshold of a new phase in international relations. *Society and Natural Resources*, *4*(3), 259–272.

Caldwell, L. (1996). *International environmental policy: From the twentieth to the twenty-first century*. 3rd ed. Durham, NC: Duke University Press.

Cohen, J. E. (1998). How many people can the earth support? *New York Review of Books*, *155*(15), 29–31.

Conlan, T. J., Riggle, J. D., and Schwartz, D. E. (1995). Deregulating federalism? The politics of mandate reform in the 104th Congress. *Publius: The Journal of Federalism*, *25*, 23–39.

deLeon, P., and Resnick-Terry, P. (1998). Comparative policy analysis: Déjà vu all over again. *Journal of Comparative Policy Analysis: Research and Practice*, *1*(1), 9–22.

Dernbach, J. C. (1998). Sustainable development as a framework for national governance. *Case Western Reserve Law Review*, *49*(1), 1–103.

Desai, U., Holden, M., and Shelley, M., eds. (1998). The politics of policy [Symposium]. *Policy Studies Journal*, *26*(3), 421–568.

De-Shalit, A. (1995). *Why posterity matters: Environmental policies and future generations*. New York: Routledge.

DiMento, J. F. (1989). Can social science explain organizational noncompliance with environmental law? *Journal of Social Issues*, *45*(1), 109–132.

Dowie, M. (1995). *Losing ground: American environmentalism at the close of the twentieth century.* Cambridge: MIT Press.

Doyle, T. (1997). *Green power: The environment movement in Australia*. Melbourne, Australia: Scribe.

Doyle, T., and McEachern, D. (1998). *Environment and politics*. London: Routledge.

Dunlap, R. E. (1991). Public opinion in the 1980s: Clear consensus, ambiguous commitment. *Environment*, *33*(8), 9–15, 32–37.

Dunlap, R. E., Gallup, G. H., Jr., and Gallup, A. M. (1992). *The health of the planet survey*. Washington, DC: George H. Gallup International Institute.

Emerson says U.S. should beware agreement to fight global warming. (1998, November). *St. Louis Post-Dispatch*, 18, A13.

Evernden, N. (1993). *The natural alien*. Toronto: University of Toronto Press.

Fischer, F., and Black, M., eds. (1995). *Greening environmental policy: The politics of sustainable future*. New York: St. Martin's Press.

Freestone, D., and Hey, E., eds. (1996). *The precautionary principle and international law: The challenges of implementation.* The Hague: Kluwer Law International.

Gallup, G., Jr., and Newport, F. (1990). Americans strongly in tune with the purpose of Earth Day 1990. *Gallup Poll Monthly, April*, 6.

Gottlieb, R. (1993). *Forcing the spring: The transformation of the American environmental movement.* Washington, DC: Island Press.

Grubb, M., Koch, M., Munson, A., Sullivan, F., and Thompson, K. (1993). *The Earth Summit agreements: A guide and assessment.* London: Earthscan.

Gundersen, A. G. (1995). *The environmental promise of democratic deliberation.* Madison: University of Wisconsin Press.

Guruswamy, L. (2000). The annihilation of sea turtles: World Trade Organization intransigence and U.S. equivocation. *Environmental Law Reporter, 30*(4), 10261–10276.

Haas, P. M. (1990). *Saving the Mediterranean: The politics of international environmental cooperation.* New York: Columbia University Press.

Haas, P. M. (2000). International institutions and social learning in the management of global environmental risks. *Policy Studies Journal, 28*(3), 558–575.

Hafbauer, G. C., Esty, D. C., Orejas, D., Rubio, L., and Schott, J. J. (2000). *NAFTA and the environment: Seven years later.* New York: Institute for International Economics.

Hampson, F. O., and Reppy, J., eds. (1996). *Earthly goods: Environmental change and social justice.* Ithaca: Cornell University Press.

Harrison, K. (1996). *Passing the buck: Federalism and Canadian environmental policy.* Vancouver: University of British Columbia Press.

Hawken, P., Lovins, A., and Lovins, L. H. (1999). *Natural capitalism: Creating the next industrial revolution.* Boston: Little, Brown.

Hays, S. P. (1987). *Beauty, health and permanence: Environmental politics of the United States, 1955–1985.* Cambridge: Cambridge University Press.

Heclo, H. (1990). The comparative history of public policy. Book review. *Political Science Quarterly, 105*(3), 479–480.

Holland, K. M. (1996). Introduction. In K. M. Holland, F. L. Morton, and B. Galligan, eds., *Federalism and the environment: Environmental policymaking in Australia, Canada, and the United States* (pp. 1–15). Westport, CT: Greenwood Press.

Holland, K. M., Morton, F. L., and Galligan, B. eds. (1996). *Federalism and the environment: Environmental policymaking in Australia, Canada, and the United States.* Westport, CT: Greenwood Press.

Hunter, L. M. (2000). A comparison of the environmental attitudes, concern, and behaviors of native-born and foreign-born U.S. residents. *Population and Environment, 21*(6), 565–580.

Hurrell, A., and Kingsbury, B. (1992). The international politics of the environment: An introduction. In A. Hurrell, and B. Kingsbury, eds., *The international politics of the environment: Actors, interests, and institutions* (pp. 1–47). Oxford: Clarendon Press.

Inglehart, R. (1977). *The silent revolution: Changing values and political styles among Western publics.* Princeton: Princeton University Press.

Inglehart, R. (1981). Post-materialism in an environment of insecurity. *American Political Science Review, 75*, 880–900.

Inglehart, R. (1982). Changing values in Japan and the West. *Comparative Political Studies, 14*, 445–479.

Inglehart, R. (1990). *Culture shift in advanced industrial society.* Princeton: Princeton University Press.

Inhofe, J., Milloy, S., Peters, P., Cohen, B. R., Avery, D., Miller, H. I., Smith, R. J., Wise, H., and Lewis, D. L. (2000). *Big government and bad science: Ten cases in regulatory abuse.* No. 30. Lewisville, TX: Institute for Policy Innovation.

Jahn, D. (1997, August). Environmental performance and policy regimes: Explaining variations in 18 OECD-countries. Paper presented at the IPSA World Congress, Seoul, South Korea.

Jobs for the future. (1995). *Environmental Action Magazine*, 27(2), 11–13.

Kamieniecki, S. (2000). Testing alternative theories of agenda setting: Forest policy change in British Columbia, Canada. *Policy Studies Journal*, 28(1), 176–189.

Keating, M. (1989). *Toward a common future: A report on sustainable development and its implications for Canada.* Ottawa: Environment Canada.

Kempton, W., Boster, J. S., and Hartley, J. A. (1995). *Environmental values in American culture.* Cambridge: MIT Press.

Keohane, R. O. (1989). *International institutions and state power.* Boulder, CO: Westview Press.

Kourous, G., and Carter, N. (2000a). Border environment policy. *Borderlines, 8*(9), 1–4, 10–11.

Kourous, G., and Carter, N. (2000b). Border environment policy. *Borderlines, 8*(11), 1–4.

Kourous, G., and Carter, N. (2001). Border environment policy. *Borderlines, 9*(1), 1–4, 12.

Kraft, M. E. (1996). *Environmental policy and politics.* New York: HarperCollins.

Kraft, M. E., and Vig, N. J. (1997). Environmental policy from the 1970s to the 1990s: An overview. In N. J. Vig, and M. E. Kraft, eds., *Environmental policy in the 1990s* (pp. 1–30). Washington, DC: Congressional Quarterly Press.

Ladd, E. C., and Bowman, K. H. (1995). *Attitudes toward the environment: Twenty-five years after Earth Day.* Washington, DC: AEI Press.

Laquatra, J., and Potter, G. L. (2000). Building a balance: Housing affordability and environmental protection in the USA. *Electronic Green Journal*, no. 12, Earth Day 2000.

Lowry, W. (1992). *The dimensions of federalism: State governments and pollution control policies*. Durham, NC: Duke University Press.

Luke, T. W. (1993). Green consumerism: Ecology and the ruse of recycling. In J. Bennett, and W. Chaloupka, eds., *In the nature of things: Language, politics, and the environment* (pp. 154–172). Minneapolis: University of Minnesota Press.

Mann, D. E., ed. (1981). *Environmental policy formation: The impact of values, ideology, and standards*. Lexington, MA: Heath.

March, J. G., and Olsen, J. P. (1989). *Rediscovering institutions: The organizational basis of politics*. New York: Free Press.

March, J. G., and Olsen, J. P. (1995). *Democratic governance*. New York: Free Press.

March, J. G., and Olsen, J. P. (1996). Institutional perspectives on political institutions. *Governance: An International Journal of Policy and Administration*, *9*(3), 247–264.

McCormick, J. (1998, April). *Regional integration and the environment*. Paper presented at the meeting of the Midwest Political Science Association, Chicago.

McKinney, J. A. (2000). *Created from NAFTA: The structure, function, and significance of the treaty's related institutions*. Armonk, NY: M. E. Sharpe.

Mertig, A. G., and Dunlap, R. E. (2001). Environmentalism, new social movements, and the new class: A cross-national investigation. *Rural Sociology*, *66*(1), 113–136.

Milbrath, L. W. (1984). *Environmentalists: Vanguard for a new society*. Albany, NY: SUNY Press.

Mitchell, R. C. (1978). The public speaks again: A new environmental survey. *Resources*, *60*, 2.

Mittelman, J. H. (2000). Environmetal resistance to globalization. *Current History*, *99*(640), 383–388.

Moffatt, I. (1992). The evolution of the sustainable development concept: A perspective from Australia. *Australian Geographical Studies*, *30*(1), 27–42.

Moffatt, I. (1996). *Sustainable development: Principles, analysis, and policies*. New York: Parthenon.

Mongillo, J., and Booth, B., eds. (2001). *Environmental activists*. Westport, CT: Greenwood Press.

Ophuls, W., and Boyan, S. S., Jr. (1992). *Ecology and the politics of scarcity revisited: The unraveling of the American dream*. New York: Freeman.

Organization for Economic Cooperation and Development. (1993). *Environmental policies and industrial competitiveness*. Paris: Organization for Economic Cooperation and Development.

Paehlke, R. (1995). Environmental values for a sustainable society: The democratic challenge. In F. Fischer, and M. Black, eds., *Greening environmental policy: The politics of a sustainable future* (pp. 129–144). New York: St. Martin's Press.

Porter, G., and Brown, J. W. (1996). *Global environmental politics*. Boulder: Westview Press.

President's Council on Sustainable Development. (1996). *Sustainable America: A new consensus for prosperity, opportunity, and a healthy environment for the future*. Washington, DC: U.S. Government Printing Office.

Punishing Polluters. (1999, April). *St. Louis Post-Dispatch*, *26*, D14.

Rabe, B. G. (1997). Power to the states: The promise and pitfalls of decentralization. In N. J. Vig, and M. E. Kraft, eds., *Environmental policy in the 1990s* (pp. 31–52). Washington, DC: Congressional Quarterly Press.

Rabe, B. G. (1999). Federalism and entrepreneurship: Explaining American and Canadian innovation in pollution prevention and regulatory integration. *Policy Studies Journal*, *27*(2), 288–306.

Rabe, B. G., and Lowry, W. R. (1999). Comparative analyses of Canadian and American environmental policy: An introduction to the symposium. *Policy Studies Journal*, *27*(2), 263–266.

Repetto, R., Rothman, D., and Faeth, P. (1997). Has environmental protection really reduced productivity growth? *Challenge*, *40*, 46–57.

Ringquist, E. (1993). *Environmental protection at the state level: Politics and progress in controlling pollution*. Armonk, NY: M. E. Sharpe.

Rose, R. (1993). *Lesson-drawing in public policy: A guide to learning across time and space*. Chatham, NJ: Chatham House.

Sabatier, P. A. (1999). The need for better theories. In P. A. Sabatier, ed., *Theories of the policy process* (pp. 3–17). Boulder: Westview Press.

Schmidheiny, S., and Zorraquin, F. (1996). *Financing change: The financial community, eco-efficiency, and sustainable development*. Cambridge: MIT Press.

Schnaiberg, A., and Gould, K. A. (1994). *Environment and society: The enduring conflict*. New York: St. Martin's Press.

Schnaiberg, A., Watts, N., and Zimmerman, K. eds. (1986). *Distributional conflicts in environmental-resource policy*. New York: St. Martin's Press.

Shabecoff, P. (1996). *A new name for peace: International environmentalism, sustainable development, and democracy*. Hanover, NH: University Press of New England.

Skogen, K. (2001). Who's afraid of the big bad wolf? Young people's responses to the conflicts over large carnivores in eastern Norway. *Rural Sociology*, *66*(2), 203–226.

Sustainable development: The U.K. strategy. (1994). London: Her Majesty's Stationery Office.

Swift, B. (2000). *How environmental laws can discourage pollution prevention: Case studies of barriers to innovation.* Washington, DC: Democratic Leadership Council Progressive Policy Institute.

Templet, P. (1995). The positive relationship between jobs, environment, and the economy: An empirical analysis and review. *Spectrum, 68*, 37–49.

Toner, G., and Doren, B. (1994). Five political and policy imperatives in green plan formation: The Canadian case. *Environmental Politics, 3*, 395–420.

Vogel, D. (1986). *National styles of regulation: Environmental policy in Great Britain and the United States.* Ithaca: Cornell University Press.

Von Moltke, K. (2000). *Trade and the environment: The linkages and the politics.* Winninpeg, Canada: International Institute for Sustainable Development.

Water Management in the next 25 years: An NGO vision. (2000). *Transnational Associations, 52*(3), 98–114.

Weaver, R. K., and Rockman, B. A. (1993). Assessing the effects of institutions. In R. K. Weaver, and B. A. Rockman, eds., *Do institutions matter? Government capabilities in the United States and abroad* (pp. 1–41). Washington, DC: Brookings Institution.

Weintraub, I. (2000). The celebration of Earth Day: Perspectives on an environmental movement. *Electronic Green Journal*, no. 12, Earth Day 2000.

World Commission on Environment and Development. (1987). *Our common future.* Oxford: Oxford University Press.

Wright, M. (1994). Conservation and development. *Conservation Issues, 1*(2), 1, 3–10.

2

Environmental Policy and Politics in the United States: Toward Environmental Sustainability?

Michael E. Kraft

Environmental policy in the United States is in the midst of a profound transition, the full consequences of which remain unclear. There has long been widespread dissatisfaction with the costs, efficiency, and effectiveness of the policies and implementation strategies of the past 30 years (Davies and Mazurek 1998; National Academy of Public Administration, 2000). As a result, support for new policy approaches has been growing within both major political parties and at all levels of government. The dominant "command-and-control" regulation is increasingly being supplemented with the use of market incentives, voluntary pollution prevention, collaborative decision-making processes, public education, and similar policy tools (Chertow and Esty 1997; Mazmanian and Kraft 1999).

Broad agreement exists as well, particularly among scientists and policy analysts, that environmental-quality goals should be redefined around the concept of sustainability or sustainable development. Proponents argue that doing so will provide a more comprehensive and holistic framework for integration of economic and environmental policies and for long-term ecosystem management and protection of public health (President's Council on Sustainable Development, 1996; National Research Council, 1999). Similar developments can be seen in other industrialized nations (Vig and Axelrod 1999).

For many reasons, a transition from the regulatory regime initiated in the 1970s to one based on environmental sustainability and new policy approaches will be ill defined, at any given time, and its effects unpredictable. Despite general agreement on long-term goals, policymakers and affected interests are likely to find the achievement of consensus for specific policy choices an elusive enterprise. The transition will also be neither smooth nor conflict free, as was demonstrated throughout the 1990s, when fierce battles erupted in the U.S. Congress over implementation of environmental policies, their impact on business and property owners, and reform of regulatory processes. Those disputes continued early in the twenty-first century.

Yet these conflicts over environmental policy have propelled a hopeful search for less intrusive, more flexible, more integrative, and more cost-effective approaches to environmental protection and natural resource management. Bureaucracies, such as the Environmental Protection Agency (EPA) and the Department of the Interior, and their counterparts within the states, have been exploring and experimenting with innovative policy mechanisms. The trends have been evident in the EPA's Project XL and Common Sense Initiative and in Interior's support for ecosystem management as a new way to handle complex issues affecting natural resources (Cortner and Moote 1999; Rosenbaum 2000). In a diversity of venues at federal, state, and local levels, terms such as *flexibility*, *incentives*, *collaboration*, and *partnership* are increasingly heard and endorsed. In comparison, the language of the rigid and adversarial command-and-control regulation, so common in the 1970s and 1980s, has been losing its appeal (Chertow and Esty 1997; Kraft and Scheberle 1998; Sexton et al. 1999; Lowry 2000; Rabe 2000).

These processes of policy change afford students of environmental policy, in the United States and elsewhere, a unique opportunity. The effects of proposed reforms and of the many pilot programs and other experiments being pursued in the early twenty-first century remain uncertain and lead to some obvious questions. Is the nation moving too quickly to abandon regulatory policies that have been moderately effective, if not always efficient, in improving environmental quality? Is there persuasive evidence, to date, that where adopted, new approaches are producing better results—or a reasonable basis for assuming they will produce comparable or greater achievements in the future? If so, what political, economic, cultural, or institutional conditions foster such policy success, and what factors are likely to inhibit it? Those are inviting subjects for scholarly inquiry (Knaap and Kim 1998; Mazmanian and Kraft 1999).

This chapter has two purposes. Consistent with the objective of this book, one is to provide a broad survey of environmental policy and politics within the United States. Thus, I review policy evolution since the late 1960s; highlight the major issues and controversies in recent years—with a special focus on mounting criticism of federal regulation and proposals for reform; and analyze the key determinants of U.S. policymaking on the environment, such as institutional structures, the role of interest groups, political culture and public opinion, and the effects of international environmental institutions and processes. I also examine selected data on environmental-quality outcomes as one basis for appraising the effectiveness of policy efforts. Like the other chapters, then, this chapter covers policy history, policy processes, and policy performance. The second purpose of the chapter is to explore, tentatively and briefly, the shifts now underway in U.S. environmental policy, as new directions for the future are formulated, debated, approved, and implemented from the local to the national level, and within both the private and public sectors. These transformations promise to significantly alter policy processes and performance as we enter a new era of sustainability-based environmental policy (Mazmanian and Kraft 1999).

The Evolution of U.S. Environmental Policy: From Consensus to Conflict

Over the past three decades, U.S. environmental policy has undergone an astonishing evolution. Prior to 1970, the federal government played a sharply limited role in policymaking on the environment. Most responsibilities lay with the 50 states and with local governments, and that institutional arrangement was considered satisfactory until the rise of the modern environmental movement in the late 1960s. The major exception to this pattern was management of public lands, primarily in the West, where, for nearly a century, Congress had set aside portions of the public domain for preservation as national parks, forests, grazing lands, recreation areas, and wildlife refuges (Clarke and McCool 1996; Davis 2001). Only slowly was environmental policy extended to control of industrial pollution and human waste, and, until 1970, that effort primarily occurred at the state and local level.

A New Environmental Policy Agenda Emerges

In an abrupt change in the national political agenda, the environment emerged as a prominent issue in the late 1960s. The evidence can be seen in opinion surveys documenting a rapid and unprecedented rise in public concern over the environment, as well as in an enormous increase in the membership of national environmental groups, such as the Sierra Club, the National Audubon Society, the Wilderness Society, and the Natural Resources Defense Council (Dunlap 1995; Bosso 2000). Between 1960 and 1970, for example, the Sierra Club saw its membership increase almost tenfold, from a mere 15,000 to over 113,000. By 1990, it had reached 630,000 (Kraft 2001, 90). Such growth translated into expanded financial resources and political influence, most notably for mainstream environmental groups active at the national level. Underlying these shifts were a greater visibility of environmental problems (thanks to increased scientific research and greater media coverage) and a shift toward "postmaterialist" or postindustrial values among an increasingly affluent and well-educated American public.

In the 1970s, environmental groups found a highly receptive and newly reformed Congress encouraged by a group of extraordinary policy entrepreneurs—including such luminaries as Senators Edmund Muskie, Henry Jackson, and Gaylord Nelson and Representatives Morris Udall, Paul Rogers, and John Blatnik. Along with other key members of Congress, they provided the leadership essential for congressional action. As astute politicians, they recognized the popularity of environmental policies, and they endorsed stringent federal programs that would force offending industries to clean up. The political climate was shaped heavily by public demand for action, strong lobbying by environmental organizations, and relatively constrained pressure from industry and trade associations. In part as a result, costs were rarely a major consideration and likely implementation difficulties received relatively little attention (Jones 1975; Kraft 1995).

An important signpost of the times was overwhelming congressional approval of the National Environmental Policy Act (NEPA) on the last day of 1969. NEPA defined as national policy the creation and maintenance of "conditions under which man and nature can exist in productive harmony." It also required detailed environmental impact statements for all major federal actions that significantly affect the environment, and it established the Council on Environmental Quality (CEQ) to advise the president and Congress on environmental issues. President Richard Nixon signed NEPA as his first official act of 1970, and he proclaimed the 1970s as the "environmental decade."

At the end of 1970, President Nixon also created the EPA by executive order, a move that transferred most of the existing pollution control programs to the agency. The EPA was established as an independent executive agency, with an administrator and other top officials nominated by the president and confirmed by the Senate. Unlike environmental ministries in other Western democracies, the EPA was not given cabinet rank. During the 1990s, Congress considered granting that status, and it may yet do so. The EPA's creation through the mechanism of executive order also means that it has no congressional charter or organic law to define its core mission and set agency priorities. That is a distinct liability, given the EPA's diversified responsibilities, tenuous political support, and inadequate fiscal resources (National Academy of Public Administration, 1995, 2000; Rosenbaum 2000).

Environmental Policy Advances: 1970 to 1990

By 1970, Congress began making significant modifications in the previously weak federal air and water pollution control laws, pushed, in part, by a White House eager to claim credit for such popular actions. Beginning with the Clean Air Act Amendments of 1970, these policy changes fell well outside the norm of incremental policy advancement said to characterize the U.S. political system. As Charles Jones (1975) has remarked, this policy escalation constituted "speculative augmentation" that would severely challenge implementing agencies that lacked the requisite technical, managerial, and political capacities to succeed. The policy changes in this early period are consistent with the idea of "punctuated equilibria" advanced by Baumgartner and Jones (1993), where major alterations can occur fairly rapidly in the political system and in public policy, despite a seeming stability in both, over time.

Under the Clean Air Act of 1970 and the Water Pollution Control Act Amendments of 1972 (now called the Clean Water Act), Congress imposed, for the first time, uniform national air- and water-quality standards to protect human health and the environment. These two laws shifted environmental policy responsibilities from the states, which had been reluctant or unable to act, to the newly established federal EPA.

Close on the heels of the Clean Air and Clean Water acts, Congress approved a wide variety of other measures intended to control the use of dangerous pesticides, protect threatened and endangered species, limit the use of hazardous and toxic chemicals, protect oceans and coastlines, promote better stewardship of public lands, and restore lands ravaged by strip mining. In 1980, Congress wound up the most productive environmental policy period in U.S. history with the creation of a "Superfund" for cleaning up toxic waste sites.

After a brief period of retrenchment in the early 1980s, during the administration of President Ronald Reagan, Congress resumed its aggressive stance by strengthening most of the major acts and adding several new ones. Table 2.1 lists the legislative achievements of the environmental decade, key statutes adopted just prior to this time, and the major amendments and related legislation enacted through 2000. Descriptions of the major environmental policies and the programs they established can be found in Kraft 2001, Portney and Stavins 2000, and Vig and Kraft 2000, among other sources.

In the area of natural resources, the most distinctive change was the approval by Congress of new requirements for management of public lands under the National Forest Management Act and the Federal Land Policy and Management Act, both adopted in 1976. The former mandated that the Forest Service prepare long-term, comprehensive plans for lands under its jurisdiction and involve the public in decision making. The latter gave the Bureau of Land Management full multiple-use powers that matched those of the Forest Service, defining *multiple-use* in a way that should encourage environmental sustainability. An emphasis is to be placed on land use that will "best meet the present and future needs of the American people," including the "long-term needs of future generations for renewable and nonrenewable resources."

As these new resource policies were implemented, political disputes arose between the federal government and Western economic interests in mining, ranching, agriculture, and logging. Traditional beneficiaries of federal subsidies became increasingly vocal about their economic

Table 2.1
Major U.S. Environmental Laws: 1964 to 2000

Year	Law
1964	Wilderness Act, PL 88-577
1968	Wild and Scenic Rivers Act, PL 90-542
1969	National Environmental Policy Act (NEPA), PL 91-190
1970	Clean Air Act Amendments, PL 91-604
1972	Federal Water Pollution Control Act Amendments (Clean Water Act), PL 92-500
	Federal Environmental Pesticides Control Act of 1972 (amended the Federal Insecticide, Fungicide and Rodenticide Act (FIFRA) of 1947), PL 92-516
	Marine Protection, Research and Sanctuaries Act of 1972, PL 92-532
	Marine Mammal Protection Act, PL 92-522
	Coastal Zone Management Act, PL 92-583
	Noise Control Act, PL 92-574
1973	Endangered Species Act, PL 93-205
1974	Safe Drinking Water Act, PL 93-523
1976	Resource Conservation and Recovery Act, PL 94-580
	Toxic Substances Control Act, PL 94-469
	Federal Land Policy and Management Act, PL 94-579
	National Forest Management Act, PL 94-588
1977	Clean Air Act Amendments, PL 95-95
	Clean Water Act, PL 95-217
	Surface Mining Control and Reclamation Act, PL 95-87
1980	Comprehensive Environmental Response, Compensation and Liability Act (Superfund), PL 96-510
1982	Nuclear Waste Policy Act, PL 97-425 (amended in 1987 by the Nuclear Waste Policy Amendments Act, PL 100-203)
1984	Hazardous and Solid Waste Amendments (RCRA amendments), PL 98-616
1986	Safe Drinking Water Act Amendments, PL 99-339
	Superfund Amendments and Reauthorization Act (SARA), PL 99-499
1987	Water Quality Act (CWA amendments), PL 100-4
1988	Ocean Dumping Act, PL 100-688
1990	Clean Air Act Amendments, PL 101-549
	Oil Pollution Act, PL 101-380
	Pollution Prevention Act, PL 101-508
1991	Intermodal Surface Transportation Efficiency Act (ISTEA), PL 102-240
1992	Energy Policy Act, PL 102-486
	The Omnibus Water Act, PL 102-575
1996	Food Quality Protection Act (amended FIFRA), PL 104-170
	Safe Drinking Water Act Amendments, PL 104-182
1998	Transportation Equity Act for the 21st Century (also called ISTEA II or TEA 21), PL 105-178.

Note: A fuller list with a summary description of key features of each act can be found in Vig and Kraft 2000, appendix 1, and in Kraft 2001, pp. 104, 162–63.

concerns, with this conflict culminating in the Sagebrush Rebellion of the late 1970s and the early 1980s. During the 1990s, a parallel set of concerns was voiced by the Wise Use and property rights movements (Switzer 1997; Davis 2001). In both instances, the affected interests sought to weaken federal control over Western land and roll back some of the environmental policies of the 1970s, which they believed unduly restricted economic growth (Clarke and McCool 1996; Davis 2001).

Somewhat different issues arose in the area of environmental protection, or pollution control. In nearly all of the major statutes, Congress chose direct regulation as the chief policy approach or tool to achieve its highly ambitious goals for improving environmental quality. Such an approach requires the setting of environmental-quality criteria (determining what kinds of pollutants are associated with adverse health or environmental effects), determination of appropriate environmental-quality standards (the acceptable quality level, or deciding how clean is clean enough), and the setting and enforcement of emission standards for both stationary and mobile sources. Statutory language often dictated that industry be forced to adopt the "best available technology" and provide an "adequate margin of safety" to protect public health, even where scientific and technical knowledge was insufficient. The laws sometimes allowed balancing of costs, benefits, and risks and sometimes did not. Moreover, historically, Congress has been reluctant to grant the EPA the discretion needed to set priorities among its expanding array of programmatic responsibilities (Rosenbaum 2000; Portney and Stavins 2000).

Partly because the policies have not always permitted such balancing and priority setting and because the EPA's flexibility in implementing the laws was greatly constrained by Congress, the cost of compliance with environmental mandates has risen steadily since the early 1970s. For the seven major environmental protection statutes implemented by the EPA, the agency estimated the compliance costs in 1994 to be about $140 billion a year. That was more than four times the $30 billion the nation spent in 1972 (using constant 1990 dollars). The EPA fully expected the upward trend to continue (U.S. Environmental Protection Agency, 1990b). Indeed, by the late 1990s, the agency estimated that public and private costs of the key environmental protection laws stood at over $180 billion per year. By one reputable calculation, private industry bears about 57 percent of such costs, local governments about 24 percent, the federal government about 15 percent, and state governments about 4 percent (Portney and Probst 1994).

Criticism of Regulation and Environmental Policy Mounts

The rising costs and burdens of new federal regulations dealing with health, safety, and the environment led business groups and conservative interests to mount a multifaceted drive to shift the political climate in their favor. Among other actions, they helped to fund a substantial growth in conservative policy research institutes, such as the American Enterprise Institute, Heritage Foundation, and Competitive Enterprise Institute. The reports, books, articles, and position papers that flowed from these organizations, and from groups with similar interests, fueled a reaction against federal regulation in general and environmental policy in particular. As a result, a new political agenda began to emerge that was far less friendly to the heritage of the environmental decade (Switzer 1997).

That new agenda was most evident in Ronald Reagan's presidency. Reagan was determined to deregulate and defund federal environmental policy and, initially, he gained the tentative support of the Congress. That support did not last long, however, as members soon learned that the 1980 election results included no public mandate to turn back the clock on environmental protection (Vig and Kraft 2000).

Political reaction to the Reagan administration's clumsy deregulatory efforts effectively derailed environmental policy reform efforts for much of the rest of the decade. It even hindered consideration of more moderate proposals supported by economists and other analysts in such policy research institutions as Resources for the Future and the Urban Institute (National Academy of Public Administration, 1995; Portney and Stavins 2000). Thus, many of these issues temporarily receded from view in American politics, to be incubated throughout the 1980s and 1990s.

Some of the reforms, however, including use of market incentives like tradable discharge permits (TDPs), were incorporated into federal legislation. The 1990 Clean Air Act Amendments used TDPs to reduce emissions of sulfur dioxide (a major contributor to acid deposition), with significant cost savings over direct regulation. Others, such as prudent use of cost-benefit analysis and comparative risk assessment, were widely endorsed by environmental policy analysts and promoted by the EPA itself during the 1980s and 1990s (Davies 1996; Andrews 2000; Freeman 2000). Nonetheless, because they were so strongly embraced by anti-environmental advocates as a way to curtail environmental regulation, cost-benefit analysis and risk assessment remained controversial. Even many conservatives and business leaders complained that politicization of environmental issues and the resulting public backlash during the 1980s precluded the adoption of sensible reforms.

As John Kingdon (1994) has noted so well, changes in the political agenda and in public policy depend on an unpredictable alignment of policy ideas, perceptions of the problem being addressed, and the political mood of the nation, the latter being shaped, in part, by the actions of policy entrepreneurs pushing new approaches. For environmental policy, the case for major changes was solidifying, but the window of opportunity to act in the early 1980s was closed prematurely. It would take another decade before an equal opportunity emerged again.

The effects of the Reagan years reverberated across U.S. environmental policies. Members of Congress grew more suspicious of executive agency officials in a period of divided government in which Republicans controlled the executive branch and Democrats dominated in at least one house of Congress and, more often, both. After 1983, Congress increasingly favored tougher environmental statutes and reduced administrative discretion for agency officials, particularly for the EPA. Polls indicated continued public concern about the environment, and Congress was prodded by a reinvigorated environmental lobby that found great success in its campaign against the Reagan administration (Dunlap 1995). One consequence was that Congress was inclined to disregard the rising costs and administrative complexity of environmental policy.

As discussed above, during the 1980s, most of the major environmental statutes were strengthened, and Congress added new, intricate, and far-reaching regulatory programs to the responsibilities the EPA already had. The 1984 renewal of the Resource Conservation and Recovery Act, the nation's chief hazardous waste law and, at the end of the decade, the 1990 expansion of the Clean Air Act are cases in point. In these and other acts, Congress chose to keep the EPA on a short leash by prescribing extremely detailed and rigid requirements for implementation to compel administrative compliance. Those statutory mandates left the EPA with little opportunity to set priorities among its myriad programs or to use more flexible regulatory strategies to achieve environmental-quality goals. They also rendered the agency incapable of responding to growing evidence that many of the most serious risks to public and environmental health, such as climate change, habitat alteration and loss of biodiversity, and indoor air quality, were not being given adequate attention (U.S. Environmental Protection Agency, 1990a).

Continuing budgetary shortfalls exacerbated the EPA's difficulties. Excluding the Superfund program, between 1980 and 1998, the agency's staff grew by only about 13 percent. Over the same time period, its operating budget in constant dollars actually decreased slightly (Vig and Kraft 2000, 397–398). Yet the EPA's responsibilities grew apace during this same period as Congress added to its already-imposing workload.

Another line of criticism of environmental policy emerged in the late 1980s. This concerned the inequitable distribution of environmental burdens on the nation's population and, especially, the way environmental problems affected poor and minority communities. A diversity of studies confirm that residents in such communities experience higher-than-normal rates of exposure to toxic chemicals and other pollutants and possibly greater adverse health effects (Ringquist 2000). A new "environmental justice" movement arose to advance these arguments and to call for governmental intervention to correct the injustices. Among other policy responses, the EPA created an Office of Environmental Justice, and President Clinton issued an executive order calling for all federal agencies to develop strategies for achieving environmental justice. As Ringquist (2000, 253) notes, it seems likely that environmental justice concerns "will continue to occupy a place next to risk assessment, federal mandates to the states, property rights, and economic incentives as the major forces reshaping the context of environmental policymaking into the twenty-first century."

Policy Proposals During the 1990s and Early Twenty-First Century: Reform or Reaction?

Throughout the 1980s and into the 1990s, critics of environmental policy continued to argue, with increasing persuasiveness, that environmental goals were often too ambitious in light of human and ecological risks being addressed and the cost of implementation and compliance. They argued, as well, that reliance on a policy strategy of command-and-control regulation to achieve those goals was too costly, inflexible, and intrusive (Mazmanian and Kraft 1999; Portney and Stavins 2000). The EPA itself supplied some of the evidence for that case in its economic studies, as discussed above. Many observers suggested that use of economic incentives and other approaches might achieve the same environmental goals at much lower cost (Davies and Mazurek 1998; Freeman 2000).

Throughout the 1990s and the early years of the twenty-first century, such concerns led to new demands at all levels of government for regulatory reform and for balancing environmental policy actions with their economic costs, both real and perceived (Vig and Kraft 2003). At the national level, Congress was deeply involved in these debates under the presidencies of George H. W. Bush, William Clinton, and George W. Bush. Indeed, regulatory reform became one of the most prominent issues in the 104th and 105th Congresses, with environmental policies a chief target. Among other actions, a powerful coalition of business groups lobbied intensely for bills requiring the EPA and other agencies to conduct elaborate risk assessments and cost-benefit analyses for proposed regulations and for legislation mandating compensation to property owners for regulatory "takings" of their property—even when property values were diminished only slightly (Andrews 2000; Kraft 2000). Those actions were favored, in part, because they promised to slow or halt the regulatory machinery of government while Congress turned its attention to detailed review of major environmental statutes for possible revisions.

The business community's concerns were genuine, yet the approach employed was not only controversial, but unlikely to succeed. Backers of such regulatory reform risked the same result that occurred in Reagan's first term: public and congressional backlash. Short-term economic relief might be gained, as it was in the early 1980s, but at the expense of the more important goal of long-term reform of environmental statutes that can promote priority setting and more effective decision making in the agencies (National Academy of Public Administration, 1995, 2000).

The Republican majority in the House easily passed regulatory reform bills in early 1995, despite expressions of outrage in the environmental community and warnings by scientists and policy analysts that the measures would not achieve their stated goals. The House also voted to slash the EPA's budget by over one-third, and the Senate by about 22 percent. There was little mistaking the new sentiment toward environmental policy. However, Congress's traditionally slow legislative pace worked against the antienvironmental forces in both houses. By the summer of 1995, when Majority Leader Robert Dole tried to bring regulatory reform bills to the Senate floor, the tide had turned. Environmental groups had mobilized their members to block such drastic changes in policy, and the nation's media belatedly began extensive coverage of congressional actions on the environment. Environmentalists and the Clinton White House successfully portrayed these events on Capitol Hill as an attempt to roll back 25 years of progress in protecting public health and the environment. Clinton also vetoed appropriation bills incorporating the sharp budget cuts and eventually forced Congress to restore most EPA funds (Kraft 2000).

These activities in the 104th Congress were threatening enough, however, to push the Clinton White House to escalate its own efforts at reinventing environmental regulation, which it announced prominently in March 1995. The result was that, by the late 1990s, the EPA, Interior Department, Food and Drug Administration, and other agencies had become far more sensitive to the costs and burdens of their regulations—although not enough to satisfy their critics. They also made a variety of adjustments in agency decision making in light of congressional criticism, quietly rolling back some regulations and softening enforcement of others, much to the dismay of the environmental community. In one of its most publicized efforts under Project XL, the administration, in November 1996, granted the Intel Corporation extraordinary flexibility to operate a huge new computer-chip factory in Arizona under simplified permits; in exchange, the company pledged to go beyond legal requirements in controlling pollution (Kraft and Scheberle 1998).

At the end of the 104th Congress, policymakers approved several significant measures, including reform of long-outmoded pesticide policy (adopted as the Food Quality Protection Act of 1996) and revision of the Safe Drinking Water Act (also in 1996). Nonetheless, most federal environmental policies that had been generally acknowledged to be at least somewhat ineffective, inefficient, or inequitable were left untouched. The 105th and 106th Congresses were no more successful in overcoming environmental gridlock throughout the rest of the 1990s. A Republican Congress continued its efforts to weaken environmental protection and natural resource policies, and a Democratic White House consistently blocked those actions.

The result of this political standoff was that little progress was made through the end of the Clinton administration in altering the nation's basic environmental policies (Kraft 2000; Vig 2000). That may have been good news to environmentalists who applaud the stringency of laws such as the Clean Water Act and the Endangered Species Act. To many who remain convinced that reform is essential, however, this outcome was a disappointment. Administrative changes put into effect by the Clinton White House under the banner of "reinvention" would help to meet some of the objections raised by industry and other critics. However, they could not substitute for statutory reform. Among other consequences, the statutory language leaves the agencies vulnerable to litigation and court-imposed administrative mandates and priorities. Eventually, Congress will have to fashion the appropriate legislative compromises to address needed changes in the laws themselves.

Determinants of U.S. Environmental Politics and Policy

These patterns of legislative outcomes in recent years make clear that policymaking on the environment in the U.S. political system is not a simple function of formal institutional arrangements, public opinion, or the balance of power among environmental and business interest groups. At any given time, it is influenced by an intricate and dynamic set of factors that make prediction of particular outcomes difficult. These include the institutional characteristics of government, particularly federal-state relations; the attitudes, motivations, and resources of key policy actors, most notably business and industry groups; media coverage of environmental issues; changes in socioeconomic conditions; and changes in science and technology, among others (Ingram and Mann 1983; Schneider and Ingram 1990; Sabatier and Jenkins-Smith 1993).

To these well-recognized determinants of U.S. policymaking, one could add two other significant factors: historical commitments to particular policy paths, and the increasingly important impact of international and multinational institutions. Each merits brief mention.

The former can be seen in earlier comments on U.S. choices during the 1970s to rely heavily on direct regulation as the fundamental tool of environmental policy. Despite extensive and ongoing criticism of regulation and the availability of alternative policy strategies that promise greater success, policymakers have largely left the original policy designs unaltered. Why? Part of the answer is simply institutional inertia. Bureaucratic policy actors and legislators are familiar with regulatory strategies and have learned to work well with the vast array of rules, regulations, and administrative procedures that such strategies have produced. Much the same is true of nongovernmental policy actors, particularly those from environmental and business interest groups. A second explanation for the extraordinary persistence of direct regulation is environmental policy gridlock. It is simply too difficult to build consensus for adoption of new policy strategies; hence the old ones remain in force (Kraft 2000).

With respect to international and multinational institutions, the past three decades have witnessed striking advances in international environmental treaties and the establishment of new institutions and policymaking regimes for their monitoring and implementation (Hempel 1996; Vig and Axelrod 1999). From the Montreal Protocol on ozone-depleting chemicals to the Kyoto Protocol on global climate change, we have, by some counts, about 240 such international accords on the environment. About two-thirds of them date from the first UN conference on the environment, held in Stockholm in 1972. The pace of international policymaking, or at least a sense of urgency, appears to have accelerated following the watershed 1992 United Nations Conference on

Environment and Development, which placed sustainable development firmly on both the national and international policy agendas.

Whether the subject is ocean pollution, trade in hazardous wastes, or protection of biological diversity, there is no question that the line between domestic and international environmental protection is much less sharp than it used to be. In particular, U.S. policymakers are now influenced by a wide array of forces external to the nation. These include requirements emanating from treaty obligations, pressures from international environmental groups and multinational corporations, and a multitude of reports and recommendations flowing from international institutions, such as the Global Environment Facility, the World Bank, and the UN Commission on Sustainable Development.

Thus, consistent with the overall framework introduced in the first chapter of this book, institutions matter—both those that exist within the United States and those that shape U.S. actions from abroad. The focus here is on the former, including the formal arrangement of power within government and between layers of government, the access to the policy process extended to nongovernmental organizations (NGOs) (both industry groups and environmentalists), and the rules and procedures that govern implementation of policies within government agencies. Examination of some of the important trends now underway in environmental policymaking can illuminate these institutional impacts and the prospects for change over the next decade as the transitions in environmental policy discussed earlier continue to unfold.

Generalizations about U.S. policymaking on the environment are difficult because environmental policy cuts an exceptionally wide swath and involves a great diversity of institutions and policy actors, from the U.S. EPA to the Interior and Energy departments. This is especially so when environmental policy is linked to such highly disparate concerns as energy use, transportation, housing, urban design, agriculture, and the protection of vital global ecological, chemical, and geophysical systems. There is no single environmental agency or minister to coordinate this multifaceted and interrelated set of activities. Indeed, as is the case in other policy areas, the U.S. political system is characterized by an unusually high degree of institutional fragmentation, or pluralism, that generally inhibits the successful pursuit of holistic or ecological policy design.

Institutional Pluralism

One of the most distinctive features of the U.S. system is the constitutional specification for a wide distribution of authority. Policymaking is shared by the legislative and executive branches and overseen by an independent judicial branch. Within the legislature, power is divided between the House of Representatives and the Senate, which often have conflicting views on environmental policy that are traceable, in part, to differences in member constituencies, terms of office, and deeply embedded institutional rules and culture. The federal government, in turn, shares authority over environmental policy with the 50 states and some 80,000 local governments.

Within the federal executive branch, the EPA has responsibility for the lion's share of environmental protection or pollution control policies, and the Interior and Agriculture departments govern most natural resource policies. However, some authority also has been given to nine other cabinet departments and several independent agencies (see figure 2.1). Even if the EPA were to be reorganized into a Department of the Environment, responsibility for environmental protection and management of natural resources would remain widely shared. That arrangement militates against comprehensive and integrated policymaking on the environment.

As the earlier review of environmental policy evolution suggests, the U.S. Congress is intimately involved in policy decisions, from program creation to annual appropriations and oversight of agency operations. Indeed, Congress has played the leading role in federal environmental policymaking since the 1960s, in both Democratic and Republican administrations (Kraft 1995). However, power over environmental policy is also widely dispersed in Congress. One recent study found that 13 committees and 31 subcommittees have responsibility for at least some oversight of EPA decision making (National Academy of Public Administration, 1995).

Critics often fault Congress for inappropriate and ineffective micromanagement of environmental policy and for hobbling the agencies with detailed and inflexible regulatory policies (Rosenbaum 2000). These actions have reflected congressional distrust of the executive branch during times of divided government, which characterizes much of the

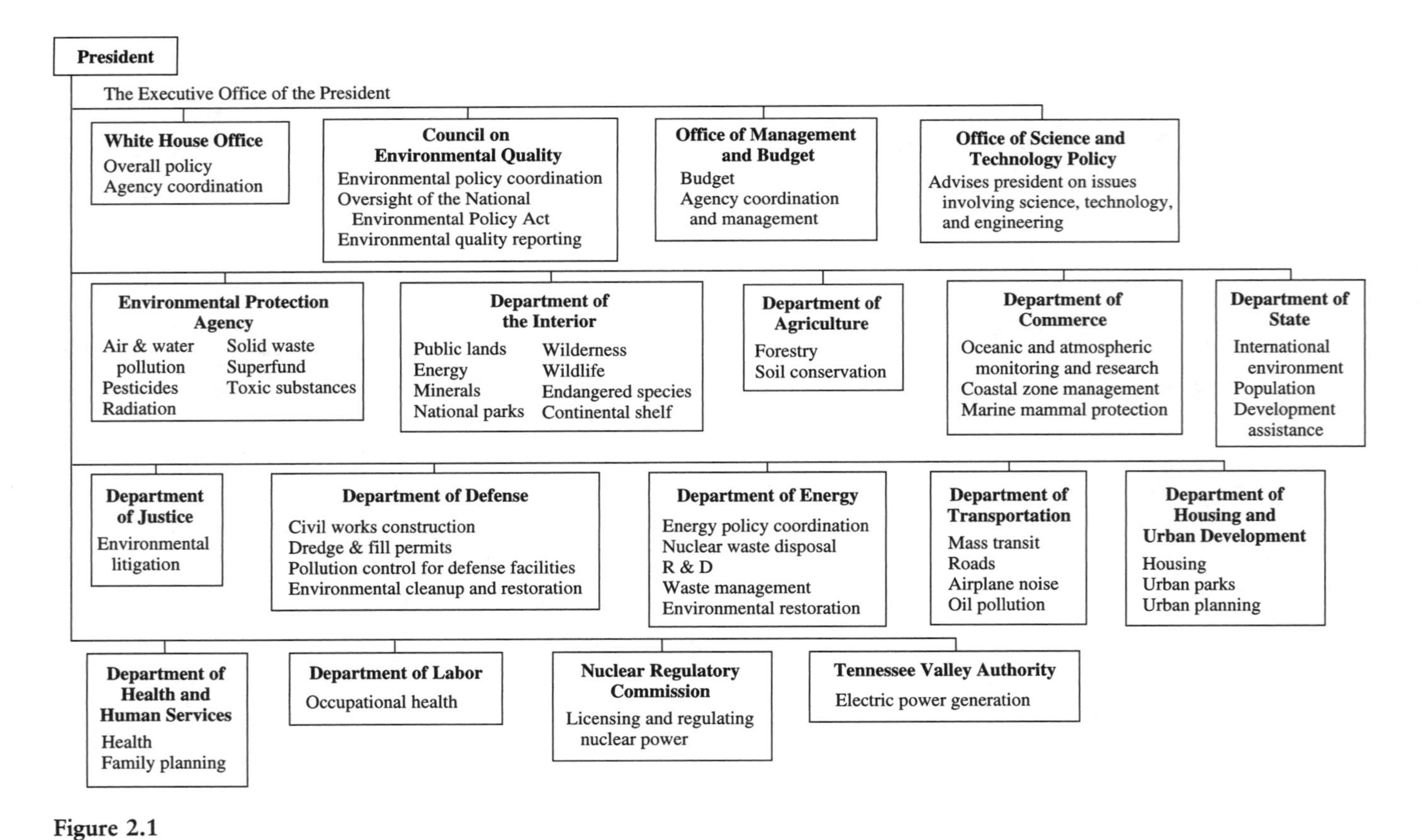

Figure 2.1
Executive Branch Agencies with Environmental Responsibilities
Source: Vig, N. J., and Kraft, M. E., eds. (2000). *Environmental policy: New directions for the twenty-first century*. 4th ed. Washington, DC: CQ Press, p. 7. Reprinted by permission.

past three decades. They also serve as a reminder that the U.S. Congress remains the strongest national legislature in the world at a time when authority in most nations has drifted to chief executives and high-level agency staff (Davidson and Oleszek 2002).

With a representative assembly so deeply involved with environmental policy, one might expect that public opinion and organized interest groups will be accorded great weight. They are, much to the consternation of many scientists and environmentalists who often bemoan the resulting policy stalemates and emphasis given to short-term impacts on members' constituencies. Internal organizational reforms during the 1970s made Congress even more open and democratic than in earlier decades, providing greater opportunities for involvement by both environmental and business groups in nearly all of its policymaking activities (Kraft 1995, 2000). The roles of both environmental and industry groups in lobbying for environmental policy are addressed below.

The courts are also unusually active and influential in U.S. environmental politics, and nothing on the horizon suggests their role is likely to change appreciably. More than 100 federal trial and appellate courts interpret federal environmental legislation and adjudicate disputes over administrative and regulatory actions. Their influence can be seen in the enormous spurt in environmental litigation over the past 30 years (O'Leary 1993; McSpadden 2000). The EPA itself has reported that, during the 1970s and 1980s, about 80 percent of its major regulatory decisions were contested in court—probably a high figure, but, nonetheless, a telling sign of the litigious context for environmental policy. Even if regulatory decisions are contested less frequently in the early twenty-first century, the result, nevertheless, can be protracted legal proceedings, long delays in implementing environmental laws, and high costs. Such results have spurred the use, in recent years, of formal regulatory negotiation (in which a committee of affected interests seeks consensus on a proposed regulation) and other efforts at environmental dispute resolution. These include various forms of collaborative decision making, endorsed by the EPA, state and local governments, and industry, which are sustained by a shared desire to avert litigation (Weber 1998).

Yet another distinctive feature of U.S. environmental politics is the extent of decentralization of both policymaking and implementation. Many areas of environmental policy have long been left to state and local governments, including most aspects of waste management, groundwater protection, coastal zone management, and land use. The Council of State Governments has estimated that about 70 percent of significant environmental legislation enacted at the state level reflects an independent judgment of policy needs; it has little to do with federal policy (Rabe 2000).

Moreover, virtually all federal environmental protection statutes of the past three decades, including air and water pollution and hazardous waste control, are implemented by the states, in cooperation with the EPA and its ten regional offices. About two-thirds of the EPA's staff is located in those regional offices to facilitate cooperation with the states. In most cases, the states have been given primary responsibility for issuing permits, monitoring environmental conditions, and carrying out enforcement activities. By the late 1990s, about 75 percent of major federal programs that could be delegated to the states had been delegated. The states also collect over 94 percent of the environmental-quality data that appears in the EPA's national databases (Brown 2000). The states are assisted by federal grants that cover an average of 20 percent of state expenditures on pollution control policies. Consistent with that spending pattern, by 1996, the states were spending twice as much as the entire EPA budget on environmental and natural resource policies. Despite these patterns of significantly increased delegation of authority to the states, even further decentralization of environmental policy is likely over the next decade. It is a shift that policymakers in both major parties tend to endorse. Nonetheless, recent studies suggest the need for careful oversight of state actions by the EPA to help ensure "accountable devolution" of environmental responsibilities, as well as provision to the states of adequate financial and technical assistance to achieve national environmental quality goals. The development, in the mid-1990s, of a National Environmental Performance Partnership System raises expectations that states with solid performance can be rewarded by the EPA with greater flexibility in program management (Kraft and Scheberle 1998; Rabe 2000).

Incentives and Policy Choices

Governmental structures affect the general character of the policymaking process and the outcomes, but they do not entirely explain the variability in patterns of policymaking in the United States. Of perhaps equal importance are the cluster of interests involved, the incentives of policy actors, and the distribution of costs and benefits for any given policy. The way these variables interact affects both environmental protection and natural resource conservation policies.

Ingram and Mann (1983) argue that the types of environmental policies adopted (for example, regulatory or distributive) reflect political variables, such as the structure of demand (conflict or consensus among interest groups), the structure of decision making (integrated or fragmented government institutions), and the structure of impacts (the actual effects on society of the policies, including costs and other burdens and on whom they fall). The result is that, in some policy areas, the nation has adopted the kind of tough regulatory policies so criticized by business interests and applauded by environmentalists. In some natural resource policy areas, however, Congress has approved distributive policies that reward narrow economic interests. Provision of generous subsidies for hard-rock mining and private logging on public lands are examples.

In a widely cited analysis of regulatory politics, James Q. Wilson (1980) suggested that such differences in policy outcomes can be explained by the perception of policy impacts, particularly the concentration or dispersal of costs and benefits and, hence, the incentives that are created for different actors to participate in the policymaking process. Economic interests (for example, ranchers, loggers, mining companies, and land development corporations) adversely affected by proposed environmental policies have a reason to organize and oppose them, as they did throughout the 1990s. The public receiving the broadly dispersed benefits of environmental protection (current and proposed) is rarely interested enough to pay much attention to or defend the policies.

What Wilson calls "entrepreneurial politics" may alter the usual logic of collective action, where the public has little material incentive to organize or actively support actions that benefit society as a whole. Policy entrepreneurs in environmental groups and Congress mobilize latent public sentiment on the issues (which is generally sympathetic to strong environmental protection), capitalize on well-publicized crises and other catalytic events, attack their opponents for endangering the public's welfare, and associate proposed legislation with widely shared values, such as clean air, clean water, and public health (Kraft and Wuertz 1996).

A different set of incentives exists when the benefits of environmental policies flow to narrow interests and the costs are broadly distributed. In this case, the beneficiaries are likely to organize and lobby to protect their interests, while the general public will usually prove hard to mobilize against natural resource subsidies or "giveaways" that are low-salience issues for the vast majority of citizens (for example, low fees for grazing of animals on public lands). Whether the result is what Wilson (1980) calls "client politics" under such circumstances depends on how visible the policies are and the extent to which the public and environmental groups are able to challenge the beneficiaries.

As these examples illustrate, the salience and complexity of the issues and the degree of conflict that exists over them are also important factors, particularly in affecting decisions by the U.S. Congress (Kingdon 1994; Kraft 1995). These qualities affect the extent to which the public and policymakers pay attention or participate actively in decision making. For example, the visibility and salience of many older natural resource policies affecting hard-rock mining and timber harvesting increased sharply in the Clinton administration as a result of Interior Secretary Bruce Babbitt's campaign to impose higher user fees and to curtail environmentally harmful practices encouraged by federal policies (Lowry 2000; Davis 2001). Those reformist efforts sparked considerable controversy over property rights and the use of public lands in the West, and stimulated aggressive efforts by antienvironmental forces, represented in the Wise Use and property rights movements (Switzer 1997).

The Impact of Political Culture: Interest-Group Activity and Public Opinion

U.S. political culture places an exceptionally high value on individual rights and liberties, the sanctity of private property, and a relatively unfettered free-market economy. Historically, it has given much less attention to the protection of collective goods, such as environmental quality and public health. Notwithstanding the development of modern health, safety, and environmental policies, particularly during the 1960s and 1970s, these beliefs continue to constrain governmental programs.

Public Opinion and Social Values Against this general backdrop of U.S. social values, recent trends affecting the resources and political influence of environmental groups, as well as public attitudes toward government and politics, are likely to shape the general patterns of policymaking noted above. The salience of environmental issues diminished in the early 1990s, the extent of public support declined slightly as well, and the leading national environmental groups struggled to cope with stagnant or declining memberships and funding. By the end of the 1990s, however, some leading groups (notably the Natural Conservancy) enjoyed additional growth in membership, while others (e.g., the Sierra Club) saw their membership fall a bit more (Bosso 2000).

Despite these alterations in the political environment, environmental values continue to be broadly endorsed by the U.S. public. Most survey data from the past two decades, including polls conducted in the late 1990s and early twenty-first century, show strong public concern about environmental problems, despite their typical low salience (Dunlap 1995; Roper Starch Worldwide 2000). Antienvironmental pressures on policymakers increased as well over the past decade, particularly from industry, property rights groups, and Western resource interests. They have had a major impact on environmental policy decisions, especially when the Republican Party controlled the Congress, the White House, or both. These groups were well represented in the administration of George W. Bush, particularly within the Interior Department.

The Role of Environmental NGOs and Business Groups The relative influence of such different constellations of interests in the future will depend on how well each is able to take advantage of both conventional and newer opportunities for public involvement to mold the political agenda. Interest groups on both sides have long attempted to affect legislative decision making through direct lobbying by their professional staffs and by education and mobilization of their members, usually called indirect or grassroots lobbying.

Several leading environmental groups, particularly the Sierra Club and the League of Conservation Voters (LCV), have placed increasing emphasis on involvement in the electoral process, primarily through campaign contributions and candidate endorsements. For both the 1996 and 1998 election campaigns, the Sierra Club spent an astonishing $7 million on voter education, issue-advocacy advertisements, and direct support of candidates for office. The LCV spent more than $2 million in each recent election (Kraft and Wuertz 1996; Kraft 2001).

Partly because of its election laws, the United States has never been fertile ground for organized Green parties, although they can be a force in selected local and state elections. During the 2000 presidential election, for example, Green Party nominee Ralph Nader won only 3 percent of the national vote. While impressive from historical perspectives, the Green vote was insufficient to qualify the party for public funding for the 2004 presidential election.

Both environmental NGOs and business groups have been active as well in presenting data and taking positions at public hearings and in commentary on proposed regulations at all levels of government. Many also have become deeply involved in local and regional public-private partnerships, in which collaborative decision making involving industry, government, and citizen groups is viewed as preferable to often adversarial regulatory proceedings (Beierle 1999; Knopman, Susman, and Landy 1999; Mazmanian and Kraft 1999).

Inventive forms of citizen activism have been spurred as well by revolutions in communications technology. In the early twenty-first century, environmental groups routinely make use of electronic mail and faxes and help to coordinate their members' responses through centralized electronic services. Increasing use is made of the Internet as well, with an enormous surge in the number of environmentally oriented Web sites. Interaction with government agencies, legislators, and executive officials has been greatly facilitated with the development of Web sites that simplify the acquisition and use of studies, technical reports, and issue documents. Beyond the activist groups, research institutes, such as Resources for the Future, and professional societies, such as the Ecological Society of America, maintain Web sites for circulation of their studies and discussion papers, as do government agencies, which commonly maintain elaborate sites providing access to thousands of reports, studies, and databases (Kraft 2001).

Industry groups, trade associations, and similar organizations use most of the same strategies and technologies that environmental NGOs do. By most accounts, they have greater economic and political resources to bring to the task, and there is little question that they significantly affect environmental policy decisions at all levels of government. This is particularly the case in the often-complex administrative and legal proceedings of regulatory agencies.

Whether representing automobile manufacturers, oil refineries, chemical companies, or the pulp-and-paper industry, business groups are able to employ the abundant talent of Washington law firms and technical consultants to help make their case in rule-making bodies, such as the EPA, and before the courts. Environmental NGOs also have considerable legal and technical expertise available to them, and they can be influential as well in these decision-making arenas, but generally not to the same degree as business groups. Whereas corporations find it essential to track and respond to nearly every potential environmental action that may affect their operations (at the state as well as national level), environmental NGOs must carefully target only the most important issues to make good use of their limited resources (Furlong 1997).

The effects of recent changes in the distribution of environmental information, citizen activism, and industry lobbying cannot be determined precisely, but they are likely to prove exceedingly important for U.S. environmental policy. Development of green software, databases, networks, and bulletin boards proceeds apace, and citizen groups are increasingly likely to seek out and use such information. Because such activism depends on citizen motivation, the salience of environmental issues will continue to be an important variable affecting the extent of activism and the relative influence of environmental and industry groups. Conventional opportunities for public involvement (for example, attending hearings and commenting on policy proposals) also remain, and various grassroots organizations (including those widely considered to be antienvironmental) are certain to use them to good effect in pursuing their interests at the local, state, and regional levels. There is every reason to expect that business and industry groups will be equally, if not more, alert to opportunities to shape environmental policy decisions from local to national and international levels.

Judging the Effectiveness of Environmental Policy

As the comments above suggest, at both national and local levels, one finds dramatically different appraisals of the effectiveness of environmental policies and the need for change. Business interests and many political conservatives question the severity of public health and environmental risks. They also argue that regulatory costs (particularly direct-compliance costs) may not be justified by what they see as the often meager benefits the nation receives from its environmental policies. Some of them are deeply concerned about perceived impacts on jobs, property rights, and personal freedoms (Switzer 1997). Environmentalists argue, in contrast, that properly valued, long-term benefits to public and environmental health can be easily justified in light of the consequences of failing to act. They also tend to favor stringent regulation to help ensure that implementation and compliance are consistent with statutory promises and not subject to the discretion of regulatory agencies they do not trust.

There is evidence to support each argument, but judging the effectiveness of environmental and resource policies is far more complex than participants in these debates assert. It is also inescapably subjective, in that some evaluative criteria must be established and judgments made on the basis of incomplete and inconsistent information. Program evaluation is clearly important to any effort to supplement or replace present policies or to foster a transition to sustainability. Nonetheless, reliable assessments of policy implementation and of environmental-quality outcomes remain rare. This is particularly true for measures of long-term changes in environmental conditions, such as air and water quality and ecosystem health (Ringquist 1995; Knaap and Kim 1998).

Selected Trends in Environmental Quality

The demographic, economic, and environmental data in table 2.2, drawn largely from government documents, can assist in evaluating trends in environmental quality. Some of the information, such as population size and growth rates, gross national product per capita, and land-use patterns, facilitates a comparison of the United States to other nations covered in this book. Perhaps most notable here is the unusually high rate of growth in the U.S. population (over 1 percent a year), the affluence of its citizens, the nation's dependence on private automobiles for transportation, and the level of fossil-fuel use. Transportation and energy use contribute importantly to the high rate of production of greenhouse gases and air pollutants.

The trends in air quality reflect some of the best data available and also permit comparison to progress made in other nations. The data show significant decreases in the major air pollutants between 1970 and 1999, as well as for more recent periods for which the data are more reliable. Reports from the U.S. EPA indicate that, over the past three decades, the nation also has significantly improved water quality and reduced emissions of toxic chemicals (Kraft 2001).

The gains in air quality are particularly striking and reflect a variety of changes since the early 1970s, including the impact of the federal Clean Air Act. The U.S. EPA reports that, between 1970 and 1999, total emissions of the six principal air pollutants decreased by some 31 percent, even while the nation's population grew by 33 percent, the gross domestic product by 147 percent, and total vehicle miles traveled by 140 percent. Between 1980 and 1999, the nation experienced a 94 percent reduction in ambient levels of lead, 57 percent in carbon monoxide, 50 percent in sulfur dioxide, 12 percent in ozone, and 25 percent in nitrogen dioxide. For fine particulates (PM_{10}), measurements from 1990 through 1999 show an 18 percent decrease in ambient air concentrations (U.S. Environmental Protection Agency, 2000b). Beyond the six principal pollutants covered by the Clean Air Act, toxic air emissions—from sources such as oil refineries, dry cleaning facilities, and chemical plants—are beginning to decrease as new federal regulations on air toxics take effect. The same is true for production and release of ozone-depleting chlorofluorocarbons, which began moving downward by the end of 1994, largely as a consequence of actions taken under the Montreal Protocol to end the production of the chemicals (Council on Environmental Quality, 1999; U.S. Environmental Protection Agency, 2000b).

Table 2.2
United States: A Statistical Profile

Land	
Total area (1,000 km^2)	9,159.0
Arable and permanent crop land (% of total) 1997	19.50%
Permanent grassland (% of total) 1997	26.10%
Forest and Woodland (% of total) 1997	32.60%
Other land (% of total) 1997	21.80%
Major protected areas (% of total) 1997	21.20%
Nitrogenous fertilizer use (t/km^2/arable land) 1997	6.20%
Wetland (acres lost/year) mid-1980s to late 1990s[f]	100,000
Population	
Total (100,000 inhabitants) 2002[d]	2,859.0
Growth rate % 1980–1998	18.7%
Population density (inhabitant/km^2) 1998	28.9
Projected population (millions) 2025[g]	346.0
Projected population (millions) 2050[g]	413.5
Gross domestic product	
Total GDP (billion U.S. dollars) 1998	$7,350
Per capita (1,000 U.S. dollars/capita) 1998	$27.2
GDP growth 1980–1998	63.6%
Current general government	
Revenue (% of GDP) 1997[b]	32.8%
Expenditure (% of GDP) 1997[b]	32.7%
Government employment (%/total) 1997[b]	13.2%
Energy	
Total supply (MTOE) 1997	2,162
% change 1980–1997	19.3%
Consumption (MTOE) 1998[c]	1,429.66
% change 1988–1998[c]	3.5%
Transport consumption (MTOE) 1998[c]	586.66
% change 1988–1998[c]	16.5%
Road vehicle stock	
Total (10,000 vehicles) 1997	21,022
% change 1980–1997	34.9%
Per capita (vehicles/100 inhabitants) 1997	79
Increase in total vehicle miles traveled 1970–1999[e]	140%

Table 2.2
(continued)

Air profile and greenhouse gases	
% change in emissions of six principal air pollutants 1970–1999[e]	–31%
% change in CO emissions 1980–1999[e]	–22%
% change in Pb (lead) emissions 1980–1999[e]	–95%
% change in NO_x emissions 1980–1999[e]	+2%
% change in VOC emissions 1980–1999[e]	–33%
% change in PM_{10} emissions 1980–1999[e]	–55%
% change in SO_2 emissions 1980–1999[e]	–50%
CO_2 emissions (tons/capita) 1998[a]	20.3
CO_2 emissions (kg/USD GDP) 1998[a]	.78
Nitrogen oxide emissions (kg/capita) 1998[a]	80.0
Nitrogen oxide emissions (kg/USD GDP) 1998	3
Sulfur oxide emissions (kg/capita) 1998[a]	69
Sulfur oxide emissions (kg/USD GDP) 1998	2.6
Pollution abatement and control	
Total (excluding household) expenditure (%/GDP) 1994	1.6%

Sources
OECD, *OECD Environmental Data Compendium 1999* (Paris: OECD, 1999), except as noted.
[a] OECD, The OECD Observer, *OECD in Figures* (Paris: OECD, June 2000).
[b] OECD, *National Accounts of OECD Countries* (Paris: OECD, 2000).
[c] IEA/OECD, *Energy Balances of OECD Countries, 1997–1998* (Paris: IEA/OECD, 2000).
[d] U.S. Census Bureau (http://www.census.gov), January 13, 2002.
[e] U.S. Environmental Protection Agency, *Latest Findings on National Air Quality: 1999 Status and Trends* (Research Triangle Park, NC: Office of Air Quality Planning and Standards, August 2000).
[f] U.S. Environmental Protection Agency, *National Water Quality Inventory: 1998 Report to Congress* (Washington, DC: U.S. EPA, Office of Water Quality, 2000).
[g] Population Reference Bureau, "2001 World Population Data Sheet" (Washington, DC: Population Reference Bureau, 2001. Also available at http://www.prb.org).

As impressive as most of these gains in air quality are, in 1999, nearly 62 million people nationwide lived in counties that failed to meet at least one of the national air-quality standards for the six major pollutants covered by the Clean Air Act. More than 53 million people resided in counties where pollution levels in 1999 exceeded federal standards for ground-level ozone, the chief ingredient of urban smog. These estimates are based on the EPA's previous 1-hour ozone standard; the numbers more than double using the new 8-hour, 0.08 parts per million standard (U.S. Environmental Protection Agency, 2000b).

The nation's water quality has also improved since passage of the Clean Water Act of 1972, although more slowly and more unevenly than has air quality. Monitoring data are less adequate for water quality than for air quality. For example, the best evidence for the state of water quality can be found in the EPA's biennial National Water Quality Inventory, which compiles data reported by each state. Yet in 1998, the states surveyed only about 25 percent of rivers and streams (but about 64 percent of those that flow year round) and only 40 percent of lakes, ponds, and reservoirs.

Based on those state inventories, for the nation as a whole, 35 percent of rivers and streams were found to be impaired to some degree, as were 45 percent of lakes, ponds, and reservoirs. Such a classification means they were not meeting or fully meeting the national minimum water-quality criteria for "designated beneficial uses," such as swimming, fishing, drinking-water supply, and support of aquatic life. Those numbers show some improvement over previous years, yet they also indicate that many problems remain, particularly siltation and nutrients that flow from urban and agricultural runoff (nonpoint sources) and the impact of persistent toxic pollutants (U.S. Environmental Protection Agency, 2000a). Prevention of further degradation of water quality, in the face of a growing population and strong economic growth, could be considered an important achievement. At the same time, water quality nationwide clearly falls short of the goals of the federal Clean Water Act.

Implications of the Trends

Assessments of this kind suggest that the United States made important advances, between 1970 and 2000, in controlling many conventional pollutants, as well as in expanding parks, wilderness areas, and other protected public lands (Council on Environmental Quality, 1999). Changes in some economic sectors, such as chemical manufacturing and pulp and paper, have been far more extensive than the figures above indicate (Press and Mazmanian 2000). Much the same is true for passenger vehicles, which emit a small fraction of the pollutants common in the early 1970s and are now far more fuel efficient.

Much debate surrounds the question of whether gains like these have been achieved at a reasonable cost. Critics of environmental policy often point to the approximately $180 billion per year spent during the late 1990s on compliance with the major pollution control programs as wasteful: Some suggest that the same environmental gains might be achieved far more cheaply with redesigned policies, in some cases, at only 50 percent of the cost (Freeman 2000). When new proposals are made for raising environmental-quality standards, the business community often voices strong objection. In the case of action by the EPA and the Clinton White House in 1997 to tighten standards for ozone and fine particulates, industry argued that too little was known about the health risks to justify the costs, which it put at $23 billion per year. The EPA countered that it was confident the new regulations would bring health care benefits of more than $120 billion, and it estimated the costs of new regulations at only $8.5 billion. Industry, in turn, questioned the validity of the EPA's economic and scientific analyses (Kriz 1997). Similar disputes arose early in the administration of George W. Bush over the allowable level of arsenic in drinking water (Vig and Kraft 2003). Inevitably, the insufficiency of environmental health and ecological knowledge feeds such conflict over the economic impact of both new rules and existing statutes and regulations.

Some economic studies have found that, on a national basis, environmental policies have been net producers of jobs and have contributed to the international competitiveness of American businesses (Porter and van der Linde 1995). Others have concluded that macroeconomic effects, such as impacts on productivity and inflation, may have been adversely affected, but generally only slightly (Tietenberg 1998). Moreover, the economic costs of environmental regulations, although often said to be a burden, typically are quite small in comparison to other business costs, such as taxes, wages, benefits, and interest rates. Hence, the other factors are more likely to affect business success. None of this means that regulatory costs are of no consequence or that some businesses are severely affected by specific environmental controls. But the evidence does help to put the costs of environmental policies into the larger context in which they must be considered.

There is another important implication to the trends in environmental quality reviewed above. Even if policies have fallen short of expectations or needs, the degree of improvement in environmental quality suggests that regulatory policies have not failed. The record of the past three decades would seem to indicate that well-designed and properly implemented regulatory programs can produce measurable progress in achieving environmental goals. Evidence from state-level studies reinforces this conclusion. Everything else being equal, states with the strongest environmental policies tend to exhibit greater improvement in environmental quality; this is particularly so in air quality. The most important variables are well-focused and well-supported administrative efforts, in combination with consistency in enforcement (Ringquist 1993).

Although federal environmental policies are clearly in need of reform to improve efficiency and effectiveness, such evidence hints at the virtue of a hybrid model that combines the advantages of regulatory policy with the greater flexibility and incentive mechanisms of approaches touted during the past decade. Nothing in the trend data indicates that much is to be gained by a wholesale abandonment of the environmental regulatory regime established since 1970. Indeed, the success of new approaches may well depend on the continuation of that regulatory apparatus. Voluntary action, public information campaigns, collaborative decision making, and public-private partnerships may work best when participants from the business community are aware that failure to achieve results will bring a return to a more formal regulatory framework that can compel action under the force of law.

It is also evident that, although advances in environmental quality continue to be achieved, further gains will be more difficult, costly, and controversial. This is largely because the easy problems have already been addressed, and marginal gains in air and water quality will cost more per unit of improvement than in the past. Moreover, second-generation environmental threats, such as toxic chemicals, hazardous wastes, and nuclear wastes, are proving to be even more difficult than was regulating bulk air and water pollutants in the 1970s. In these cases, substantial progress may not be evident for years to come, and it is likely to be expensive. In addition, it will involve far greater impact on individual lifestyles.

Similarly, dealing effectively with the third generation of ecological problems, such as global climate change and protection of biodiversity, will be costly and difficult. The costs imposed are often tangible, visible, and immediate, yet the benefits may be uncertain, intangible, obscure, and remote. Solutions will require an unprecedented degree of cooperation among nations and substantial improvement in institutional capacity for research, data collection, and analysis, as well as policy development and implementation. They will also demand greater cooperation from citizens that may not be easily forthcoming.

The imperative to resolve conflict in all of these issue areas comes at a time of continuing public distrust of government and other institutions, stagnant or declining budgetary resources, and the need to act in the face of scientific uncertainty over risks to public and environmental health. As discussed above, these conditions have inspired a wide search for effective and acceptable solutions. That search is setting a new environmental policy agenda for the twenty-first century. Policy analysts and officials now face the intriguing task of determining which policy approaches to use in any given circumstances (Sexton et al. 1999). Only further study, experimentation (in part at state and local levels), and assessment of results through formal environmental program evaluations can answer those questions satisfactorily (John 1994; Knaap and Kim 1998; Mazmanian and Kraft 1999). Rethinking the logic of policy design should help as well, because it forces policymakers to consider how regulatory requirements or incentives affect both those who implement policy and those who are affected by it (Schneider and Ingram 1990). In the meantime, there is at least some agreement on the longer-term agenda of sustainable development that may help to elucidate the most promising policy paths.

Toward Environmental Sustainability?

At least since its endorsement by the World Commission on Environment and Development (1987), the concept of sustainable development has been promoted as a unifying concept that can help to integrate economic and environmental values and guide public policy. The 1992 United Nations Conference on Environment and Development put that concept at the center of its deliberations and in the plan of action delegates endorsed, Agenda 21. Within the United States, the concept has emerged as a key element in political dialogue on the environment.

President George H. W. Bush, for example, appointed a President's Commission on Environmental Quality, a committee of corporate, foundation, and environmental-group executives, to seek new approaches to environmental problems. In early 1993, it issued a report calling for the creation of a national council on sustainable development that would promote appropriate management practices in the private sector. President Clinton responded by appointing a new President's Council on Sustainable Development. Consisting of 25 leaders from industry, government, and the environmental community, the council was charged with formulating a sustainable development plan for the United States.

With remarkable unanimity, the council's members called for maintaining the present regulatory system, but improving it through adoption of a new generation of flexible, consensual environmental policies. The council believed such a strategy could maximize economic welfare, while achieving more effective and efficient environmental protection. The United States, they said, "must change by moving from conflict to collaboration and adopting stewardship and individual responsibility as tenets by which to live" (President's Council on Sustainable Development, 1996, 1). These kinds of recommendations have been echoed in scores of similar reports (e.g., National Commission on the Environment, 1993; National Research Council, 1999). This strong embrace of the idea of sustainable development is an important signpost for the 1990s, even if it remains a vague term that can mask serious economic and political conflicts.

The movement toward sustainability at the community level captures many of the most important trends in U.S. environmental policy today.

It encourages a holistic and long-term assessment of environmental problems and a search for new mechanisms and institutions, at local and regional levels, to better balance the needs of human populations and natural systems. Cities and regions are beginning to modify a diversity of practices, such as building construction, water and energy use, recreation and preservation of open space, and municipal transportation, to emphasize life-cycle analysis and full cost accounting to help ensure that environmental impacts of decisions are taken into account. Many are also designing creative ways for citizens to become involved in local decision making to move beyond the inherent limitations of electoral participation and public hearings.

To be successful, such initiatives will have to attract broader recognition and support than they have to date, and they will doubtless be modified after periods of trial and error. Moreover, the scientific and other information and data management needs for such decision making will be prodigious and likely involve substantial investments in data gathering, computer modeling, and collaborative decision making. Yet such an integrated approach to environmental, economic, and social problems promises to assist the United States, over time, in moving from an era of conflict over environmental policy to one based far more on a search for common ground and mutual interests (Mazmanian and Kraft 1999).

Conclusions

This overview of environmental policy and politics in the United States makes clear the intimate interrelationship between the political process and policy actions and outcomes. Environmental policies reflect the institutional arrangements of the U.S. political system, the way key policy actors define the problems and appraise proposed solutions, the extent and kind of media coverage of disputes, and, especially, the relative influence of opposing interest groups. Environmental science and policy analysis contribute as well, though, primarily as they are filtered through the eyes of the lead players and serve their strategic purposes.

The policies produced as a result of these various forces are undergoing a transformation, even if the outcomes cannot yet be known. The changes being made reflect the mixed record of success of environmental and natural resource policies over the past 30 years and diverse judgments about their cost, intrusiveness, effectiveness, and efficiency. The clearest trend is a movement away from conventional command-and-control regulation and toward a hybrid model that supplements such policies with elements of market-based incentives, public information campaigns, voluntary pollution prevention initiatives, decentralization of authority to states and localities, and greater flexibility in the implementation of existing regulatory statutes. In its twenty-fifth anniversary report, the Council on Environmental Quality (1997) referred to this transition as "forging a new paradigm" that can complement traditional approaches and build on proven means to protect human health and the environment.

It is too early, however, to judge the success of any of these endeavors, singly or in combination. All merit further scrutiny. Students of environmental policy need to determine which policy approaches produce what kinds of outcomes and which tools in what combinations are most appropriate for various environmental problems. Some assessments of this kind have already begun (Mazmanian and Kraft 1999; Sexton, et al. 1999), and similar policy analyses would improve our understanding of precisely how such reforms are likely to affect environmental programs and the agencies that administer them. The same could be said for the U.S. role in international environmental policy development and implementation. As was made clear in the introductory chapter and above, international organizations and globalization of the economy affect U.S. environmental actions (most notably on climate-change issues), while U.S. domestic politics will affect the way the nation responds to international challenges (Hempel 1996; Vig and Axelrod 1999; DeSombre 2000).

Despite these uncertainties, perhaps the most promising development in recent years is endorsement by many business and political leaders of environmental sustainability as a long-term goal that can help to integrate a range of current policies in a more holistic or ecological manner and to reconcile conflicts between environmental and economic values. Only a naive observer of environmental politics, however, would assume that the current movement toward sustainable development will be as unimpeded as its advocates hope. Experience with the stridently antienvironmental 104th and 105th Congresses serves as a useful reminder that intense conflicts continue to exist over environmental values, programs, and national priorities. Such conflicts were evident, once more, in 2001, at the beginning of George W. Bush's administration, as the president sought to reverse many environmental commitments, both domestic and international, of the Clinton-Gore era (Vig and Kraft 2003).

As these examples suggest, even if agreement on the broadest environmental values is achieved and maintained, there is no guarantee that consensus on policy choices will come easily when those decisions impose significant short-term costs. There will always be much to fight about, and hence a role for politics as a process for resolving conflicts. These same forces ensure that policy change, for the most part, will occur incrementally and slowly, as the nation tries new approaches and learns from the experience.

On a positive note, even the 104th Congress eventually agreed on major redirection in pesticide control and drinking-water policy that had suffered from years of confrontation and bitter partisan gridlock. There were unique reasons for legislative success in these cases. Yet with polls continuing to show strong public support for environmental protection, lawmakers in both parties were eager to enact environmental measures in an election year. That experience should be a hopeful sign for the future as environmental policy goes through a period of much-needed reassessment and redirection.

References

Andrews, R. N. L. (2000). Risk-based decisionmaking. In N. J. Vig, and M. E. Kraft, eds., *Environmental policy*. 4th ed. Washington, DC: CQ Press.

Baumgartner, F. R., and Jones, B. D. (1993). *Agendas and instability in American politics*. Chicago: University of Chicago Press.

Beierle, T. C. (1999). Using social goals to evaluate public participation in environmental decisions. *Policy Studies Review*, *16*(Fall/Winter), 75–103.

Bosso, C. J. (2000). Environmental groups and the new political landscape. In N. J. Vig, and M. E. Kraft, eds., *Environmental policy*. 4th ed. Washington, DC: CQ Press.

Brown, R. S. (2000). The states protect the environment. Web page of Environmental Council of the States (http://www.sso.org/ecos). Appeared originally in *ECOStates*, Summer 1999.

Chertow, M. R., and Esty, D. C., eds. (1997). *Thinking ecologically: The next generation of environmental policy*. New Haven: Yale University Press.

Clarke, J. N., and McCool, D. C. (1996). *Staking out the terrain: Power and performance among natural resource agencies*. 2nd ed. Albany: State University of New York Press.

Cortner, H. J., and Moote, M. A. (1999). *The politics of ecosystem management*. Washington, DC: Island Press.

Council on Environmental Quality. (1997). *Environmental quality: 25th anniversary report of the Council on Environmental Quality*. Washington, DC: Government Printing Office.

Council on Environmental Quality. (1999). *Environmental quality: The 1997 report of the Council on Environmental Quality*. Washington, DC: Council on Environmental Quality. (Available at http://ceq.eh.doe.gov/nepa/reports/1997/index.html)

Davidson, R. H., and Oleszek, W. J. (2002). *Congress and its members*. Washington, DC: CQ Press.

Davies, J. C., ed. (1996). *Comparing environmental risks: Tools for setting government priorities*. Washington, DC: Resources for the Future.

Davies, J. C., and Mazurek, J. (1998). *Pollution control in the United States: Evaluating the system*. Washington, DC: Resources for the Future.

Davis, C., ed. (2001). *Western public lands and environmental politics*. 2nd ed. Boulder: Westview Press.

DeSombre, E. (2000). *Domestic sources of international environmental policy: Industry, environmentalists, and U.S. power*. Cambridge: MIT Press.

Dunlap, R. E. (1995). Public opinion and environmental policy. In J. P. Lester, ed., *Environmental politics and policy: Theories and evidence*. Durham, NC: Duke University Press.

Freeman, A. M. (2000). Economics, incentives, and environmental regulation. In N. J. Vig, and M. E. Kraft, eds., *Environmental policy*. 4th ed. Washington, DC: CQ Press.

Furlong, S. R. (1997). Interest group influence on rulemaking. *Administration and Society*, *29*(July), 325–347.

Hempel, L. C. (1996). *Environmental governance: The global challenge*. Washington, DC: Island Press.

Ingram, H. M., and Mann, D. E. (1983). Environmental protection policy. In S. S. Nagel, ed., *Encyclopedia of policy studies* (pp. 687–725). New York: Marcel Dekker.

John, D. (1994). *Civic environmentalism: Alternatives to regulation in states and communities*. Washington, DC: CQ Press.

Jones, C. O. (1975). *Clean air: The policies and politics of pollution control*. Pittsburgh: University of Pittsburgh Press.

Kingdon, J. W. (1994). *Agendas, alternatives, and public policies*. 2nd ed. New York: HarperCollins.

Knaap, G., and Kim, T. J., eds. (1998). *Environmental program evaluation: A primer*. Champaign: University of Illinois Press.

Knopman, D., Susman, M. M., and Landy, M. K. (1999). Civic environmentalism: Tackling tough land-use problems with innovative governance. *Environment, 41*(December), 24–32.

Kraft, M. E. (1995). Congress and environmental policy. In J. P. Lester, ed., *Environmental politics and policy*. 2nd ed. Durham, NC: Duke University Press.

Kraft, M. E. (2000). Environmental policy in Congress: From consensus to gridlock. In N. J. Vig, and M. E. Kraft, eds., *Environmental policy*. 4th ed. Washington, DC: CQ Press.

Kraft, M. E. (2001). *Environmental policy and politics*. 2nd ed. New York: Addison Wesley Longman.

Kraft, M. E., and Scheberle, D. (1998). Environmental federalism at decade's end: New approaches and strategies. *Publius: The Journal of Federalism, 28*(Winter), 131–146.

Kraft, M. E., and Wuertz, D. (1996). Environmental advocacy in the corridors of government. In J. G. Cantrill, and C. Oravec, eds., *The symbolic Earth: Discourse and our creation of the environment*. Lexington: University Press of Kentucky.

Kriz, M. (1997, January 1). Heavy breathing. *National Journal*, pp. 8–12.

Lowry, W. R. (2000). Natural resource policies in the twenty-first century. In N. J. Vig, and M. E. Kraft, eds., *Environmental policy*. 4th ed. Washington, DC: CQ Press.

Mazmanian, D. A., and Kraft, M. E., eds. (1999). *Toward sustainable communities: Transition and transformations in environmental policy*. Cambridge: MIT Press.

McSpadden, L. (2000). Environmental policy in the courts. In N. J. Vig, and M. E. Kraft, eds., *Environmental policy*. 4th ed. Washington, DC: CQ Press.

National Academy of Public Administration. (1995). *Setting priorities, getting results: A new direction for EPA*. Washington, DC: National Academy of Public Administration.

National Academy of Public Administration. (2000, November). *Environment gov.: Transforming environmental protection for the 21st century*. Washington, DC: National Academy of Public Administration.

National Commission on the Environment. (1993). *Choosing a sustainable future: The report of the National Commission on the Environment*. Washington, DC: Island Press.

National Research Council. (1999). *Our common journey: A transition toward sustainability*. Washington, DC: National Academy Press.

O'Leary, R. (1993). *Environmental change: Federal courts and the EPA*. Philadelphia: Temple University Press.

Porter, M. E., and van der Linde, C. (1995). Green and competitive: Ending the stalemate. *Harvard Business Review*, *73*(September–October), 120–134.

Portney, P. R., and Probst, K. N. (1994). Cleaning up Superfund. *Resources*, *114*(Winter), 2–5.

Portney, P. R., and Stavins, R. N., eds. (2000). *Public policies for environmental protection*. Washington, DC: Resources for the Future.

President's Council on Sustainable Development. (1996, February). *Sustainable America: A new consensus for prosperity, opportunity, and a healthy environment*. Washington, DC: President's Council on Sustainable Development.

Press, D., and Mazmanian, D. A. (2000). Understanding the transition to a sustainable economy. In N. J. Vig, and M. E. Kraft, eds., *Environmental policy*. 4th ed. Washington, DC: CQ Press.

Rabe, B. G. (2000). Power to the states: The promise and pitfalls of decentralization. In N. J. Vig, and M. E. Kraft, eds., *Environmental policy*. 4th ed. Washington, DC: CQ Press.

Ringquist, E. J. (1993). *Environmental protection at the state level: Politics and progress in controlling pollution*. Armonk, NY: M. E. Sharpe.

Ringquist, E. J. (2000). Environmental justice: Normative concerns and empirical evidence. In N. J. Vig, and M. E. Kraft, eds., *Environmental policy*. 4th ed. Washington, DC: CQ Press.

Roper Starch Worldwide. (2000, November). Roper Green Gauge 2000: Rising concern. New York: Roper Starch Worldwide.

Rosenbaum, W. A. (2000). Escaping the "battered agency syndrome": EPA's gamble with regulatory reinvention. In N. J. Vig, and M. E. Kraft, eds., *Environmental policy*. 4th ed. Washington, DC: CQ Press.

Sabatier, P. A., and Jenkins-Smith, H. C. (1993). *Policy change and learning: An advocacy coalition approach*. Boulder: Westview Press.

Schneider, A. L., and Ingram, H. (1990). Policy design: Elements, premises, and strategies. In S. S. Nagel, ed., *Policy theory and policy evaluation: Concepts, knowledge, causes, and norms*. Westport, CT: Greenwood Press.

Sexton, K., Marcus, A. A., Easter, K. W., and Burkhardt, T. D., eds. (1999). *Better environmental decisions: Strategies for governments, businesses, and communities*. Washington, DC: Island Press.

Switzer, J. V. (1997). *Green backlash: The history and politics of environmental opposition in the U.S.* Boulder: Lynne Rienner.

Tietenberg, T. (1998). *Environmental economics and policy*. 2nd ed. Reading, MA: Addison-Wesley.

U.S. Environmental Protection Agency. (1990a, September). *Reducing risk: Setting priorities and strategies for environmental protection*. Washington, DC: U.S. Environmental Protection Agency, Science Advisory Board.

U.S. Environmental Protection Agency. (1990b, November). *Environmental investments: The cost of a clean environment: Report of the administrator of the Environmental Protection Agency to the Congress of the United States.* Washington, DC: U.S. Environmental Protection Agency.

U.S. Environmental Protection Agency. (2000a). *National Water Quality Inventory: 1998 Report to Congress.* Washington, DC: Office of Water. (Available at the EPA Web site: http://www.epa.gov/305b/98report/)

U.S. Environmental Protection Agency. (2000b, August). *Latest findings on national air quality: 1999 status and trends.* Research Triangle Park, NC: Office of Air Quality Planning and Standards. (Available at the EPA Web site: http://www.epa.gov/airtrends/)

Vig, N. J. (2000). Presidential leadership and the environment: From Reagan to Clinton. In N. J. Vig, and M. E. Kraft, eds., *Environmental policy.* 4th ed. Washington, DC: CQ Press.

Vig, N. J., and Axelrod, R. S. (1999). *The global environment: Institutions, law, and policy.* Washington, DC: CQ Press.

Vig, N. J., and Kraft, M. E., eds. (2000). *Environmental policy: New directions for the twenty-first century.* 4th ed. Washington, DC: CQ Press.

Vig, N. J., and Kraft, M. E., eds. (2003). *Environmental policy.* 5th ed. Washington, DC: CQ Press.

Weber, E. P. (1998). *Pluralism by the rules: Conflict and cooperation in environmental regulation.* Washington, DC: Georgetown University Press.

Wilson, J. Q. (1980). The politics of regulation. In J. Q. Wilson, ed., *The politics of regulation.* New York: Basic Books.

World Commission on Environment and Development. (1987). *Our common future.* New York: Oxford University Press.

3

Contesting the Green: Canadian Environmental Policy at the Turn of the Century

Glen Toner

This chapter analyzes five major policy initiatives that dominated federal environmental politics of the 1990s. Combined, these five areas continue to comprise the bulk of the Canadian government's environmental policy activity in the early twenty-first century. The policies include the Canadian Environmental Protection Act (CEPA), the institutionalization of sustainable development practices, the Canadian Environmental Assessment Act (CEAA), and the Canadian response to the 1992 Rio Earth Summit Climate Change and Biodiversity conventions. The policy history of each initiative will be explored to reveal the dynamics of the policy process and to assess policy performance.

The contested and multifaceted nature of Canadian environmental politics reflects both the inherent complexity of environment-economy policy issues and the institutional arrangements that govern the policy field. But it also reflects the importance of the environment to a country that has relied on its natural resource endowment to generate a high standard of living, while at the same time valuing the beauty of the landscape as a central feature of its national identity.[1]

Federal Departments

Environment Canada (EC) is the lead federal department (executive agency) for environmental issues, though the programs and responsibilities of many other departments, such as Transport, Industry, Natural Resources, Fisheries and Oceans, Agriculture and Agri-food, Public Works and Government Services, and Finance, also impact on the environment. This vertical division of authority within the Canadian state, at both the federal and provincial levels, between departments charged with protecting the environment and those responsible for resource and economic development is a principal feature of Canadian environmental politics. The vertical division of state authority means that policy relationships among departments of the same government are often as important and problematic as relationships between different governments and among governments, industry, and environmental non-governmental agencies (ENGOs) (Hessing and Howlett 1997; Doern and Conway 1994; Harrison 1996; Toner 2000). Indeed, the Commissioner of the Environment and Sustainable Development (CESD) identified the lack of horizontal coordination and integration across federal departments as one of the three major constraints hobbling the federal government's environmental policy performance (Commissioner of the Environment and Sustainable Development, 1997, 11).

Canada is a constitutional monarchy and has a federal system of government in which both the federal and provincial governments exercise authority over the environment. In fact, the Canadian Constitution makes no direct reference to the environment. Jurisdiction over the environment is complicated by a distribution of powers between the two levels that touches on numerous fields of power relating to environmental protection and sustainable development. Section 91 of the Constitution Act identifies federal powers over the seacoast and inland fisheries; navigation and shipping; federal lands and waters; and peace, order, and good government. These powers have enabled the federal government to pass legislation such as the Fisheries Act, the Navigable Waters Protection Act, the Arctic Waters Pollution Prevention Act, and CEPA. International obligations are a federal authority, and this power has enabled the federal government to pass the Migratory Birds Convention Act and the International River Improvements Act, as well as to sign international conventions and protocols, such as the Biodiversity Convention and the Montreal Protocol on Ozone Depleting Substances.

Section 92 outlines a number of provincial powers, such as local works and undertakings, property and civil rights within the province, matters of a local or private nature, and authority over provincially owned lands and resources. The latter provision is particularly important in Canada, where most lands outside of the northern territories are provincially owned. In a 1982 amendment to the Constitution Act, the provinces were assigned exclusive jurisdiction over the development, conservation, and management of nonrenewable resources in the province, including forestry and hydroelectric facilities. Hence, both levels have substantial constitutional authority to govern in this policy area (Vanderzwagg and Duncan 1992). Such jurisdictional overlap has led to much uncertainty, frustration, and conflict over the years. The federal, provincial, and territorial governments created the Canadian Council of Ministers of the Environment (CCME) to provide a forum for coordinating intergovernmental actions on the environment (Skogstad and Kopas 1992).[2]

Industry

Canada is a trading nation, and its success in trade is heavily dependent on natural resources. Indeed, the Canadian economy retains a higher level of dependence on natural resource extraction and export than most Organization for Economic Cooperation and Development (OECD) countries. Almost 20 percent of Canada's gross domestic product (GDP) is earned through export or processing for export of renewable and nonrenewable natural resources. Forestry and products alone account for 3.5 percent of the nation's GDP. One in 16 workers rely on forestry for their livelihood. Commercial production from the Atlantic, Pacific, and freshwater fisheries totaled $3.1 billion in 1990. However, the collapse of the Atlantic cod fishery in 1993 dramatically reduced both employment and the value of the fish harvest. The Pacific salmon fishery has subsequently experienced similar sustainability problems associated with overcapacity and distribution of the catch. Agriculture and food processing contribute 4 percent to the country's GDP. Sixty percent of Canada's electricity is hydro generated, and the energy sector accounts for over 7 percent of Canada's GDP. The minerals industry is responsible for 4.3 percent of GDP and 2.1 percent of national employment (Canada, Statistics Canada, 1997a; Organization for Economic Cooperation and Development, 2000).

So, while Canada retains a heavy economic dependence on its natural resources, like other modern industrial economies, it has developed a diversified and increasingly knowledge-based industrial sector (Newton and Besley 2001). Throughout the 1970s and recessionary early 1980s, Canadian industry made marginal and grudging adjustments to its practices for environmental reasons. However, since the late 1980s, a number of industrial sectors have put in place programs and strategies to reduce the impact of their activity on the environment. Individual companies have led their sectors in introducing environmental audits, environmental management systems, environmental corporate reporting, and other techniques to improve corporate decision making on environment-economy issues. Industry associations are well organized and have been active in promoting environmental codes of practice and environmental management systems to their membership. As in other advanced industrial economies, Canadian industry has recognized that environmental protection can be both profitable and a job creator (Five Winds International, 2000), and the environmental industries sector has been among the fastest-growing segments of the economy. An extensive "culture of consultation" has emerged around the Canadian environmental policy process that has required industrialists to engage extensively with other members of the policy community, both governmental and nongovernmental.

International Organizations and Agreements

Throughout the 1980s, there was growing concern with global issues, such as ocean- and air-carried toxic pollution, ozone depletion, a loss of biological diversity, and climate change. The recognition of these biophysical changes was the result of advances in environmental science that shed light on both the nature of the problems and the directions of solutions. Scientific evidence, together with advances in telecommunications and the rise of trade and investment-liberalization agreements, have coalesced to globalize the environmental policy framework. Thus, in the environmental policy field, globalization is a biophysical phenomenon that is influenced by technical and economic factors (Parsons 2000; Toner and Conway 1996).

Canada has a strong internationalist tradition and has been an enthusiastic joiner of international organizations and agreements (Doern, Pal, and Tomlin 1996). This has been true for United Nations environment agencies (United Nations Environment Program, United Nations Economic Commission for Europe, United Nations Commission on Sustainable Development) and initiatives such as the World Commission on Environment and Development (Brundtland Commission). Canada also supports various international science and standards-based organizations (World Meteorological Organization, International Union for the Conservation of Nature, World Health Organization, and International Standards Organization). The International Joint Commission and the North American Commission on Environmental Cooperation are two key regional bodies (Hufbauer et al. 2000). Indeed, several of Canada's bilateral environmental problems with the United States have been addressed through formal negotiated agreements, some of which are part of broader multilateral agreements (Hoberg 1998).

In addition to the UN Conventions on Biodiversity and Climate Change, Canada has signed on to a number of international agreements, including the Declaration on Protection of the Arctic Environment, the Basel Convention on the Control of Transboundary Movements of Hazardous Wastes and Their Disposal, the Montreal Protocol on Substances That Deplete the Ozone Layer, and the Convention on Long-Range Transboundary Air Pollution. In 2001, Canada signed the Treaty on Persistent Organic Pollutants. This latest treaty will attempt to reduce pressures on the once "pristine" Arctic ecosystem, which is polluted by acid rain, soot, PCBs, pesticides, heavy metal radionuclides, and other contaminants that originate in industrialized regions of Europe, Asia, and North America, thousands of miles away (Jaimet 2001).

The Politics of Canadian Environmental Policy: Policy History, Policy Process, and Policy Performance in Five Key Areas

The Canadian Environmental Protection Act

CEPA was introduced in 1988 to modernize and coordinate the federal government's approach to managing toxic chemicals. It embraced several new ideas that had been percolating up through the environmental debate over the previous few years. CEPA was required because the existing piecemeal and uncoordinated regulatory effort had proven inadequate—partly as a result of problems with the existing federal legislation and partly as a result of the power exercised by industry and the provinces.

In the 1970s, total prohibitions against all emissions were replaced by legislation that allowed for "control regulatory regimes," which essentially recognized that some level of pollution was bound to occur from all human production and consumption. Thus "the realistic goal was to set standards and 'control' this pollution within technological limits and within the capabilities of the natural environment to cleanse the pollution from its system" (Doern and Conway 1994, 212). CEPA was intended to represent a conceptual departure from this "react-and-cure" approach to a proactive "anticipate-and-prevent" approach to managing toxic chemicals, and the legislation employed the "life-cycle" and "ecosystem" language.

The first wave of environmental consciousness in the late 1960s and early 1970s led to the creation of a generation of environmental protection legislation. The century-old Fisheries Act was amended in 1970. The Canada Water Act became law in 1970 and the Clean Air Act in 1971. As a result of serious pollution problems in the Great Lakes from sewage, industrial wastes, and farm runoff, the International Joint Commission was urging the Canadian and American governments to act, and the Canada–United States Great Lakes Water Quality Agreement was passed in 1972. Media-based regulations were passed under these various authorities. In 1975, the Environmental Contaminants Act was created to deal with the new generation of toxic chemicals coming into the market at an ever-increasing pace, and the Ocean Dumping Control Act was passed. Even with these various new legal authorities, enforcement of regulations throughout the 1970s was minimal and was constrained by EC's limited capacity. In fact, the department faced major reductions in its fiscal resources throughout the late 1970s (Whittington 1980; Brown 1992).

The vacillation of the various Pierre Trudeau–led Liberal governments on the enforcement of its environmental protection laws reflected a lack of political will to challenge (1) provincial claims that federal initiatives were a major infringement on their jurisdiction, and (2) threats by industry that the regulatory burden represented by environmental laws would cost jobs and investment. Often industry would be supported in its charges by provincial governments. The "lack of stomach" for major battles with the provinces or industry, combined with the massive reduction in political and financial support for EC, meant that little real progress was made under the Liberals in the last half of the 1970s or the first half of the 1980s. (For detailed analyses of the pre-CEPA period of toxics management policy, see Webb 1986; Schrecker 1984; Ilgen 1985.)

In fact, despite a series of high-profile national and international environmental disasters that heightened public anxiety about the human health implications of toxic chemicals, the Conservatives, under Brian Mulroney, undertook another major cut in EC's resources in 1985. However, in 1986, when the Conservative Environment Minister realized he had minimal capacity to respond to a "toxic blob" leaked from the Dow Chemical plant into the St. Clair River at Sarnia, Ontario, he responded by stating his intention to get tough with polluters through a new generation of legislation.

To take advantage of increasingly supportive public opinion and a proactive minister, senior managers in EC launched a major public consultation process to review the draft CEPA legislation. In a novel experiment, EC funded the participation of ENGO representatives. The formal involvement of environmentalists in the policy formulation process created a greater degree of balance in an arena normally dominated by assertive industry and provincial representatives and strengthened EC's hand. As a result of the consultations, several substantive changes were made to strengthen the legislation.

Both political will and bureaucratic commitment are crucial elements for the successful implementation of regulations (Pal 2001), and there were real doubts about the political, legal, scientific, and economic capacities of EC in the CEPA-era regulatory process (Doern 2000). EC had always faced shortages of these strengths, which often had left it at a considerable disadvantage in negotiations and political struggles with industry and other government departments and agencies (Brown 1992; Harrison 1999).

CEPA is a complex piece of legislation that is administered by both EC and Health Canada (HC). The legislation stipulated that Parliament review CEPA five years after its enactment. In preparation for the review by the Parliamentary Standing Committee on Environment and Sustainable Development, CEPA was subjected to a formal program evaluation in 1993. That evaluation gives the government some credit:

> The federal government has allocated considerable resources towards CEPA's implementation: about $360 million over the past five years. During this period, it has established a number of new management and co-ordination structures, recruited staff, released 10 Priority Substance List assessments (and completed the scientific work on most of the remaining 34), refined the assessment process, developed 12 new regulations, replaced 16 regulations developed under previous legislation, improved its compliance and enforcement capacity, liaised extensively with the provinces, published the State of the Environment Report, and made important changes in the application of the ocean dumping provisions. (Resource Futures International, 1993, 141)

The evaluation noted that "some evidence of direct health and environmental impacts can be discerned, including reductions in emissions of ozone-depleting substances, lead in gasoline, phosphates in detergents, and dioxin and furan levels in fish downstream of pulp mills in British Columbia. The publicity surrounding the introduction of CEPA may also have induced a number of provinces to strengthen their environmental legislation" (Resource Futures International, 1993, 141–142). The report, however, went on to identify a series of problems with both the substance and the administration of CEPA. Not surprisingly, the hearings held by the Standing Committee on revising CEPA were the venue for much conflicting advice from industry associations and environmental groups.

The Liberal Party's electoral manifesto for the 1993 election stated that toxic substances would be a central feature of Liberal environmental policy (Liberal Party of Canada, 1993, 66–67). The document went on to stress that the focus would be on reducing pollution at source through technological innovation and retooling production processes, rather than at the point of discharge. This focus on pollution prevention was increasingly perceived as the smart way to combine competitiveness and innovation with environmental protection. In the summer of 1995, the government released two policy documents: *Pollution Prevention: A Federal Strategy for Action* and *Toxic Substances Management Policy*. In 1996, Environment Canada announced that, starting with the 1997 reporting year, businesses would have to report pollution prevention measures through the National Pollutant Release Inventory.

The Standing Committee's Report on CEPA was released in June 1995 and called for a much more interventionist, regulation-oriented, enforcement-driven approach, including faster toxicity assessments, regulation of more substances, and greater enforcement powers, as well as a more prominent federal role in pollution management across the country (Canada, House of Commons Standing Committee on Environment and Sustainable Development, 1995a). It also endorsed the government's shift toward pollution prevention. The report was praised by environmentalists and condemned by industry (Toner 1996, 107). While industry groups supported the idea of pollution prevention, they rejected its implementation through regulation. They denounced the proposed approach as too bureaucratic and heavy-handed, which would lead to unprecedented intrusions into corporate operations. Both environmental and industry groups vigorously lobbied Environment Canada in an attempt to shape the government's response. Industry was assisted in its efforts by a rigorous intragovernmental lobby led by Natural Resources Canada and Industry Canada (Leiss 1996).

The CESD's 1999 report dedicated two chapters to assessing the government's management system for toxic substances under the original CEPA. The report was highly critical of the pace of assessment of toxic substances on the Priority Substances List, arguing that "by 2000, after ten years of effort, fewer than 70 substances will have been assessed by the CEPA Priority Substances Assessment Program" (Commissioner of the Environment and Sustainable Development, 1999a, 3–19). Resource constraints, after several years of budget cuts, along with potential new demands under the renewed CEPA, were presenting a major challenge to EC and HC, which were required to categorize an additional 23,000 substances on the Domestic Substances List.

Eighteen months after the Standing Committee report, in December 1996, the government introduced a bill in the House of Commons proposing a renewed CEPA. The revised act integrated the objectives of the 1995 toxic substance management policy by requiring the virtual elimination of the toxic substances found to be persistent and bioaccumulative. The government proposed changes to the toxicity assessment process to accelerate the evaluation of priority substances by modifying assessment criteria and by making greater use of the data developed in other OECD countries. EC inspectors would be given greater powers and citizens granted the right to sue for damage to the environment when the government fails to enforce the new law. The act did propose mandatory pollution prevention planning, though in a more flexible manner than the committee proposed.

The CEPA bill seemed to please no one. Industry groups charged that, in its haste to regulate more chemicals, the government was sidestepping science and consultation. They argued that mandatory pollution prevention measures and expanded regulatory powers, over areas such as fuel efficiency and international pollution, implied that the government was reverting to an old-style command-and-control approach, not giving sufficient credit to voluntary measures. Environmentalists were also critical, arguing that the new criteria for virtual elimination were so restrictive that few substances would actually be targeted. Thus, the debate over pollution prevention, which it was hoped would reconcile economic competitiveness and environmental protection, had resulted in a 1970s-style confrontational environmental politics in which environmentalists demanded tougher regulations, while industry sought less government involvement in their daily operations (Juillet and Toner 1997, 181–184). The government's middle-ground position pleased neither side, and the ongoing conflict both within and outside government meant the government was unable to pass the act prior to Parliament being dissolved for the May 1997 national election. The government promised, in its successful reelection bid, to reintroduce the CEPA legislation early in its next mandate (Liberal Party of Canada, 1997).

They did reintroduce the CEPA bill in May 1998. Some additional changes were made to the legislation, including the introduction of a CEPA Registry that would improve citizens' access to information, increased recognition of voluntary efforts by industry, and enhanced coordination with other levels of government (Environment Canada, 1998). The Standing Committee review of the CEPA bill was one of the most difficult, time-consuming, and acrimonious battles in Canadian parliamentary history. Literally hundreds of amendments to the CEPA bill were voted on, many sponsored by the government's own backbenchers on the committee. Indeed, when the amended bill was voted on in the House of Commons, three government backbench members of the committee, including the committee chair, voted against the amended act. This is a highly unusual occurrence in Canadian parliamentary practice and reveals just how divisive the debate was.

The renewed CEPA received Royal Assent in September 1999 and became law in April 2000, almost seven years after the initial review of original CEPA was launched. The new Environment Minister, who had inherited the bill partway through the committee process, announced that an additional $72 million would be provided to help implement the renewed CEPA. Time will tell if the new CEPA and the additional resources provided to it advance the government's management of toxic substances. Given the tone and content of the CESD's 1999 assessment, the government has its work cut out for it:

> Overall, we conclude that the federal government is not adequately managing the risks to the public that toxic substances and pesticides create. We are deeply concerned by the degree of conflict between departments, their inertia toward implementing government policies, and the lack of rigour in existing voluntary initiatives. We are also concerned . . . about inadequate tracking of releases of toxic substances and pesticides into the environment. We believe the federal government is not doing its part to effectively manage the risks posed by toxic substances (Commissioner of the Environment and Sustainable Development, 1999a, 4–31).

Institutionalization of Sustainable Development

The May 1986 World Commission on Environment and Development visit to Canada had an institutional impact, when the Canadian Council of Resource and Environment Ministers created the National Task Force on Environment and Economy. The task force brought together federal, provincial, and territorial ministers and senior members of the corporate, environmental, and academic communities. One of its most far-reaching recommendations was its proposal to institutionalize its innovative method of multisectoral collaboration by having each government establish a Round Table on Environment and Economy. The idea was to have a body of influential sectoral leaders reporting directly to the prime minister and the premiers. And, indeed, governments did adopt this recommendation in the heady days of the early 1990s, and

Table 3.1
Canada: A Statistical Profile

Land	
Total area (1,000 km^2)	9,215.00
Arable and permanent crop land (% of total) 1997	4.5%
Permanent grassland (% of total) 1997	2.9%
Forest and Woodland (% of total) 1997	45.30%
Other land (% of total) 1997	47.30%
Major protected areas (% of total) 1997	9.6%
Nitrogenous fertilizer use (t/km^2/arable land) 1997	4.10%
Population	
Total population (100,000 inhabitants) 1998	306.00
Growth rate % 1980–1998	24.40%
Population density (inhabitant/km^2) 1998	3.10
Gross domestic product	
Total GDP (billion U.S. dollars) 1998	$637.00
Per capita (1,000 U.S. dollars/capita) 1998	$20.80
GDP growth 1980–1998	55.30%
Current general government	
Revenue (%/GDP) 1997[b]	44.10%
Expenditure (%/GDP) 1997[b]	42.80%
Government employment (%/total) 1997[b]	18.90%
Energy	
Total supply (MTOE) 1997	238.00
% change 1980–1997	23.30%
Consumption (MTOE) 1998[c]	182.54
% change 1988–1998[c]	13.10%
Road-vehicle stock	
Total (10,000 vehicles) 1997	1,786.00
% change 1980–1997	35.20%
Per capita (vehicles/100 inhabitants)	59.00
Air profile and greenhouse gases	
CO_2 emissions (kg/1990 USD GDP) 1998[a]	0.72
CO_2 emissions (Tons/capita) 1998[a]	15.84
Chlorofluorocarbon CFC_{12} (parts/trillion by volume) 1992	503.00
$HCFC_{22}$ (CFC substitute—parts/trillion by volume) 1992	105.00
Halons—CF_4 (a pentafluorocarbon) 1992	70.00
Nitrogen oxide emissions (kg/capita) 1998[a]	68.00
Nitrogen oxide emissions (kg/USD GDP) 1998	3.40

Table 3.1
(continued)

NO emissions (million tons) 1980	1.959
NO emissions (million tons) 1990	2.106
Change (increase/millions tons) 1980–1990	0.147
Nonmethane volatile organic compounds (tons/year) 1992[c]	2.00–2.50
Sulfur oxide emissions (kg/capita) 1998[a]	90.00
Sulfur oxide emissions (kg/USD GDP) 1998	4.40
Pollution abatement and control	
Total expenditure (%/GDP) 1995	1.10%
Household expenditures excluded	

Sources: OECD, *OECD Environmental Data, Compendium 1999* (Paris: OECD, 1999), except as noted.
[a] OECD, The OECD Observer, *OECD in Figures* (Paris: OECD, June 2000).
[b] OECD, *National Accounts of OECD Countries* (Paris: OECD, 2000).
[c] IEA/OECD, *Energy Balances of OECD Countries, 1997–1998* (Paris: IEA/OECD, 2000).

round tables were created at the national, provincial, and municipal levels. By 2001, there had been some retrenchment in the use of round tables, particularly at the provincial level. However, the National Round Table on the Economy and Environment (NRTEE) has been granted its own statutory base and continues to work on a number of important sustainable development issues.

In December 1990, the Mulroney Conservative government introduced a five-year, $3 billion Green Plan (GP). It was an early effort by an OECD government to deal with environment-economy issues in a comprehensive manner. Indeed, one analyst has called it "arguably the 'mother' of green planning" (Dalal-Clayton 1996, 21). The GP was the sustainable development strategy the Canadian government took to UNCED in 1992. It began with a broad commitment to sustainable development, calling it no less than an effort at "planning for life." This commitment was then linked through the ecosystem approach to the natural environment and the human decisions and actions that impact on it. While much of the spending had to do with cleaning up past mistakes, there was also an emphasis on introducing industrial technologies and practices to promote pollution prevention and sustainable development. The second focus was directed at programs that would contribute to sustainable development by addressing normative principles that shape decision-making systems in government and society (Toner and Doern 1994).

The early bureaucratic drafts of the GP were actually much closer to a sustainable development strategy than the version that ultimately emerged from the cabinet process. Indeed, the drafts written in the autumn of 1989 identified the societal and economic decision-making systems as the "root cause" of environmental degradation. Those early drafts envisioned the GP as representing a turning point in the Canadian discourse by moving the conceptual basis of environmental policy away from resource management and environmental cleanup to pollution prevention and sustainable development. When the politicians on the Cabinet Committee on the Environment undertook their detailed review of the draft GP in the autumn of 1990, they imposed a traditional "distributive politics" template on the document by moving the expensive environmental cleanup programs to the front and burying the chapter on the need to change societal decision making in the back (Toner 1994). As a result, it looked less like a novel sustainable development strategy and more like just another environmental protection program. Even then, it eschewed greater reliance on the traditional regulatory approach (Hoberg and Harrison 1994).

In accordance with the UNCED Secretary General's Guidelines for the Preparation of National Reports, the federal government created a multistakeholder participatory process around the writing of Canada's Report for Rio. Specifically, a National Report Steering Committee was created to assist the government in the preparation of the report. Business and labor organizations, the provincial governments, the round tables, aboriginal organizations, and the NGO community participated in the Steering Committee under the general coordination of Environment Canada (Canada, Environment Canada, 1991). Each morning during the Earth Summit, the Canadian delegation, along with all other Canadians participating in the other Rio meetings, were invited to a briefing with Environment Minister Jean Charest. These sessions could have as many as 200 attendees and became dubbed the "Team Canada" briefings. The previous day's activities would be reviewed and Canada's response would be discussed. The current day's agenda would also be discussed (Canada, Supply and Services Canada, 1992). This provided Canadian NGO and industry representatives participating in the other events with special insight into developments at the official conference. This was an extraordinarily inclusive and open process for an international, intergovernmental meeting.

The initial plan was to use the GP as a follow-up to Rio, signaling the development of a GP II. This never happened, however, because the Conservatives lost interest in the environment and sustainable development policy area in their last year in government, which coincided with the year following Rio. While the Conservatives found the political will to launch the GP, they had difficulty sustaining their commitment since it was not based on any strong ideological or emotional foundation in their party. Their commitment was simply poll driven. As the economic recession deepened in the 1992–93 post-Rio period, environmental issues declined in the "top-of-mind" public opinion surveys and the Conservatives backtracked from their commitment to the GP without ever explicitly or publicly renouncing it.

At Rio, the Canadian government outlined a six-part "quick-start agenda" and challenged other countries to take immediate action. One of these agenda items included the development of a national report on plans and policies related to the conference's objectives. Agenda 21 encouraged countries to adopt national sustainable development strategies (NSDSs) as a central mechanism for implementing the actions and accords agreed to at the Earth Summit (Canada, Supply and Services Canada, 1992, 3).

There was pressure, however, to go beyond updating the GP to try to maintain the momentum generated by the "Team Canada" multistakeholder approach developed around the Rio process. In a speech to Parliament in November 1992, Charest proposed a national response to the commitments of Rio and to the challenge of sustainable development. Later that month, a national stakeholder meeting with representatives of over 40 sectors of Canadian society agreed to launch a "Projet de société." The Projet was to analyze Canadian responses to Rio and to draft a concept paper on sustainability planning. It was not intended to be a representative assembly, though by its conclusion in 1995, representatives of over 100 sectors of society had participated in its work (Projet de société, 1995). The Projet was more of a coalition of networks working together to generate a national strategy. As might be expected of such a broad-based initiative, the Projet had both organizational and conceptual difficulties. Tensions arose between participants attracted to developing strategic plans and those inclined to do more specific projects. The Projet also lost political momentum quickly. As government support slowly dissipated and as the difficulties of writing a document through an open-ended, volunteer-driven process became obvious, the NRTEE became increasingly involved in actually writing the document. The final draft of the document was published in May 1995, but by then there was no longer even an illusion of government support or involvement.

The Projet strove to meet Agenda 21's standard for NSDSs by integrating economic, environmental, and social objectives; by involving the widest possible participation; and by providing a thorough assessment of the current situation. While the broad and inclusive nature of the process was its great attraction, its unofficial—that is, nongovernmental—status proved, over time, to be its greatest weakness (Dalal-Clayton 1996, 106). The NRTEE withdrew its support for the Projet in early 1996 and it expired. Canada, to this day, has no NSDS, official or unofficial.

The Conservative government was defeated in October 1993, halfway through the GP's intended life span. The new Liberal government dedicated a chapter of its 1993 electoral manifesto to sustainable development and initially claimed it would not dump the GP simply because it was introduced by its partisan rival. However, it soon began to ignore the title "Green Plan" when discussing initiatives that had been developed under the Green Plan's authority or budget, and within a year or so after taking power, the Liberals eliminated all reference to it in official government documents.

In opposition, the Liberals had been highly critical of the Conservatives' GP, arguing that it did not go far enough in changing the decision-making system to institutionalize a sustainable development framework. Their electoral manifesto called for a "fundamental shift in values and public policy," arguing that "sustainable development—integrating economic with environmental goals—fits the Liberal tradition of social investment as sound economic policy" (Liberal Party of Canada, 1993, 63). Several dimensions of the Liberals' sustainable development program surpassed Conservative promises in the GP and challenged the bureaucratic forces within the federal departments, which, often successfully, had resisted the institutionalization of new sustainable development administrative practices and policy priorities under the Conservatives. After coming to power, the Liberals created, for the first time, a Parliamentary Standing Committee with Sustainable Development in its title. The Liberals' vision of aggressive activism was, however, blunted by the contact with the hard realities of Canadian politics in the 1990s, particularly fiscal reality.

The first Liberal budget, in February 1994, announced that a multistakeholder task force would be established to undertake a major campaign commitment to review federal taxes, grants, and subsidies in order to identify barriers and disincentives to sound environmental practices. A 40-member multistakeholder task force was established in July 1994, with a membership consisting of industry representatives, environmentalists, academics, and government officials. The task force reported in November 1994, proposing a series of immediate options for the 1995 budget and recommending market-based instruments that could be developed and implemented over a longer time frame (Task Force on Economic Instruments and Disincentives to Sound Environmental Practices, 1994). Further work on ecological fiscal reform was not undertaken for the rest of the 1990s, however, because the Department of Finance argued that the most egregious environmental barriers and disincentives were removed by extensive cuts to business subsidies that were undertaken as part of a broader deficit-fighting program introduced with the 1995 budget. As a consequence, the systematic review of the use and impact of fiscal instruments for sustainable development lay fallow for six years before reemerging in 2000–01, led by the NRTEE (National Round Table on the Environment and the Economy, 2000b).

In June 1995, the Liberals released a vision document called *A Guide to Green Government*. It was signed by the prime minister and all cabinet members and represented a government-wide commitment to sustain-able development. It argued that "achieving sustainable development requires an approach to public policy that is comprehensive, integrated, open and accountable. It should also embody a commitment to continuous improvement" (Canada, Supply and Services Canada, 1995b, 1). To institutionalize this approach, the Liberal government established a legal requirement that departments develop and implement Sustain-able Development Strategies (SDSs) and created the Office of Commissioner of the Environment and Sustainable Development (CESD) to monitor departments' performance. The CESD's first report was released in May 1997. It identified three key weaknesses in the federal government's management of sustainable development issues: (1) a gap between policy commitments and concrete action, (2) a lack of coordination among departments and across jurisdictions, and (3) the inadequate review of performance and provision of information to Parliament.

The CESD's second report, in May 1998, had chapters on environmental assessment, performance measurement for SDSs, advances in environmental accounting, development of a strategic approach to sustainable development, meeting Canada's international environmental commitments, climate-change implementation, and biodiversity-strategy implementation. The second report also assessed the first-generation SDSs. As the Brundtland Commission indicated, sustainable development is not a fixed state, but rather a process of change (World Commission on Environment and Development, 1987, 9). The SDSs were intended to encourage this process of change, challenging departments' entrenched normative assumptions about their roles and mandates by encouraging them to think about the sustainable development impacts of their policies and administrative practices. To ensure openness and accountability, the departments were required to seek stakeholders' views on departmental priorities for sustainable development and plans for achieving them. In his assessment, the CESD noted that "for the first time, we have a picture of how each department views sustainable development and the actions each department plans to take to promote it. Preparing their strategies has also raised awareness of sustainable development issues within departments" (Commissioner of the Environment and Sustainable Development, 1998, 15). However, he also identified two fundamental weaknesses in the strategies. First, almost all departments failed to set clear targets that could be used to judge whether the strategy is being successfully implemented. Second, many of the strategies were more a restatement of the status quo than a commitment to change.

The third report, in May 1999, assessed the first annual progress reports to Parliament on sustainable development submitted by departments. It concluded that "the links between the large number of actions that departments reported and the objectives set out in their strategies are frequently too abstract to provide insights about progress. As a result, beyond tallying the activities reported accomplished by departments, we are unable to conclude whether the strategies are on track or whether corrective action is required" (Commissioner of the Environment and Sustainable Development, 1999a, 5). The CESD again underscored the importance of departments putting in place management systems and training programs to build departmental capacity to get the implementation job done. As part of the CESD's commitment to help departments build sustainable development capacity, the report included chapters on how 17 North American and European organizations are building sustainable development considerations into the way they do business.

Not surprisingly, the CESD cited performance review as a serious problem both within departments and horizontally across departments. As part of the continuous-learning approach, he launched a research program to assist departments with the development of key indicators. Internationally, Canadians continue to contribute to the debate. The International Institute for Sustainable Development helped sponsor the development of the ten Bellagio (Italy) Principles for the "Practical Assessment of Progress Toward Sustainable Development" (International Institute for Sustainable Development, 1996). The North American Commission for Environmental Cooperation (NACEC) launched its own

State of the Environment (SoE) reporting activities. It recognized that, while the governments of Mexico, Canada, and the United States have published reports in the past, budget cuts in all three countries have reduced the resources available for SoE reporting. The NACEC reports provide important information on the North American region by analyzing the interactions among economic, social, and institutional change in the region and the environment.

Departments are legally required to submit new SDSs to Parliament every three years. The commissioner was quite critical of departmental efforts in the first round and, in December 1999, laid out his expectations for improvements in the quality of the second SDSs in a report titled *Moving Up the Learning Curve: The Second Generation of Sustainable Development Strategies* (Commissioner of the Environment and Sustainable Development, 1999b). He expected departments to assess their first strategies to determine what had been achieved, what had changed, and what needed to be done differently. Departments were expected to make these assessments available to participants in the multisectoral consultations that are part of the strategy development process. The CESD argued that departments had to strengthen the planning of strategies by drawing clear links between the department's activities and the significant impacts of those activities and priorities for action. Finally, he wanted to see departments accelerate the development of management systems that were needed to turn the strategies from talk into action, including evidence of the support, involvement, and commitment of senior management (Carley and Christie 2000), which had often been missing in the first generation of strategies (Commissioner of the Environment and Sustainable Development, 1999b, 5).

The CESD's fourth report, in May 2000, assessed the performance of Canadian governments in reducing smog. While the Canadian Council of Ministers of the Environment endorsed a plan to reduce ground-level ozone in 1990, with the goal of "fully resolving" the ozone problem by 2005, they never implemented the plan. Hence, the smog problem is far from resolved. While the report found that the federal government did most of what it said it would do, it failed to lead a national effort, including getting the provinces to do their part. The CESD's thesis was that, for governments in Canada, entering an agreement is too often seen as an end itself, rather than as a means of delivering results in an efficient and effective manner (Commissioner of the Environment and Sustainable Development, 2000, 5). The report was not all bad news with respect to partnerships, however, and identified several successful working relationships that combine credible reporting, effective accountability mechanisms, transparent processes, and protection of the public interest.

The report also raised questions about the government's performance in demonstrating leadership in the greening of government operations. Even though the government had been making commitments to green its operations since the 1990 GP, the CESD found that a decade later there was only rudimentary information available about the government's operations and their environmental consequences. In his view, it was unacceptable that the government does not have complete and accurate data on the annual costs of running its buildings and on the environmental impacts of its operations. A lack of common performance indicators and a lack of a leading organization within the federal government with responsibility to establish a coherent approach meant it was impossible to report fully on the government's performance. This led to the statement that "if the government cannot demonstrate that it can effectively manage and report on its operational performance, how can it deal with the larger, more complex task of integrating sustainable development into decision making for government policies and programs?" (Commissioner of the Environment and Sustainable Development, 2000, 2–23). The CESD stated that he expected the government to be able to produce a government-wide performance report on greening operations by 2002, five years after departments embarked on this journey with their first SDSs.

A common theme of the CESD's first four reports is that the support, involvement, and commitment of senior management is critical to changing attitudes within the department and moving up the learning curve. Another recurring theme is that most of the pressing issues cut across departmental mandates and political jurisdictions. Since effective coordination across these mandates is essential for meeting Canada's environmental and sustainable development objectives, he stated that he expected to see departments working together more effectively. For the first time, the CESD echoed the sentiment of many participants in the first round of consultations who argued that, while it was important to have a decentralized departmental strategy approach, it was also necessary to have a "Government of Canada" SD strategy, which would require departments to work together under a more coherent and common framework (Commissioner of the Environment and Sustainable Development, 1999b, 4).

While the government has not yet chosen to develop a Government of Canada SDS, an important initiative to engage senior officials with the strategy development process was the April 4, 2000, "Leaders Forum on Sustainable Development," which brought together 65 leaders from inside and outside government. The goal was to generate advice for the federal departments on the formulation of their second SDS and to provide feedback on eight horizontal issues facing all departments. Canada's senior public servant, Mel Cappe, Clerk of the Privy Council, noted, "We in government, generally, can benefit from all the help we can get in terms of rethinking or thinking further about SDSs, and how we can prepare to improve the role of government as steward of the economy and society and the environment" (National Round Table on the Environment and the Economy, 2000a).

Several coordinating committees of deputy ministers and assistant deputy ministers have been established to overcome the departmental divisions on horizontal issues, and it appears that senior managers have been much more deeply engaged in the development of the second generation of SDSs, which were tabled in the House of Commons on February 14, 2001 (Industry Canada, 2001; Natural Resources Canada, 2001).

The Canadian Environmental Assessment Act

In a country where the history of economic development has been synonymous with natural resource exploitation, it should come as no surprise that the assessment of major industrial projects for their impact on the environment has been the source of much controversy. Indeed, the history of the federal environmental assessment (EA) reflects a clash between the ideas of prevention, or ensuring that damaging environmental impacts are considered and minimized at the beginning of a development's decision-making process, and remediation, in which "react-and-cure" strategies attempt to repair damage after the project is underway or completed. The debate over how to change behavior to ensure that environmental considerations were taken into account when projects are planned separated those who believed it was necessary to codify behavior through legislation from those who believed it was possible to rely on less formal methods. This 30-year-long debate has been the intellectual crucible within which politicians, bureaucrats, and interest groups contested their varying visions of social change (Hazell 1999; Doern and Conway 1994). In the end, it was the courts that finally forced a legislative base for EA, in 1990. The U.S. experience with a legislated process in the 1970s cast a large shadow over the Canadian approach. More specifically, Canada's EA process was both inspired by and hobbled by the earlier American experience. (For broader assessments of U.S. influence on Canadian environmental policy, see Moffet 1994; Howlett 1994; Hoberg 1991.)

In 1969, the U.S. National Environmental Policy Act (NEPA) established the legal basis for environmental impact assessment (EIA), which was strengthened by the Council on Environmental Quality in 1978. Two features of the American approach influenced Canadian officials as they created the Canadian process. First, EIAs under NEPA were based on a legal requirement. Individuals and groups were able to use the requirement that government initiatives that have potentially "significant" environmental impacts must enter the EIA process to stall or prevent developments that they believed were environmentally damaging. This requirement led to a plethora of expensive and time-consuming court proceedings and generated tremendous resistance from other U.S. government agencies and successive administrations. Second, the scope of NEPA encompassed not only environmental effects but also related aesthetic, historic, cultural, economic, social, and health effects and applied to legislative proposals and major programs, as well as projects. Thus "[EC] officials involved in developing the Canadian process were chastened by these early US results. They were not at all certain about what an alternative process should be, but 'the US experience was far more significant to us than any experience we had here in Canada'" (Doern and Conway 1994, 193). Based on the American experience, senior

Canadian officials in the 1970s vigorously promoted a cautious, non-legislative approach.

The retreat from the idea of legislating EA was reflected in the December 1973 cabinet decision to create the process. It established that EA was to be based on self-assessment, with departments developing their own screening procedures and applying them to their own proposals. Crown corporations and key regulatory agencies would only be "invited" to participate, and cabinet policy would not be subject to EIA. The idea defended by the opponents of legislation was that ministerial discretion should be maximized and judicial and public involvement minimized.

Since the cabinet decision in 1973, there has been an ongoing struggle between EC and its environmental assessment agencies and the resource departments who sponsored, supported, or regulated various economic developments. For most of its life, EC lacked resources and political support in comparison with the economic departments. Given the constant pressure for economic development and the pro-development mandates of the economic departments and the provinces, EA, more often than not, appeared problematic (Doern and Conway 1994).

Because the constitutional authorities were contested and often unclear, EA has been the locus of a great deal of federal-provincial conflict. The search for "federal-provincial peace" is an ongoing theme in Canadian politics and public policy. The 1970s were a period of both economic stagnation and increased provincial assertiveness (Pratt and Richards 1979), and the Trudeau Liberals were searching for ways to accommodate provincial demands. Trudeau did not want EA to become "regulatory," a "burden" on the economy, or a federal-provincial "irritant." Thus, the federal-provincial conflict was minimized in the 1970s by the federal reluctance to be more assertive in exercising its responsibilities for EA. Harrison has characterized this as a period of "Federal Retreat." These artificially constrained conflicts would dramatically reemerge with the assertion of the federal role in the late 1980s (Harrison 1996, 81–161).

Despite continued hostility to strengthening EA from the economic departments, the environment minister was, in 1984, finally able to formalize the largely unwritten and vague EA process and have it codified as an administrative code of practice. In fact, the Trudeau cabinet passed an Environmental Assessment Review Process (EARP) Guidelines Order at its last meeting, in June 1984. In the subsequent election, the victorious Conservative platform had called for the legislation of EARP (Fenge and Smith 1986). While further discussions, consultations, and promises about introducing legislation followed over the next five years, nothing concrete happened until two precedent-setting court cases in 1989 and 1990. Environmental groups used the courts to challenge provincially sponsored dam projects in the provinces of Saskatchewan and Alberta. In both cases, the courts decided that because federal responsibilities for, among other things, fisheries, transboundary waters, and navigable water were involved, all federal departments and ministers were subject to the Guidelines Order. These court decisions gave the Guidelines Order the status of administrative law, which dramatically reduced opposition within the federal government to the development of legislation. The controversy surrounding these court cases also led to a dramatic escalation in federal-provincial conflict over EA as a process and raised the public profile of potential environmental harm resulting from major developments (Hood 1994). The provinces apparently "received encouragement from the business community, which was anxious about the climate of uncertainty surrounding new projects" (Harrison 1996, 137). The legal uncertainty, federal-provincial hostility, rising level of public concern, and new stature of the Guidelines Order combined to convince the Mulroney Conservative Government to introduce the Canadian Environmental Assessment Act in June 1990 (Gibson 1992).

Much of the political struggle over what to do about EA has taken place at the most senior levels of government, and ministers and cabinets have been prominent players. This was also the case with the CEAA. Indeed, between 1990 and 1995, five ministers were involved in the passage of the legislation. It took more than two years to get the legislation through Parliament, and even this did not guarantee smooth implementation. In fact, pressure from industry and provincial governments and continued bureaucratic resistance from federal economic development departments delayed its proclamation. Consequently, after the legislation received Royal Assent in June 1992, another two and a half years passed before CEAA was proclaimed in January 1995. The reason for this delay was that once the enabling legislation was passed, regulations still had to be developed to determine, among other things, which federal legislative and regulatory authorities would trigger an EA (Hazell 1999, 67–88).

While the government had established a multistakeholder committee of environmentalists and industry representatives to determine which laws and regulations would initiate an assessment, some industrial sectors were pressuring officials in the Industry department to work behind the scenes to minimize this list. So, even though CEAA had passed Parliament, some officials in the economic departments within the policy sector continued to play the role of internal opposition to EA legislation, articulating, within the state, the interests of some of the more strident industrial sectors (Toner 1994, 250).

In opposition, the Liberals had been very critical of the Conservatives for failing to proclaim CEAA. Indeed, in their 1993 electoral manifesto, they charged that

> the gap between rhetoric and action under Conservative rule has been most visible in the area of environmental assessment. All too often, the Conservatives have ignored the solid recommendations for environmental protection offered by public review panels. Under a Liberal government, the Canadian Environmental Assessment Act will be amended to shift decision-making powers to an independent Canadian Environmental Assessment Agency, subject to appeal to the Cabinet. (Liberal Party of Canada, 1993, 64)

To allow for the involvement of the Canadian public without financial or legal restriction, CEAA was amended to guarantee intervenor funding as an integral component of the assessment process. However, the independent status of the Canadian Environmental Assessment Agency never materialized. Instead, the Liberals shifted the final authorization for projects examined by review panels from the responsible ministers alone to the cabinet as a whole. Rather than simply consider appeals of decisions rendered by an independent agency, the cabinet would make the decisions in the first instance.

Given the history of EA within the bureaucracy, it is easy to understand the reluctance of the government to grant more decision powers to an independent agency. Ongoing, strong criticism from resource ministers about the bureaucratic delays created by the new assessment process made the allocation of greater power and independence to the agency politically impossible. By 1997, concerns about the dismissal of public review-panel recommendations under the Conservatives had been replaced by concerns about the reluctance of the Liberal government to establish public review panels in the first place (Juillet and Toner 1997, 184). Despite their disagreement with certain aspects of the legislation, environmental groups still had great expectations about CEAA. Eighteen months after its proclamation, one group complained that the act has become "nearly invisible" (Sierra Club of Canada, 1996, 11–12).

In his 1998 report, the CESD undertook a detailed review of the government's implementation of CEAA. While numerous shortcomings were identified, none were "universal or catastrophic" (Commissioner of the Environment and Sustainable Development, 1998, 6–18). The most salient weaknesses were related to the quality and quantity of screenings and to the usefulness of the information made available to Parliament and the public regarding the environmental effects that had been considered. While the act had been applied to around 5,000 projects a year, 99.7 percent were screenings, or the most limited form of EA. These are self-directed assessments undertaken by departments. Only 0.1 percent, or ten assessments, reached the most demanding stage of being submitted to independent-panel review.

There is a recognition that project-based EA, despite design improvements, focuses rather late in the development process once the project has already been planned. Reflecting the growing influence of the sustainable development orientation, there was an emerging recognition that EA could have even more influence on the development process if it was applied at the policy- and program-planning stage (Rosario Partidario and Clark 2000). This form of EA is known as *strategic environmental assessment* (SEA). In June 1990, as part of its effort to institutionalize sustainable development practices, the Conservative Government introduced a nonstatutory requirement for the SEA in the form of a cabinet directive. The Commissioner found that, overall, the implementation effort was woeful. This reflected a lack of effort by senior management who control departmental initiatives related to the cabinet process. The CESD was concerned that, without proper SEA of programs and policies, federal departments will be unable to implement the government's sustainable development objectives.

The Environmental Assessment Agency had come to a similar conclusion about the application of SEA and produced an updated cabinet directive on SEA in 1999. It is designed to strengthen the role of SEA and clarify the obligations of departments and agencies. While the updated directive is welcomed, some critics continue to ask if a nonlegal approach can work. While legalizing the cabinet process would be tricky, some argue that "a legal framework may be desirable from the perspective of improving compliance with SEA rules across government departments, as well as improving the quality of SEAs conducted under those rules" (Hazell and Benevides 2000, 66). While the rigorous application of SEA will be a challenge for departments, many have pledged to strengthen SEA as part of their 2001 SDSs.

In late 1999, the minister launched a review of CEAA, as was required by the legislation, as a first step toward amending the act. There was considerable debate during the lengthy consultation process as to whether CEAA was achieving its goals. As is the norm now in Canada, the public is invited to participate in consultation exercises through meetings and interactive Web sites. Twelve hundred Canadians took part in organized consultation meetings, and the Web site received over 40,000 visits. In March 2001, the minister tabled a bill proposing amendments to CEAA. The minister's three key goals for renewing the federal process include the following: providing a greater degree of certainty, predictability, and timeliness to all participants; enhancing the quality of assessments; and ensuring more meaningful public participation. Changes would include focusing on projects with a greater likelihood of adverse environmental effects, improving coordination among federal departments when several are involved in the same assessment, enhancing cooperation with other governments, strengthening the role of follow-up to ensure that sound environmental protection measures are in place for the project, improving the consideration of cumulative effects, and improving access to reports and other assessment information. The government committed $51 million over five years to implement these changes (Canadian Environmental Assessment Agency, 2001).

Climate Change

Climate change epitomizes the challenge of sustainable development. It is not just an environmental issue. It has important dimensions related to the economy, including trade and competitiveness considerations, as well as social aspects. It also raises concerns about equity between generations and among Canadian provinces and economic sectors, as well as nations and regions of the world. These considerations have to be taken into account in deciding the policy response (Commissioner of the Environment and Sustainable Development, 1998, 3–15). At Rio, the Conservatives committed Canada to stabilize CO_2 emissions at 1990 levels by the year 2000 in accordance with the terms of the Framework Convention. In their 1993 electoral Red Book, the Liberals "raised" the Conservatives by stating that they would work with provincial and urban governments to improve energy efficiency and increase the use of renewable energies, with the goal of cutting CO_2 emissions by 20 percent from 1988 levels by 2005.

Because of its size and location, Canada is projected to experience greater temperature changes than most regions of the world. As a coastal and northern country and as a renewable-resource producer in the forestry, agriculture, and fisheries sectors, Canada is vulnerable to damage from climate change (Canada, Environment Canada, 1997b). Fossil-fuel use poses a special problem: "In 1995, approximately 89 percent of total greenhouse-gas emissions in Canada were attributable to transportation and fossil-fuel production and consumption. Reducing fossil-fuel use in Canada is a challenge, due in part to our large landmass, cold climate, an increasing population, and a growing economy" (Canada, Environment Canada, 1997c, 5). Despite the increasing certainty of the science, the climate-change debate in Canada has been divisive along sectoral, regional, ideological, and policy lines. Canada has major oil-, gas-, and coal-producing industries in Western Canada, centered in the province of Alberta, and they have resisted action by challenging both the science and the economics of climate change (Dotto 1999).

The climate-change issue has bedeviled the Liberals throughout their tenure. An extensive multistakeholder consultative process during 1993 and 1994 failed to come to agreement on a national strategy. As a result,

Canada attended the first Conference of the Parties in Berlin, in May 1995, noting that it was on course to be 13 percent above the target 1990 emissions level by 2000, but still committed to meeting the stabilization goal by 2000. By the second conference in Geneva, in July 1996, Canadian ministers admitted that Canada's emissions had increased by over 9 percent since 1990 and that Canada would not meet the target (Russell 1997).

The first Liberal environment minister adopted an aggressive stance on climate change, which placed her in a confrontation with the oil and gas industry, the Conservative government of Alberta, and her cabinet colleague, the Minister of Natural Resources Canada (NRCan), who was the Alberta representative in the federal cabinet. The Minister of Natural Resources used her cabinet colleagues' anxiety about jobs and growth to gain acceptance of a cautious go-slow approach to climate change. As a result of these dynamics, open conflict between the federal environment and natural resource ministers and departments was a prominent feature of climate-change politics under the Liberals during their first mandate. This open warfare within the cabinet reflected the total lack of leadership or even engagement by Prime Minister Chrétien between 1993 and late 1997 (Toner 2000).

Despite efforts by some industry groups and their allies in the business press and the right-wing Reform Party to cast doubt on the science, there is really no longer any serious dispute on that front (Watson 2000). The real conflict has moved to the debate over the policy instruments required to reduce emissions. Alberta and industry groups advocate voluntary emission-reduction initiatives. Alberta is strongly opposed to the use of regulatory or fiscal instruments, such as a carbon tax, that could reduce consumption of its oil, gas, and coal resources, while environmentalists have taken increasingly strong positions in favor of both. To date, the fossil-fuel sector, the government of Alberta, and NRCan have created a formidable juggernaut against regulating economic activities to achieve reductions in greenhouse gases. They have been successful in promoting a National Action Plan on Climate Change that consists primarily of a voluntary challenge and registry (VCR) initiative. The VCR involves individual companies from the major greenhouse-gas-emitting industrial sectors (electrical utilities, manufacturing, energy, transportation and commercial, forestry, pulp and paper, agriculture, mining) submitting action plans detailing the measures they will take to reduce greenhouse-gas emissions. Over 700 companies responsible for more than 50 percent of Canada's total greenhouse-gas emissions have signed on to the Registry, but with little impact to date (Hornung and Bramley 2000).

EC, the scientific community, and environmentalists share a much greater sense of urgency and question the effectiveness of voluntary initiatives to fully meet the targets. The Climate Action Network in Canada, made up of more than 80 environmental and other nongovernmental organizations, argued for a portfolio of measures that combine voluntary, regulatory, and economic instruments. Economic instruments would include higher excise taxes on gasoline and a carbon charge (called an *atmospheric user charge*), which would be compensated for, in part, by a reduction in the federal Goods and Services (value-added) Tax. Existing regulatory and incentive initiatives would be strengthened to enhance fuel-economy standards for vehicles, encourage commercial and residential building retrofit measures, and increase industrial energy-use efficiency. Recognizing the importance of the "jobs agenda," environmentalists have emphasized the job-creation potential of a major energy-efficiency initiative. Indeed, they argue that climate change provides Canada with an employment-generating and technology-advancing opportunity (Climate Action Network, 1997). As part of its response, the federal government launched a multivolume scientific research project called the *Canada Country Study*, which was released in 1997. This study undertook the first nationally integrated assessment of the social, biological, and economic impacts of climate change in Canada (Canada, Environment Canada, 1997b).

Chrétien was a late entrant into the buildup to the third Conference of the Parties in Kyoto in December 1997. He finally engaged with the issue after discussions with European leaders during a trip to Europe in October 1997. He only entered the domestic debate, which had been tearing his cabinet apart, after President Bill Clinton released the U.S. position in the third week in October. Chrétien called from Europe, ordering his officials to develop a position that would "beat the Americans." His first formal statement was only four weeks before

Kyoto, on November 3. Canada was the last G7 country to state its position going into Kyoto.

A coalition of industrialists exerted extensive pressure in October–November, including intensive backroom lobbying of government officials and full-page newspaper advertisements. For instance, the Canadian coal industry employed a highly emotional, fear-based campaign characterizing action on Rio as "Ritual Suicide by Honour—Economic Suicide by Ignorance." Business research organizations released reports supporting the go-slow approach. Environmentalists countered with their own full-page ads, research studies, and opinion polls showing Canadians supported a serious effort on climate change and lamented the loss of international leadership by Canada. Just as Clinton had renounced energy taxes under pressure from the Senate and industrialists, Chrétien had rejected a carbon tax under pressure from Alberta. Yet in his November 1997 speech, he skewered those skeptical of climate-change science, comparing them to the tobacco industry and their allies who, for decades, denied that smoking causes lung cancer. He went on to argue that Canada should get international credit when Canadian natural gas exported to the United States reduces the use of coal and oil there and when Candu nuclear reactors exported to China reduce coal consumption there. He also dismissed the horror stories being spread by the fossil-fuel industry and its allies that climate-change action would cause massive reductions in the gross national product (GNP) and jobs, arguing that Canadian exports of environmental and energy technologies would benefit from a global consensus for action (Chrétien 1997).

In the Kyoto Protocol, Canada agreed to reduce emissions of greenhouse gases to 6 percent below 1990 levels by the commitment period 2008–2012. In his second report, the Commissioner audited the federal implementation effort for the period between Rio and Kyoto. The audit found a totally inadequate implementation effort, characterized by a lack of coordination among federal departments, a lack of federal-provincial cooperation, and an overall management structure that lacked accountability (Commissioner of the Environment and Sustainable Development, 1998). In a meeting immediately following Kyoto, federal, provincial, and territorial leaders agreed to renew the implementation effort. To that end, the federal government created a Federal Climate Change

Secretariat and committed $150 million over three years for further research and public education. The Secretariat managed another major consultation effort, built around the work of 16 multisectoral "issue tables," to examine the impacts, costs, and benefits of the Protocol, with the goal of determining both immediate and longer-term actions to provide sustained reductions in emissions.

In late 2000, this national consultation process produced a Climate Change Action Plan. A National Business Plan on Climate Change was agreed to by the federal government and all the provinces and territories except Ontario. The Business Plan adopts a "phased approach," with three-year business plans to be updated regularly. The plan continues to emphasize awareness programs, and further investments in science, modeling, and other knowledge-building tools, along with promotion of technology development and innovation. Recent federal budgets have allocated substantial financial resources to support further science and awareness-building programs. Yet the federal and provincial governments have continued to refuse to use fiscal tools, such as a carbon tax, and a comprehensive regulatory approach to reduce greenhouse-gas emissions.

A number of leading firms, both domestically and internationally, have begun to engage the climate-change issue seriously, and some have publicly taken on commitments to reduce their internal greenhouse-gas emissions by 10 percent below 1990 levels—a commitment that far exceeds goals set for Canada and other industrialized countries under the Kyoto Protocol. In the environmental policy area, firms tend to change their behavior in response to regulation or in anticipation of future regulation, or in response to fiscal tools, which influence the price of commodities. While the federal government has shown a commitment to reinvest significant funds in climate-change initiatives (as much as $500 million in the 2000 budget) and to encourage voluntary behavioral change, it has steadfastly refused, to date, to use fiscal or regulatory tools in support of emission reductions (Wilkinson 2000).

In early 2001, the Bush administration rejected the Kyoto Protocol and reduced commitments to energy-efficiency measures and renewable-energy sources introduced by the Clinton administration. The Canadian government has reiterated its commitment to the Kyoto target, but the

Bush administration's vision of a "continental energy deal," in which Canadian and Mexican oil and gas resources are developed for use by the United States, would undoubtedly increase Canada's already-rising greenhouse-gas emissions, because the production and transmission of energy is itself an energy-intensive activity. Hence, the prospects of Canada meeting the reduction goal by the commitment period of 2008–2012 are, at present, unimaginable.

Biodiversity

The United Nations Convention on the Conservation of Biological Diversity came into effect in December 1993. Canada was active in negotiating it and was the first industrialized country to ratify the Convention. *Creating Opportunity* included a general commitment to the protection of biodiversity and the goal of maintaining the GP commitment to complete the national parks system by 2000. It also stated that the federal government would work with the provinces to protect, in its natural state, a representative sample of each of the country's natural regions, amounting to 12 percent of Canada. To this end, Chrétien announced the creation of two new national parks in October 1996.

In 1995, the government introduced the Canadian Endangered Species Protection Act (CESPA) as a legislative proposal. A multistakeholder task force was established to provide advice on the drafting of the bill. CESPA would enshrine into law the existing administrative process used to list endangered species, with some modifications, and would make it illegal to harm or capture a member of a listed species or to damage its residence. Subsequent to its listing as endangered, the government would have one year to submit a plan stating how it intends to protect a species and assist in its recovery. One contentious point was that the final decision about the listing of species was to be made by the cabinet, on the advice of a scientific body created by the law. Environmentalists and scientists strongly opposed this change, arguing that cabinet decisions will be based on politics instead of science when there are economic pressures against listing a species. Thus, notwithstanding the consensus previously achieved by the multistakeholder task force, the government's bill quickly became the object of harsh criticism from virtually all sides (Juillet and Toner 1997).

Another main point of contention was the limited scope of the legislation. While overharvesting remains a threat to some species, the vast majority of endangered species are threatened by the destruction or contamination of their natural habitat through commercial activity (industrial pollution, forest clear-cutting, mining, and farming practices) or urban sprawl. The protection of habitat involves the regulation and voluntary modification of a wide array of activities on federal, provincial, and private lands. As such, it is a formidable challenge that requires extensive interjurisdictional cooperation.

While recognizing the requirement for such cooperation and acknowledging that Canada has a strong tradition of cooperation on wildlife management issues (Bocking 2001; Gauthier and Wiken 2001), many environmental groups have accused the federal government of refusing to fully occupy its jurisdiction regarding endangered species. They believed that the federal government possessed much more extensive jurisdiction than that proposed under CESPA, which only protected species found on federal lands (while they remain on federal lands). In addition, CESPA only applied to "federally managed" species (those covered by the Migratory Birds Convention or the Fisheries Act). The legislation also contained provisions enabling, but not requiring, the Environment Minister to make regulations for species crossing international borders. In total, CESPA would cover only about 40 percent of the species currently found on the national endangered species list.

The federal government must count on provincial cooperation to ensure adequate protection across the country. In October 1996, the national government and the provinces signed an agreement, the National Accord for the Protection of Species at Risk, committing the signatories to adopt complementary legislation. While five provinces developed legislation to protect endangered species, others were reluctant to do so. Despite polls that consistently show strong public support for federal legislation for the protection of endangered species, the Liberals have been unwilling to confront industry and landowners by establishing stringent habitat protection regulations. This federal reluctance to be confrontational reflects, in part, the tradition of multisectoral cooperation that has characterized the wildlife policy area (Organization for Economic Cooperation and Development, 1995). Industrial associations and federal resource departments have opposed more stringent habitat provisions. Indeed, they were already unhappy with the current limited provisions, arguing that the act should not apply to private lands and that producers should participate directly in the drafting of recovery plans. They also maintained that the draft act did not rely sufficiently on voluntary measures and that the regulations against the destruction of species' residences should not apply to habitat. Federal departments, like Agriculture and Agri-Food, Transport, and Fisheries and Oceans, also argued for limited regulatory measures, so that habitat protection would not hold up commercial activity. All this opposition from the provinces, the private sector, and federal departments slowed progress through the parliamentary process and, consequently, CESPA died when Parliament was dissolved for the June 1997 general election. Thus, the Liberals failed to secure the major legislative basis for fulfilling the UNCED commitments.

On the nonlegislative front, the government released the *Canadian Biodiversity Strategy: Canada's Response to the Convention on Biological Diversity*, in 1995, after a lengthy consultation process. The Biodiversity Strategy is a voluntary agreement among Canadian governments to improve citizens' understanding of the value of biological resources and to develop incentives and legislation to support their conservation and sustainable use. Internationally, one stated goal was to put in place a regime to share equitably the benefits that derive from the utilization of genetic resources between the developing countries that provide them and the industrial sectors that utilize them (Canada, Supply and Services Canada, 1995c). At the Second Conference to the Parties, in 1995, Canada competed with Switzerland, Kenya, and Spain for the right to be the seat of the Convention Office. Canada won, and the office was opened in Montreal in 1996. The federal government also created a Biodiversity Convention Office.

An important amendment to the Income Tax Act, in June 1996, encouraged Canadian landowners to participate in the preservation of biodiversity and wildlife habitats by donating ecologically sensitive land for conservation purposes. Fifteen additional national parks have been added to the national system since 1970, to bring the total number of parks in the system to 38. Provincial governments have also been adding additional protected areas. The problem with this positive news is that most of the newly protected spaces are in the North, while the ecosystems in the more heavily populated South continue to be stressed by growth pressures and inefficient models of urban development. In total, protected areas in Canada account for about 8 percent of the country, while the goal is 12. Only 24 of Canada's 31 natural regions are currently represented by national parks or park reserves (Canada, Public Works and Government Services, 1997d, 7). Drastic budget cuts to Parks Canada meant that the government missed its goal of completing the terrestrial park system by 2000 (Sierra Club of Canada, 2000, 12).

In his 1998 report, the CESD audited the federal government's effort at implementing the Biodiversity Strategy and gave the effort poor marks. Only two of eight federal departmental biodiversity implementation plans had been completed by early 1998. Even these lacked time frames, resource allocations, expected results, or performance indicators. The strategy requires an overall implementation plan that has targets and time frames, both to achieve national goals and to measure Canada's performance against its international commitments. The CESD concluded that even though the Convention has been in place for six years, progress in Canada has been slower than projected and deadlines have been missed. A growing hostility to the legislative protection of lands or species by the provinces in the face of industrial pressures, combined with significant budgetary cuts by both levels of government, led the CESD to conclude that the "present level of resources dedicated to biodiversity is inadequate for the magnitude of the task at hand" (Commissioner of the Environment and Sustainable Development, 1998, 4–11). Even the federal government's own *Report to the Biodiversity Convention* acknowledged that the eroding national scientific and monitoring capacity will slow Canada's implementation effort (Canada, Environment Canada, 1998).

In April 2000, the Liberals introduced the legislative sequel to CESPA, the Species At Risk Act (SARA). Critics of the federal legislation claimed that it did not make substantial improvements to the flaws of CESPA. SARA's proposal for mandatory habitat protection applied only to federal lands, there was still little protection for endangered species that migrated across international borders, and the decision to list a species as endangered was still left to the cabinet. Provisions for compensating landowners who suffer losses as a result of extraordinary impacts of regulations imposed under SARA were considered problematic, and the federal "habitat safety net," which would allow the federal government to intervene if territories or provinces were unwilling to protect habitat within their own jurisdictions, was considered politically unworkable (Sierra Club of Canada, 2000, 10–11). In any event, the legislation died when Parliament was dissolved for the December 2000 national general election. SARA was reintroduced in the new Parliament in February 2001 after the Liberals won reelection. It is largely identical to the previous bill. After seven years of debate and dialogue, the positions of most policy actors are locked in and predictable (Amos, Harrison, and Hoberg 2001). While ongoing programs allow the government to list species at risk and to develop recovery plans, the effort to create a legislative base for species and their habitat nears conclusion.

After a difficult period before the Parliamentary Standing on Environment and Sustainable Development in which several hundred amendments were proposed, SARA cleared the House of Commons in late 2001 and was sent to the Senate. Royal Assent is expected in the summer of 2002 with proclamation following within a year, after the development and approval of several sets of enabling regulations. In anticipation of a successful conclusion to the legislation, the 2000 budget provided $180 million in new money to support the bill.

Conclusion

The analysis presented in this chapter reinforces the thesis of this book that "institutions matter" and that "policy history is important." The chapter has applied an institutional approach in a historical context to review key developments in five areas of federal government environmental policy. The goal of the chapter, in keeping with the mission of the book, has been to provide a narrative account of policy change over time. The emphasis has been on developments in the 1980s and 1990s and the first few months of the twenty-first century, while reaching back into the 1970s for policy issues that got their launch after the first wave of environmental consciousness in the late 1960s. The main focus has been on policy process—describing how policy decisions came about, how they were implemented, and how they changed. The roles of industrial interests, interdepartmental and intergovernmental actors, and international institutions and agreements were highlighted in shaping policy choices and influencing policy performance.

As Kraft and Vig (1997, 19) argue, it is "difficult, both conceptually and empirically, to measure the success or failure of environmental policies," and as Jahn (1997, 3) notes, "estimating environmental performance has become a highly contested field." This is certainly true in the Canadian context. It is possible to find analyses that condemn Canadian environmental policy performance as among the worst in the OECD and vilify Canadians for "talking the talk of environmental protection, but not walking the walk" (Boyd 2001). But others claim that environmental quality is improving across the board, so don't worry be happy (Jones 1999).

Two recent, major international studies, one by Columbia and Yale universities in the United States and the other published by Oxford University Press in the United Kingdom, assessed Canada's environment and sustainable development performance against that of other countries. The Columbia and Yale Study compared 122 countries, with Canada ranked third overall, following Finland and Norway. The rankings were based on the Environmental Sustainability Index (ESI), which identifies 22 major factors that contribute to environmental sustainability, including urban air quality, overall public health, and environmental regulation. It measures these factors using 67 different variables, such as levels of sulfur dioxide in urban air, deaths from diseases associated with poor sanitation, and percentage of land protected from development. The study was released at the January 2001 Davos Economic Forum Annual Meeting of world leaders. The ESI is the most comprehensive global report comparing environmental conditions and environmental performance across nations. It was created to satisfy a critical need for substantive, impartial data for national and global environmental decision making. Comparable to the GDP, a central indicator for health of a country's economy, the ESI distills the health of a country's environment into a single number ranging from 0 to 100. This number represents a country's environmental success: its ability to sustain human life through food resources and a safe environment, to cope with environmental challenges, and to cooperate with other countries in the management and improvement of common environmental problems. The top country, Finland, registered 80.5 and the bottom country, Haiti, scored 24.7 (Columbia University, 2001).

The Oxford University Press study is one of the first major comparative studies of sustainable development implementation in high-consumption societies. It evaluated the performance of a number of OECD countries and ranked Canada at the top of the middle category just behind the leading group, which included the Netherlands, Norway, and Sweden. The study had the following to say about Canada's overall performance:

> Over the time-frame covered here the Canadian government deployed the most systematic response, and would appear to lie closest to the group of "enthusiasts." Canada's innovative Roundtables on the environment and economy were stimulated by the Brundtland report, and the Green Plan represented an early ambitious attempt to tackle environmental issues in a more comprehensive manner. Canada also played a pusher role in the UNCED process, helping to secure US acceptance of the climate change convention at Rio. Relatively inclusive forms of participation have also been associated with the Canadian profile, and the establishment of the Parliamentary Commissioner for the Environment and Sustainable Development and the process of preparing and reviewing departmental strategies represents a unique attempt to integrate sustainable development into the work of the government as a whole.
>
> Yet there has also been much inconsistency in the Canadian experience. Both the Green Plan and the Project de société were, for example, seriously compromised. . . . After Rio, Canada adopted a "wait and see" attitude on climate change, and there has been little movement on ecological fiscal reform. Legislation on environmental impact assessment and species protection has remained stalled or ineffectual, and budget trimming in the mid-1990s weakened environmental monitoring. (Lafferty and Meadowcroft 2000, 416–418)

Interestingly, one of the institutional innovations whose birth is itself a key outcome of the policy process under review, the Commissioner of the Environment and Sustainable Development, has developed a well-respected capacity over the past five years for evaluating federal environmental policies and programs. We have used the evaluations offered in the CESD's annual reports to help assess the policy performance in all five of the areas analyzed above. The 1997 Report to Parliament offered a broader observation on the issue of assessing policy performance:

> Although progress has been made in a number of areas, it has not been uniform. Many environmental and sustainable development issues are, by their very nature, difficult to manage. They present governments with significant challenges. They are often scientifically complex, involve long time frames and do not fit neatly within a single department's or government's mandate or jurisdiction. The global nature of environmental issues has also increased the complexity of problem solving. It is no longer enough to focus on environmental problems in our backyard, although that remains important.

Notwithstanding these complicating factors, the CESD went on to argue, first, that the federal government's performance in managing individual environmental policy issues is often characterized by an "implementation gap" in which performance falls short of its stated objectives, reflecting the failure to translate policy direction into effective action. Second, there tends to be "a lack of co-ordination and integration" on horizontal issues that cut across departmental mandates and political jurisdictions. This broad sharing of responsibility and lack of coordination often leads to the question of "Who's minding the store?" (Commissioner of the Environment and Sustainable Development, 1997, 10–11).

The history of Canadian environmental policy analyzed here supports the salience of the CESD's "implementation gap" and "lack of coordination" critique. Policy choices about the substantive policy instruments to be employed and policy processes to be applied in the CEPA, CEAA, climate-change, and biodiversity cases reflected the often-conflicting interests of different government departments, provincial governments, and industrial interests. As the CESD noted, in each of the five policy areas, international agreements were important drivers of policy change. Multilateral agreements focused on the management of a number of different types of contaminants have helped shape the policy choices involved in the evolution of CEPA. In the cases of both CEPA and CEAA, it was the American experience with toxics management and environmental assessment that motivated federal engagement with the issues and shaped distinctive Canadian policy choices. The WCED report, *Our Common Future*, triggered a series of domestic efforts by both the Conservatives and the Liberals to implement policy initiatives to institutionalize sustainable development—from the Green Plan to the SDSs. The UN Biodiversity and Climate Change conventions have been primary drivers of domestic policy choices and changes to existing approaches.

When it was motivated to act, the federal government employed its political resources—leadership, ideas, money, legislation, and scientific knowledge—to overcome internal as well as provincial opposition to its plans. With varying degrees of success, the federal government strengthened its capacity to intervene in the Canadian economy and society to support environmental and sustainable development objectives. Yet, in these five cases, the federal government was often internally divided with respect to policy choice and instrument selection. Provincial governments exercised their institutional strengths and were invariably opposed to the extension of federal power and capacity. Such pressures often sapped the political will required to see initiatives through the implementation stage, resulting in the "implementation gap." Moreover, business and ENGO interests exercised what power they could, either through multistakeholder negotiations, closed-door bargaining, or litigation, to influence policy choices and to shape policy outcomes.

Perhaps the best that can be said is that, at the federal level, Canada continues to lurch along with some advances and some setbacks, far from an ideal state, but with some renewed momentum due to the mandated requirement that federal departments develop and implement sustainable development strategies. The second generation of strategies is expected to be much stronger than the first, and the federal government has developed cross-departmental processes to address eight horizontal issues. There is no question that during the latter half of the 1990s federal and provincial governments were preoccupied by economic and fiscal matters and that environmental issues did not receive the attention they deserved. Hence, even essential activities, such as government-sponsored monitoring of environmental changes, have significantly decreased. However, Canadians' complacency over the quality of their environment has been shaken by drinking-water contamination tragedies in Ontario and Saskatchewan, as well as by increasing understanding of the health impacts of urban air quality. As the NRTEE has argued, four major challenges (accumulation of toxic contaminants, loss of natural spaces, deterioration of urban environments, and global economic changes) "will jeopardise our status within the decade, unless we begin mapping a response now" (National Round Table on the Environment and the Economy, 2001, 1).

Arguably, there is a growing appreciation that Canada's vaunted quality of life and economic prosperity are inherently linked to the quality of the environment. Hence, caring for the environment is one of the best investments Canadians can make to ensure a healthy, sustainable economy. The salience of the issue is rising in the public opinion polls again, and there are signs that the federal government is willing to reinvest in environmental initiatives in the postdeficit period. Two budgetary initiatives in 2000 committed over $1.4 billion over five years. Many of these initiatives were reinforced in the Liberals' 2000 electoral manifesto (Liberal Party of Canada, 2000). Proponents of such initiatives will have to argue their case with the Finance department, among others, because the policy history shows contending voices will emerge, with respect to both policy choices and instrument selection. All of the policy actors identified in this chapter will be part of future policy debates, driven by their contending interpretations of past policy performance. Indeed, virtually no major, or even routine, initiative or alteration to a federal program, policy, or regulation can even be contemplated without formally integrating policy "stakeholders" through some consultative mechanism. Hence, the memories and assessment of past policy choices and performance will continue to shape future policy choices.

Notes

1. Canada has a population of 30 million and a huge landmass of just under 10 million square kilometers. The country borders on three oceans and encompasses fifteen ecozones and six time zones. Canada's history was shaped by the exploitation of natural resources, and resource development continues to be a major feature of the economic base. Forest covers almost half of Canada and represents 10 percent of the world's total forest cover. Twenty-one percent of the world trade in forest products originates in Canada. It possesses 9 billion barrels of proven oil reserves and 95 trillion cubic feet of natural gas reserves. Over 30 percent of the world's nickel, 8 percent of its iron ore, and 20 percent of its zinc are produced in Canada. Canada's coastal fishing zones have traditionally been among the most productive in the world, and Canada possesses 20 percent of the world's freshwater reserves.

2. Several of the most controversial issues in Canadian environmental policy are under provincial jurisdiction in the resource (forestry, agriculture, mining, oil, and gas) and land-use management sectors. The municipal level of

government is also an important player in Canadian environmental policy. A number of Canadian cities have larger populations than several of the provinces and they have to deal with issues related to waste management, urban sprawl/land-use planning, water and sewage, urban transportation, and air quality. Many are struggling to address aging infrastructure as well. However, because the municipal governments are constitutionally under provincial jurisdiction, they have little direct interaction with the federal government, though the federal government is able to use shared cost funding to support municipal initiatives.

References

Amos, W., Harrison, K., and Hoberg, G. (2001). In search of a minimum winning coalition: The politics of species—at risk legislation in Canada. In K. Beazley and R. Boardman, eds., *Politics of the wild: Canada and endangered species* (pp. 137–166). Toronto: Oxford University Press.

Bocking, S. (2001). The politics of endangered species: A historical perspective. In K. Beazley and R. Boardman, eds., *Politics of the wild: Canada and endangered species* (pp. 117–136). Toronto: Oxford University Press.

Boyd, D. (2001). *Canada vs. the OECD: An environmental comparison.* Victoria: University of Victoria.

Brown, P. (1992). Organizational design as policy instrument: Environment Canada in the Canadian bureaucracy. In R. Boardman, ed., *Canadian environmental policy: Ecosystems, politics and process* (pp. 24–42). Toronto: Oxford University Press.

Canada, Supply and Services Canada. (1990). *Canada's Green Plan—for a healthy environment.* Ottawa: Supply and Services Canada.

Canada, Environment Canada. (1991). *Canada's National Report to UNCED.* Ottawa: Environment Canada.

Canada, Supply and Services Canada. (1992). *Canada's Green Plan and the Earth Summit.* Ottawa: Supply and Services Canada.

Canada, House of Commons Standing Committee on Environment and Sustainable Development. (1995a). *It's about our health!: Towards pollution prevention.* Ottawa: House of Commons Standing Committee on Environment and Sustainable Development.

Canada, Supply and Services Canada. (1995b). *A guide to Green government.* Ottawa: Supply and Services Canada.

Canada, Supply and Services Canada. (1995c). *Canadian biodiversity strategy: Canada's response to the Convention on Biological Diversity.* Ottawa: Supply and Services Canada.

Canada, Statistics Canada. (1997a). *Canada Year Book 1997.* Ottawa: Statistics Canada.

Canada, Environment Canada. (1997b). *The Canada Country Study: Climate impacts and adaptation.* Ottawa: Environment Canada.

Canada, Environment Canada. (1997c). *Building momentum: Sustainable development in Canada.* Ottawa: Environment Canada.

Canada, Public Works and Government Services. (1997d). *Securing our heritage: The sustainable development strategy of the Department of Canadian Heritage.* Ottawa: Public Works and Government Services.

Canada, Environment Canada. (1998). *Caring for Canada's biodiversity: Canada's First National Report to the Conference of the Parties to the Convention on Biological Diversity.* Ottawa: Environment Canada.

Canadian Environmental Assessment Agency. (2001, March 20). Minister Anderson tables bill proposing changes to strengthen environmental assessment for Canadians. Ottawa: Canadian Environmental Assessment Agency.

Carley, M., and Christie, I. (2000). *Managing sustainable development.* London: Earthscan.

Chrétien, J. (1997, November 3). *Notes for an address by Prime Minister Jean Chrétien on the subject of global warming.* Ottawa.

Climate Action Network. (1997). *Rational energy program: Analysis of the impact of national measures to the year 2010.* Ottawa: Climate Action Network.

Columbia University. (2001). (Available at http://www.ciesin.columbia.edu/indicators/ESI)

Commissioner of the Environment and Sustainable Development. (1997). *Report of the Commissioner of the Environment and Sustainable Development to the House of Commons.* Ottawa: Commissioner of the Environment and Sustainable Development.

Commissioner of the Environment and Sustainable Development. (1998). *Report of the Commissioner of the Environment and Sustainable Development to the House of Commons.* Ottawa: Commissioner of the Environment and Sustainable Development.

Commissioner of the Environment and Sustainable Development. (1999a). *Report of the Commissioner of the Environment and Sustainable Development to the House of Commons.* Ottawa: Commissioner of the Environment and Sustainable Development.

Commissioner of the Environment and Sustainable Development. (1999b). *Moving Up the Learning Curve: The Second Generation of Sustainable Development Strategies.* Ottawa: Commissioner of the Environment and Sustainable Development.

Commissioner of the Environment and Sustainable Development. (2000). *Report of the Commissioner of Environment and Sustainable Development to the House of Commons.* Ottawa: Commissioner of the Environment and Sustainable Development.

Dalal-Clayton, B. (1996). *Getting to grips with green plans: National level experience in industrialized countries*. London: Earthscan.

Doern, G. B. (2000). Patient science versus science on demand: The stretching of green science at Environment Canada. In G. B. Doern and T. Reed, eds., *Risky business: Canada's changing science-based policy and regulatory regime* (pp. 286–306). Toronto: University of Toronto Press.

Doern, G. B., and Conway, T. (1994). *The greening of Canada: Federal institutions and decisions*. Toronto: University of Toronto Press.

Doern, G. B., Pal, L. A., and Tomlin, B. W. (1996). *Border crossings: The internationalization of Canadian public policy*. Toronto: Oxford University Press.

Dotto, L. (1999). *Storm warning: Gambling with the climate of our planet*. Toronto: Doubleday.

Environment Canada. (1998). *Strengthening environmental health protection in Canada*. Ottawa: Environment Canada.

Fenge, T., and Smith, G. (1986). Reforming the federal environmental assessment and review process. *Canadian Public Policy, 12*, 596–605.

Five Winds International. (2000). *The role of eco-efficiency: Global challenges and opportunities in the 21st century*. Ottawa: Five Winds International.

Gauthier, D., and Wiken, E. (2001). Avoiding the endangerment of species: The importance of habitats and ecosystems. In K. Beazley and R. Boardman, eds., *Politics of the wild: Canada and endangered species* (pp. 49–74). Toronto: Oxford University Press.

Gibson, R. B. (1992). The new Canadian Environmental Assessment Act: Possible responses to its main deficiencies. *Journal of Environmental Law and Policy*, 2, 223–225.

Harrison, K. (1996). *Passing the buck: Federalism and Canadian environmental policy*. Vancouver: University of British Columbia Press.

Harrison, K. (1999). Retreat from regulation: The evolution of the Canadian environmental regulatory regime. In G. B. Doern, ed., *Changing the rules: Canadian regulatory regimes and institutions* (pp. 112–142). Toronto: University of Toronto Press.

Hazell, S. (1999). *CANADA v. the environment: Federal environmental assessment 1984–98*. Toronto: Canadian Environmental Defence Fund.

Hazell, S., and Benevides, H. (2000). Towards a legal framework for SEA in Canada. In M. Rosario Partidario and R. Clark, eds., *Perspectives on strategic environmental assessment*. New York: Lewis.

Hessing, M., and Howlett, M. (1997). *Canadian natural resource and environmental policy*. Vancouver: University of British Columbia Press.

Hoberg, G. (1991). Sleeping with an elephant: The American influence on Canadian environmental regulation. *Journal of Public Policy, 11*, 107–131.

Hoberg, G. (1998). North American environmental regulation. In G. B. Doern, ed., *Changing regulatory institutions in Britain and North America* (pp. 305–327). Toronto: University of Toronto Press.

Hoberg, G., and Harrison, K. (1994). It isn't easy being green: The politics of Canada's Green Plan. *Canadian Public Policy*, *20*, 119–137.

Hood, G. (1994). *Against the flow: Rafferty-Alameda and the politics of the environment*. Saskatoon: Fifth House Publishers.

Hornung, R., and Bramley, M. (2000). *Five years of failure: Federal and provincial government inaction on climate change during a period of rising industrial emissions*. Drayton Valley, Alberta: Pembina Institute.

Howlett, M. (1994). The judicialization of Canadian environmental policy, 1980–1990: A test of the Canada–United States convergence thesis. *Canadian Journal of Political Science*, *27*(1), 99–127.

Hufbauer, G., Estey, D., Orejas, D., Rubio, L., and Schott J. (2000). *NAFTA and the environment: Seven years later*. Washington, DC: Institute for International Economics.

Ilgen, T. (1985). Between Europe and America, Ottawa and the Provinces: Regulating toxic substances in Canada. *Canadian Public Policy*, *11*, 578–590.

Industry Canada. (2000). *Sustainable Development Strategy 2000–2003*. Ottawa: Industry Canada.

International Institute for Sustainable Development. (1996). *The Bellagio Principles*. Winnipeg: International Institute for Sustainable Development.

Jahn, D. (1997, August). Environmental performance and policy regimes: Explaining variations in 18 OECD-countries. Paper presented at the IPSA World Congress, Seoul, South Korea.

Jaimet, K. (2001, May 24). Canada signs, ratifies treaty to ban "Dirty Dozen." *Ottawa Citizen*, p. A4.

Jones, L. (1999). *Crying wolf: Public policy on endangered species*. Vancouver: Fraser Institute.

Juillet, L., and Toner, G. (1997). From great leaps to baby steps: Environment and sustainable development policy under the Liberals. In G. Swimmer, ed., *How Ottawa spends 1997–98: Seeing red—A Liberal report card* (pp. 179–210). Ottawa: Carleton University Press.

Kraft, M. E., and Vig, N. J. (1997). Environmental policy from the 1970s to the 1990s: An overview. In N. J. Vig and M. E. Kraft, eds., *Environmental policy in the 1990s* (pp. 1–30). Washington, DC: Congressional Quarterly Press.

Lafferty, W., and Meadowcroft J. (2000). Patterns of governmental engagement. In W. Lafferty and J. Meadowcroft, eds., *Implementing sustainable development: Strategies and initiatives in high consumption societies* (pp. 337–421). Oxford: Oxford University Press.

Leiss, W. (1996). Governance and the environment. In T. Courchene, ed., *Policy frameworks for a knowledge economy*. Kingston: John Deutsch Institute.

Liberal Party of Canada. (1993). *Creating opportunity: The Liberal Plan for Canada*. Ottawa: Liberal Party of Canada.

Liberal Party of Canada. (1997). *Securing our future together: Preparing Canada for the 21st Century*. Ottawa: Liberal Party of Canada.

Liberal Party of Canada. (2000). *Opportunity for all: The Liberal Plan for the future of Canada*. Ottawa: Liberal Party of Canada.

Moffet, J. (1994). Judicial review and environmental policy: Lessons for Canada from the United States. *Canadian Public Administration*, *37*(1), 140–167.

National Round Table on the Environment and the Economy. (2000a). *Leaders' Forum on Sustainable Development*. Ottawa: National Round Table on the Environment and the Economy.

National Round Table on the Environment and the Economy. (2000b). *NRTEE Economic Instruments Program: Ecological fiscal reform*. Ottawa: National Round Table on the Environment and the Economy.

National Round Table on the Environment and the Economy. (Spring 2001). Achieving a balance: NRTEE identifies new environmental challenges. *In Review*. Ottawa: National Round Table on the Environment and the Economy.

Natural Resources Canada. (2001). *Sustainable development strategy: Now and for the future*. Ottawa: Natural Resources Canada.

Newton, K., and Besley, J. (2001). Developing sustainability in the KBE: Prospects and potential. Paper presented to the CRUISE Conference, Building Canadian Capacity: Sustainable Production and the Knowledge Economy, Ottawa.

Organization for Economic Cooperation and Development. (2000). *OECD Economic Surveys 1999–2000: Canada*. Paris: Organization for Economic Cooperation and Development.

Pal, L. (2000). *Beyond policy analysis: Public issue management in turbulent times*. 2nd ed. Toronto: Nelson.

Parsons, E. A. (2000). Environmental trends and environmental governance in Canada. *Canadian Public Policy*, *26*(2), S123–S143.

Pratt, L., and Richards, J. (1979). *Prairie capitalism: Power and influence in the New West*. Toronto: McClelland and Stewart.

Projet de société. (1995). *Planning for a sustainable future: Canadian choices for transitions to sustainability*. Ottawa: Supply and Services Canada.

Resource Futures International. (1993). *Evaluation of the Canadian Environmental Protection Act*. Ottawa: Resource Futures International.

Rosario Partidario, M., and Clark, R. (2000). *Perspectives on strategic environmental assessment*. New York: Lewis.

Russell, D. (1997). *Keeping Canada competitive: Comparing Canada's climate change performance to other countries.* Vancouver: David Suzuki Foundation.

Schrecker, T. F. (1984). *The political economy of environmental hazards.* Ottawa: Law Reform Commission of Canada.

Sierra Club of Canada. (1996). *1996 Rio Report Card: Report on commitments made by federal and provincial governments at the United Nations Conference on Environment and Development.* Ottawa: Sierra Club of Canada.

Sierra Club of Canada. (2000). *Eighth Annual Rio Report Card 2000: Grading the federal government and the provinces on their environmental commitments.* Ottawa: Sierra Club of Canada.

Skogstad, G., and Kopas, P. (1992). Environmental policy in a federal system: Ottawa and the Provinces. In R. Boardman, ed., *Canadian environmental policy: Ecosystems, politics and process* (pp. 43–59). Toronto: Oxford University Press.

Task Force on Economic Instruments and Disincentives to Sound Environmental Practices. (1994). *Final Report.* Ottawa: Task Force on Economic Instruments and Disincentives to Sound Environmental Practices.

Toner, G. (1994). The Green Plan: From great expectations to eco-backtracking . . . to revitalization? In S. D. Phillips, ed., *How Ottawa spends 1994–95: Making change* (pp. 229–260). Ottawa: Carleton University Press.

Toner, G. (1996). Environment Canada's continuing roller coaster ride. In G. Swimmer, ed., *How Ottawa spends 1996–97: Life under the knife* (pp. 99–132). Ottawa: Carleton University Press.

Toner, G. (2000). Canada: From early frontrunner to plodding anchorman. In W. Lafferty and J. Meadowcroft, eds., *Implementing sustainable development: Strategies and initiatives in high consumption societies* (pp. 53–84). Oxford: Oxford University Press.

Toner, G., and Conway, T. (1996). Environmental policy. In G. B. Doern, L. A. Pal, and B. W. Tomlin, eds., *Border crossings: The internationalization of Canadian public policy* (pp. 108–144). Toronto: Oxford University Press.

Toner, G., and Doern, G. B. (1994). Five imperatives in the formulation and implementation of Green Plans: The Canadian case. *Environmental Politics, 3*(3), 395–420.

Vanderzwagg, D., and Duncan, L. (1992). Canada and environmental protection: Confident political faces, uncertain legal hands. In R. Boardman, ed., *Canadian environmental policy: Ecosystems, politics and process* (pp. 3–23). Toronto: Oxford University Press.

Watson, R. (2000). Presentation of the Intergovernmental Panel on Climate Change at the Sixth Conference of the Parties to the United Nations Framework Convention on Climate Change. (Available at www.unfcc.de/)

Webb, K. (1986). *Pollution control in Canada: The regulatory approach in the 1980s.* Ottawa: Law Reform Commission of Canada.

Whittington, M. (1980). The Department of the Environment. In G. B. Doern, ed., *Spending tax dollars: Federal expenditures 1980–81* (pp. 99–118). Ottawa: Carleton University Press.

Wilkinson, C. (2000). *Negotiating the climate: Canada and international politics of global warming*. Vancouver: David Suzuki Foundation.

World Commission on Environment and Development. (1987). *Our common future*. Oxford: Oxford University Press.

4

Environmental Policy in Britain

John McCormick

The British approach to environmental policy is notable for its contradictions. Britain has one of the oldest bodies of environmental law in the world, was one of the first countries to establish government agencies to address environmental issues, has one of the biggest and most professional environmental lobbies in the world, and has a populace with high levels of environmental awareness. Yet central government in Britain has responded to environmental problems only reluctantly and haphazardly, and recent changes in the structure of environmental policy institutions—rather than clarifying and streamlining the policy process—have perpetuated a policymaking system in which coordination and direction are both deficient.

Britain passed what may have been the world's first piece of antipollution legislation in 1273, created the first government environmental agency in 1863 (the Alkali Inspectorate), saw the creation of the first private environmental group in 1865, introduced a comprehensive system for managing land use with the Town and Country Planning Act of 1947, passed the first comprehensive air pollution control act in 1956, and created the first cabinet-level environment department in 1970. The popularity of natural history as a pastime has combined with the premium placed by Britons on wildlife, city parks, country estates, gardens, and the often spectacular beauty of the countryside to produce many great naturalists, including Gilbert White, Alfred Russel Wallace, Charles Darwin, Julian Huxley, and Peter Scott.

On the other hand, the environment—by almost every measure—has long been a relatively minor issue on the British political agenda, driven more by the concerns of successive governments to appease vested interests (notably industry, farmers, and rural landowners) than by a desire to develop a rational and integrated response to environmental management. Postwar experiments in economic and social policy have combined with Britain's changing place in the world to ensure that most British political debates have focused on issues such as unemployment, health care, welfare, education, and foreign policy, while touching only rarely on the environment. Ever cautious and pragmatic, British governments have been slow to recognize the environment as a distinct policy area and unwilling to provide environmental agencies with adequate power or funding.

Britain's position on acid pollution provides a good example of the contradictions. The problem was first identified by British scientist Robert Angus Smith in the 1850s (Smith 1872). Battersea power station in London (opened in 1929) was the first power station in the world to fit antipollution scrubbers, and London was one of the first cities in the world to respond successfully to smog problems. Yet when acid pollution was finally recognized as an international problem in the early 1980s, and when the Scandinavians and the Germans began taking action to curb acidifying emissions, Britain (or at least the Thatcher administration) demurred. It was only in 1988—long after all its European neighbors had agreed to act, and then mainly as a result of the requirements of European Union (EU) law—that Britain finally agreed to reduce its emissions of the sulfur dioxide and nitrogen oxides that are implicated in acid pollution.

Having said all this, however, there are signs that the environment has moved more squarely in recent years onto the national policy agenda, even if public interest has moved faster than political interest. In a famous off-the-cuff remark at the height of the Falklands war in 1982, Margaret Thatcher noted how exciting it was to have a "real crisis" on her hands as a change from "humdrum" issues, such as the environment (Young 1990, 372). By 1988–89, green consumerism was achieving a new prominence, the Green Party won 15 percent of the vote in the 1989 European Parliament elections, opinion polls regularly placed the environment among the three most important issues on the policy agenda, there was a wide-ranging political debate about environmental problems, and even Thatcher acknowledged that protecting the balance of nature was one of the "great challenges of the late twentieth century" (see McCormick 1991, 60).

The 1990 Environmental Protection Act covered a broader range of issues than any previous piece of environmental law and attempted to bring some order to a system that had become very complex. The prominent role played by Prime Minister John Major at the June 1992 Earth Summit, in Rio de Janeiro, gave a brief boost to political interest in the environment. When the Blair government took office in May 1997, it promised to be "the first truly green government ever." It presided over another reorganization of the institutional machinery for environmental management, its environment ministers won plaudits from environmental interest groups, it took a leading role in the debate over climate change, and it moved faster than its predecessors on such issues as encouraging less polluting energy. However, it fell out with environmentalists over the use of genetically modified crops, was accused of failing to provide leadership on the environment, and Tony Blair himself said and did little to suggest any personal interest in the environment, waiting more than three years to make his first major speech on the subject. Following another major speech in early 2001 (this time on renewable energy), opinion still seemed to be mixed about the extent to which his government really cared about the environment.

This chapter sets out to explain the priorities of British environmental policy, to describe and assess the major political actors involved in the development and implementation of that policy, and to draw some general conclusions about the efficacy of that policy. It argues that British governments have long taken a reactive, rather than a proactive, approach to environmental issues and that the most important policy initiatives in recent years have been taken more as a result of the requirements of EU law than of a genuine desire for change on the part of government. It concludes that Britain lacks an integrated environmental policy and that it lags some distance behind several of its European counterparts.

Policy History

Until the late 1980s, studies of British environmental policy were replete with such words as *consultation*, *consensus*, *flexibility*, *practicability*, *pragmatism*, *informality*, *incrementalism*, and *fragmentation*. Macrory (1987, 87) summed up the situation neatly: "Discretion and practicability might be described as the hallmarks of British environmental law and policy, with a degree of satisfied isolationalism and administrative complacency running closely behind." At the same time, more recent studies have emphasized the impact of EU law and policy styles on those in Britain, which has moved steadily toward the formal, legalistic, and adversarial approaches that tend to characterize both EU law and the policies pursued at the national level by most of Britain's EU partners. Those approaches include fixed emission and environmental quality standards, timetables, and concepts such as the precautionary approach, as well as the idea that polluters should pay.

The characteristics of the British environmental policy process until the late 1980s are summed up by Lowe and Ward (1998, 7–9) and by Carter and Lowe (1998, 21–28): the environment was on the periphery of both political and administrative interests; the organization and implementation of policy tended to be devolved to local authorities, quasi-governmental organizations, and interest groups; environmental policy lacked overall coherence and was driven, instead, by pragmatic responses to particular problems; technical specialists were given considerable scope over the development of policies and regulations; policymakers preferred voluntary procedures and self-regulation to standards and quality objectives; and policymakers worked closely with affected interests in the development of laws and the implementation of policies.

These principles were not unique to the environment because, in general, public policy in Britain traditionally has been worked out through a process that emphasizes consensus and consultation with affected interests; it verges, in some places, on neocorporatism. As noted by Jordan and Richardson (1987, ix), "most political activity is bargained in private worlds by special interests and interested specialists." Because British policymakers often think of themselves as custodians of the public interest and feel that they can understand the best interests of the public with minimal reference to the public itself, environmental policy was long made in closed policy communities. Throughout the 1980s, one of the strongest such communities was that consisting of farmers (represented mainly by the National Farmers Union or NFU) and the Ministry of Agriculture, Fisheries and Food (MAFF). So close was the relationship between the two that their representatives had almost daily meetings to consult and discuss policy. However, partly due to the success of attempts by environmental interest groups to draw attention to threats posed by modern farming to the countryside, the NFU-MAFF relationship weakened subsequently.

The habit of consultation between the regulators and the regulated was exemplified by the 1974 Control of Pollution Act, which was mainly shaped by industry and local government, despite more than 150 amendments tabled by environmental interest groups (O'Riordan 1988, 39–44). Similarly, the 1981 Wildlife and Countryside Act, one of the most controversial and influential pieces of legislation affecting rural scenery and seminatural wildlife habitats, was largely shaped by the powerful farming and landowning lobbies, despite more than 2,300 amendments tabled by wildlife and landscape interest groups, among others. Although the traditional power of farming and industrial lobbies has been challenged in recent years by an increasingly confident environmental lobby, consultation with affected interests had the effect of contributing to what critics of the British system of government have labeled "directionless consensus" and "pluralistic stagnation" (Kavanagh 1990, 311).

The emphasis on consultation was illustrated by British approaches to pollution control. As the first industrial state in the world, Britain has had longer experience of industrial pollution and has lived longer than any other country with the kind of mercantilist policies that discourage government regulation. While the United States and Germany have developed pollution policies based around the setting of ambient air quality standards and quantitative targets, Britain traditionally relied almost entirely on encouraging industry to comply voluntarily with "decent" standards of behavior. Pollution control laws tended to be broad and discretionary, and regulatory agencies were usually given wide scope to establish and enforce environmental objectives.

Table 4.1
Key environmental policy institutions in Britain

Institution	Year established	Main responsibilities
Countryside Agency	1949 (reorganized 1999)	Amenity in England
Department for Environment, Food and, Rural Affairs (DEFRA)	1970 (reorganized 2001)	Local government issues and limited environmental interests
Drinking Water Inspectorate	1990	Drinking-water quality in England and Wales
English Nature Scottish Natural Heritage Countryside Council for Wales Joint Nature Conservation Committee	1991	Quasi-governmental agencies responsible for rural management and nature conservation
Environment Agency (England and Wales) Scottish Environment Protection Agency	1995–96	Air and water pollution, waste management
Forestry Commission	1919	Forest management and timber supply throughout Britain
Maritime and Coastguard Agency	1998	Marine pollution
Ministry of Agriculture, Fisheries and Food	Integrated with DEFRA, 2001	Environmental impact of agriculture, with limited interests in water pollution, biodiversity, rural issues
Royal Commission on Environmental Pollution	1970	Advisory body on pollution

These values have been reflected in the institutional structure of policymaking and implementation. At first glance, Britain appears to have an impressive community of government agencies with responsibilities for environmental management (see table 4.1) and an impressive body of law to back up their work (see table 4.2). In truth, the work of many of these agencies is decentralized and divided; much of the responsibility for environmental policy has been left in the hands of quasi-

Table 4.2
Key pieces of environmental law in Britain

1853	Smoke Nuisance Abatement (Metropolitan) Act
1863	Alkali Act
1876	Rivers Pollution Prevention Act
1947	Town and Country Planning Act
1949	National Parks and Access to the Countryside Act
1953	Navigable Waters Act
1954	Protection of Birds Act
1955	Rural Water Supplies and Sewage Act
1956	Clean Air Act
1960	Radioactive Substances Act
1963	Water Resources Act
1968	Countryside Act
	Clean Air Act
1971	Prevention of Oil Pollution Act
1972	Deposit of Poisonous Wastes Act
	Road Traffic Act
1973	Water Act 1974
	Control of Pollution Act
	Dumping at Sea Act
	Road Traffic Act
1975	Conservation of Wild Creatures and Wild Plants Act
1976	Endangered Species (Import and Export) Act
1981	Wildlife and Countryside Act
1986	Housing and Planning Act
	Agriculture Act
	Food and Environment Protection Act
1988	Agriculture Act
1989	Water Act
1990	Environmental Protection Act
	Town and Country Planning Act
1991	Water Resources Act
	Planning and Compensation Act
1993	Radioactive Substances Act
1995	Environment Act

Table 4.3
United Kingdom: A Statistical Profile

Land	
Total area (1,000 km^2)	240.00
Arable and permanent crop land (% of total) 1997	26.70%
Permanent grassland (% of total) 1997	45.70%
Forest and Woodland (% of total) 1997	10.50%
Major protected areas (% of total) 1997	20.40%
Nitrogenous fertilizer use (t/km^2/arable land) 1997	19.50%
Population	
Total (100,000 inhabitants) 1998	591.00
Growth rate % 1980–1998	4.90%
Population density (inhabitant/km^2)	241.30
Gross domestic product	
Total GDP (billion U.S. dollars) 1998	$1,093.00
Per capita (1,000 U.S. dollars/capita) 1998	$18.50
GDP growth 1980–1998	53.00%
Current general government	
Revenue (% of GDP) 1997[b]	38.10%
Expenditure (% of GDP) 1997[b]	41.40%
Government employment (%/total) 1997[b]	14.40%
Energy	
Total supply (MTOE) 1997	228.00
% change 1980–1997	13.30%
Consumption (MTOE) 1998[c]	158.97
% change 1988–1998[c]	8.80%
Transport consumption (MTOE) 1998[c]	51.43
% change 1988–1998[c]	17.80%
Road-vehicle stock	
Total (10,000 vehicles) 1997	2,982.00
% change 1980–1997	71.80%
Per capita (vehicles/100 inhabitants)	51.00
Air	
CO_2 emissions (tons/capita) 1998[a]	9.57
CO_2 emissions (kg/USD GDP) 1998[a]	0.50
Nitrogen oxide emissions (kg/capita) 1998[a]	35.00
Nitrogen oxide emissions (kg/USD GDP) 1998	2.00
Sulfur oxide emissions (kg/capita) 1998[a]	34.00
Sulfur oxide emissions (kg/USD GDP) 1998	2.00
Pollution abatement and control	
Total (excluding household) expenditure (%/GDP) 1990	1.00%

Sources: OECD, *OECD Environmental Data, Compendium 1999* (Paris: OECD, 1999), except as noted.

[a] OECD, The OECD Observer, *OECD in Figures* (Paris: OECD, June 2000).

[b] OECD, *National Accounts of OECD Countries* (Paris: OECD, 2000).

[c] IEA/OECD, *Energy Balances of OECD Countries, 1997–1998* (Paris: IEA/OECD, 2000).

governmental agencies, rather than being addressed directly by central government; and much of the responsibility for promoting public awareness, generating pressure for legislative change, and overseeing policy implementation has fallen to the environmental lobby.

Until 1970, there was no single or comprehensive authority for environmental regulation in Britain. Responsibility for air pollution control, for example, was divided among the ministries of transport, housing, local government, technology, and agriculture; the Department of Social Services; the Board of Trade; and the secretaries of state for Scotland and Wales. Despite a belief that the environment might be an election issue for the first time in 1970, the closeness of the fight encouraged both major parties to focus on more familiar issues. Following the Conservative victory, the word *environment* appeared for the first time in the Queen's Speech in July 1970. In October, the Heath administration—with its predeliction for institutional reform and the creation of new "super agencies" aimed at promoting central control—announced that the ministries of housing and local government, public building and works, and transport were to be amalgamated into a new Department of the Environment (DOE).

Despite its name, the creation of the DOE was more a reorganization of government machinery than the creation of a new department with new powers. Its creation might have brought the word *environment* more centrally into cabinet discussions, but many key environmental concerns were left with other departments and with local authorities. For example, responsibility for energy supply remained with the Department of Energy and the Central Electricity Generating Board (both now defunct), while pollution issues were addressed, in part, by a new advisory Royal Commission on Environmental Pollution (RCEP), established in 1970. In 1976, the DOE's transport portfolio was hived off into a new Department of Transport.

The very title "Department of the Environment" was misleading, because only 10 percent of DOE staff dealt with environmental protection and planning (Organization for Economic Cooperation and Development, 1994, 23), while the bulk of the department's resources were devoted to local government, housing issues, building construction, and the regeneration of inner cities (Department of the Environment,

1995). The role of the DOE was further weakened by the limited interest shown by most environment secretaries in environmental issues (rather than local government) and by the responsibilities of other government departments for policy areas with critical environmental implications (for example, transport, energy supply, housing, and trade).

Among the most influential of those departments was the Ministry of Agriculture, Fisheries and Food (MAFF), which—despite its name—had a central role in British environmental policy, particularly in regard to wildlife and countryside issues. Among its aims: to "sustain and enhance" the rural environment, to promote forestry, and to manage fish stocks so as to "secure a sustainable future" for the fishing industry (Ministry of Agriculture, Fisheries and Food, 2001). It was also responsible for implementing the EU's Common Agricultural Policy, which—by guaranteeing European farmers a price for their products—prompted many of those farmers to use chemical fertilizers and herbicides and to increase the "efficiency" of their farming techniques. Whether the promotion of "efficient" agriculture is compatible with the objective of "enhancing" the rural environment is questionable. MAFF—now integrated into the new Department for Environment, Food and Rural Affairs—was repeatedly criticized by environmental groups for pursuing enviromentally destructive policies.

Rural management and nature conservation issues have, meanwhile, come under the aegis of a cluster of specialized quasi-governmental agencies (QGAs) set up by Parliament, funded mainly by the Treasury, governed by boards appointed by the sponsoring government department, and with goals defined by Acts of Parliament and the objectives of their sponsoring departments (Young 1993, 57). Foremost among these, for many years, were the Nature Conservancy Council (NCC, created in 1949) and the Countryside Commissions for Scotland and for England and Wales (created in 1968). The NCC and the Countryside Commissions were often criticized by environmental groups as ineffective, powerless, and secretive, for being slow to speak out against the destruction of sites of natural interest and value by farming and forestry activities, and for having governing councils dominated by career civil servants, foresters, farmers, and industrialists.

Relations between the environmental lobby and the state agencies improved in the late 1980s, with both the NCC and the commissions being more critical of government policy and more willing to resist that policy, notably the conversion of natural habitat to farmland. However, as they won more support from environmentalists, they aroused the ire of government ministers and landowners (Young 1993, 58). When plans were proposed by the Thatcher administration in 1989 to dismember the state agencies and to create separate bodies for Scotland and Wales, the opposition of the environmental lobby was instant, unanimous, and vocal; it seemed clear that the changes were politically motivated and aimed at weakening the powers of the NCC and the CC. Despite the opposition, the NCC was broken up in 1991 into English Nature, Scottish Natural Heritage, and the Countryside Council for Wales, and a Joint Nature Conservation Committee was created to promote coordination and to deal with UK-wide issues.

Britain's state forests, meanwhile, come under the jurisdiction of the Forestry Commission, which is responsible to Parliament in England, to the Scottish Ministers in Scotland, and to the Welsh National Assembly in Wales. In addition to promoting "the interests of forestry," the development of afforestation, and the supply of timber and forest products, the commission is responsible for overseeing forestry and forestry research. It, too, has been criticized by some environmental groups for suggesting that the interests of commercial forestry and wildlife conservation are compatible, for providing grant aid for the removal of ancient woodland, and for promoting fast-growing, commercial conifer plantations.

Tony Blair made much of the importance of sustainable development while he was in opposition (1994–1997), but Jordan (2000, 275) argues that, following its victory in the 1997 general election, Labour "found it increasingly difficult to coordinate policy across the many strands of social, environmental and economic activity in pursuit of sustainability. . . . Labour is publicly committed to greening government, but so far sustainable development has made headway only when political and economic circumstances have permitted." At least part of the explanation for this may lie in the priorities given by the Blair government, during its first term, to constitutional reform and to addressing problems in education, health care, and law and order. Following its second election victory, in June 2001, and once Blair's interest in action on international terrorism had waned, it was conceivable that Labour would turn its attention to issues previously marginalized, such as the environment.

The Policy Process

Institutional arrangements for air and water pollution control have undergone several changes in recent years. These changes have occurred not because the government has recognized the weakness of the existing system and the need for integration (advocated by the RCEP as early as 1978), but because of a combination of the requirements of EU law, changes in economic policy, legal concerns over the arrangements made for controlling pollution from Britain's newly privatized energy- and water-supply industries, and the fallout from the creation of regional assemblies in Scotland, Wales, and Northern Ireland in 1997–1999. The first change came in 1987 with the establishment of Her Majesty's Inspectorate of Pollution (HMIP), created in an attempt to develop a unified approach to pollution abatement in England and Wales (Scotland came under a separate HM Industrial Pollution Inspectorate, or HMIPI). The goal of HMIP was to ensure that every large emission source was regulated by a single inspector and to find the best practicable environmental option (BPEO) for the disposal of effluent. In 1988, prompted mainly by the requirements of EU law, the DOE finally made the case for a system of integrated pollution control, a recommendation that formed the basis of the 1990 Environmental Protection Act.

Meanwhile, plans in the late 1980s to privatize the water- and energy-supply industries further complicated the picture. The Thatcher administration had originally proposed allowing the privatized water industry to regulate its own pollution emissions, but this was found to be illegal under EU law. Consequently, a new National Rivers Authority (NRA) was created in 1989 to monitor the quality of river water, and a Drinking Water Inspectorate was created to monitor standards in drinking water. Similarly, when the electricity-supply industry was privatized in 1989, a new Office of Electricity Regulation (Offer) was created to act as a watchdog over the two new generation companies, PowerGen and National Power; its responsibilities included environmental pollution.

(Offer has since been merged into the Office of Gas and Electricity Markets, or Ofgem.) Meanwhile, the Department of Energy (created during the energy crisis of 1973) was abolished in 1992.

Two developments in 1990 were heralded by the Thatcher government as watershed policy developments, but neither contained much substance. The first was the publication, in September, of *This Common Inheritance*, a White Paper listing environmental measures already in place, but providing little in the way of new commitments for the future. The second was the passage of the Environmental Protection Act, which extended a number of existing policy initiatives, and either confirmed or introduced new institutional changes. The most significant outcomes of the act included further promotion of the concept of integrated pollution control and reorganization of waste disposal procedures, involving the creation of Waste Regional Authorities and Waste Disposal Authorities.

Both the Labour Party and the Liberal Democrats had begun calling in the early 1980s for the creation of an Environmental Protection Agency that would streamline and rationalize the responsibilities then held by a complex web of state agencies. These calls were renewed in the early 1990s, and the Major government, while arguing that this would create too much upheaval coming so soon after the creation of HMIP, nonetheless undertook a new reorganization in 1995, passing an Environment Act that replaced the eight-year-old HMIP, the six-year-old NRA, the five-year-old Waste Regulatory Authorities, and parts of the DOE with a new Environment Agency for England and Wales and a Scottish Environment Protection Agency. (Northern Ireland is largely treated separately from the rest of the United Kingdom—it has its own Department of the Environment, Department of Agriculture and Rural Development, and Environment and Heritage Service.)

The Environment Agency has responsibilities that are paid for partly by government and partly by industry; it concerns itself mainly with air and water pollution control, waste management, and radioactive waste disposal. Eleven of the fifteen members of its founding board were transferred directly from the NRA (including its first chief executive)—suggesting that it would develop the same kind of decentralized management structure as the NRA had before it. Concerns were raised early about its curious arrangements for managing rivers: while boundaries for water management were based on river catchments, those for pollution control followed the political boundaries of local districts (Environmental Data Services, 1995, 3). Critics suggested that the creation of the Environment Agency simply followed the British tradition of disjointed and incremental institutional change.

The Blair government added to the long history of changes to the structure of institutions by bringing the environment and transport back together again, in the form of a new Department of the Environment, Transport and the Regions (DETR), charge over which was given to deputy prime minister John Prescott. The objective behind this was to provide a more integrated approach to policy on the environment, but many of the old problems remained. Not only did the DETR continue to deal with many issues that had little or no impact on the quality of the environment (such as the promotion of road safety, of "social cohesion," of health and safety in the workplace, and of responsive local government), but responsibilities over key areas of environmental concern came under the aegis of other institutions. More changes came following the 2001 election, when the department was renamed the Department for Environment, Food and Rural Affairs (DEFRA). The change reflected the concerns that had arisen about the impact of mad cow disease and foot-and-mouth disease on the British countryside.

Environmental issues are touched on—and factor into—many different activities, so a single government agency with universal responsibilities for such issues would be bigger than almost any other agency and would have to possess an awesome array of powers. Nonetheless, the institutional structure for environmental policy in Britain remains fragmented and confusing, despite—or perhaps because of—the repeated attempts to make it subject to some kind of order. Not only have a broad group of different bodies been given responsibility for different issues, but these bodies have different levels of seniority in the administrative system (ranging from full-fledged cabinet departments to quasi-governmental agencies). They also cover different geographic areas (some are responsible for the whole of Britain, some for England and Wales, and some for England or Scotland or Wales alone), the more junior agencies are responsible to a confusing variety of superior bodies (including DEFRA, MAFF, Scottish Ministers, and the regional assemblies), and several have overlapping responsibilities.

The European Union

During his term as Britain's environment secretary (1993–1997), John Gummer announced that more than 80 percent of British environmental legislation was derived from the requirements of EU law (quoted in Morphet 1998, 139). While the EU—and the European Economic Community before it—had been steadily developing a body of European laws and policies on the environment since the 1960s, it moved into a new level of activity following the passage of the 1987 Single European Act. This not only gave the European Commission (the executive arm of the EU) new powers to introduce and develop environmental law, but also promoted common environmental standards as a critical element in the process of harmonization needed to create a single European market. The result is that some of the most important recent policy initiatives in Britain have come less as a result of domestic political pressure than of European regulatory pressure. For example:

- The controversial 1981 Wildlife and Countryside Act—which brought fundamental change to Britain's approach to wildlife protection—was a required response to the 1981 EU directive on wild birds.
- Britain has increasingly adopted environmental impact assessment procedures as a result of a 1985 EU directive; the European Commission, in 1991, called a halt to seven major construction projects in Britain on the grounds that they did not comply with the directive.
- The traditional decentralization of British policy has been reversed as a result of the requirements built into EU law that the governments of the member states report back to the European Commission on implementation (Haigh and Lanigan 1995).
- All the major pieces of British domestic law on air pollution since 1973 have come as a result of the requirements of EU law. Among the air pollution standards that have been set in response to the demands of EU law (rather than on British domestic initiative) are those on smoke, sulfur dioxide, nitrogen oxides, lead, and motor-vehicle emissions.

• Britain has followed the lead of the EU on all its water-quality standards, established most notably by directives on drinking water (1975, 1980), bathing water (1976), freshwater quality (1978), groundwater (1980), and urban wastewater treatment (1991).

The effect of EU law has been particularly obvious in pollution control, where Milton (1991, 42) argued as long ago as 1991 that "the most striking feature of the government's policy on pollution is the extent to which it is dictated by [EU] directives." After many years of intransigence and opposition to action on acid pollution, for example, the British government was finally obliged to take action by the 1987 EU directive on large combustion plants, which required Britain and its EU partners to make substantial reductions in emissions of sulfur dioxide and nitrogen oxides. The change of policy was notable, partly for the fact that it came after many years of pressure on Britain from the Scandinavian countries that received much of Britain's acid pollution and partly for the fact that it obliged the Thatcher administration—long opposed to acid pollution regulation—to make a policy U-turn. In short, EU law was able to achieve what many years of domestic and foreign pressure on the British government had failed to achieve.

Britain has tended to lag well behind some of the more progressive EU member states in reforming and tightening environmental regulation and so has found itself increasingly responding to the lead of Germany or the Netherlands, rather than acting independently on environmental policy. In a sense, environmental policy in the EU has been federalized, with member states losing their powers over policy development and being obliged to meet the same standards as other member states and with national government agencies increasingly switching their emphasis to responsibility for the enforcement and implementation of EU law.

The Environmental Lobby

Besides the government agencies and the EU, the most consistent source of pressure for change in British environmental policy has come from interest groups. Britain has what may well be the oldest, best-organized, and biggest (per capita) community of environmental interest groups in the world, which—by one estimate—had a combined membership in

2001 of 3.2 million (Royal Society for the Protection of Birds, 2001). Much of the growth in its size and levels of activity has come since the late 1980s, and the groups with the fastest growth have been those that are either more activist (such as Greenpeace and Friends of the Earth) or that have international interests (such as the World Wide Fund for Nature—formerly World Wildlife Fund—or WWF).

The growth of groups has paralleled a decline in party identification and membership and in voter turnout, which, at the 2001 elections, reached its lowest level since 1925. More environmental groups have also become more overtly politically active and more adept at working the political system and have changed their tactics accordingly. While concerns about their charitable status made many reluctant to become politically active in the 1980s, even the more traditional groups are more politically influential today, while (ironically) the more activist groups of the 1980s, such as Greenpeace, have become relatively more conservative and more centrally a part of the "establishment" environmental lobby. There has been a tendency for groups to move away from complaint and criticism and toward research-based appeals to policymakers, industry, and the public and toward the provision of services and solutions.

Environmental groups have built on their traditional ad hoc coordination and cooperation and issued more joint statements on government policy. They are increasingly working toward more-or-less preagreed sets of goals and are complementing each other. The work of parliamentary committees has also provided groups with opportunities to influence parliament and political parties, and several groups have, accordingly, appointed parliamentary liaison officers. With the realization that an increasing amount of British environmental policy and law is now driven by the requirements of EU law, there has also been a tendency for groups to either increase their direct representation at the EU level or to pay more attention to EU legislation and its implications for Britain. Friends of the Earth, Greenpeace, and WWF all appointed Brussels lobbyists for the first time in the 1980s (Long 1998).

Another clear change in tactics has involved environmental groups paying more attention to public attitudes and behavior. Concerns about the state of the environment tend to be highest in middle-class communities enjoying the fruits of affluence and concerned about threats to that affluence. The British middle class has slowly expanded as the working class has diminished, private home ownership has more than doubled (from 31 percent in 1951 to 67 percent today), a continuing shift in employment away from industry and toward the service sector has occurred, and the consumer society has grown. Such changes have combined to encourage more Britons to think more actively about threats to the quality of their lives, notably in the form of changes to the countryside, the physical quality of the urban environment, chemicals in food, and pollution of water and air. In times of growing affluence—such as the 1960s and the 1980s—middle-class support for tighter environmental controls has grown, while it has diminished in times of recession, such as the 1970s and the early 1990s. (A similar cycle was evident in the levels of attention paid by the government to the environment.)

One of the notable developments of the last decade in Britain has been the rise of the green consumer movement—consumers paying more attention to the ingredients of processed food and drink, switching in unprecedented numbers to organically grown produce, purchasing environmentally friendly products, and teaching themselves more about the environmental impact of the modern consumer lifestyle. The growth of green consumerism has been, at least in part, an outcome of the public-awareness activities of the environmental lobby. As green consumerism has grown, groups have found a new and potentially fruitful means of influencing public policy through encouraging changes in consumer demands. Not only are groups exerting pressure on the British government through the EU, but—by building an environmentally educated consumer population—they are exerting further pressure for policy change.

British government has long worked on the basis that elected officials will be helped, advised, and criticized in various ways by interest groups of various kinds. As Kavanagh (1990, 152) puts it, "consultation with affected and recognized interests is a cultural norm in British government. . . . A group expects to be consulted, almost as of right and certainly as a courtesy, about the details of any forthcoming government legislation and administrative change that is likely to affect it." This was traditionally true of trade unions and business and economic interests, but has become true of environmental groups as well, which have been increasingly accepted as "responsible" and so worthy of consultation on a regular basis. The importance of groups is emphasized by the extent to which they carry out roles that would more likely be the responsibility of government agencies in other industrialized countries. For example, the National Trust (2001) does much more than the British government in acquiring and maintaining historic buildings and scenic landscape. The Trust, the Royal Society for the Protection of Birds (2001), and county Wildlife Trusts (2001) together own and manage just over 1 million acres (419,000 hectares) of protected land. By contrast, state-run National Nature Reserves cover just 343,000 acres (139,000 hectares).

It took some time for activity to be translated into political influence. Writing in the early 1980s, Lowe and Goyder (1983, 58) argued that although the environmental movement was larger than either the consumer or the women's movements in the 1970s, it had won fewer institutional reforms: "Perhaps the greatest failing of environmental groups in the 1970s was their inability to translate their massive numerical support into an appreciable political force." Similar conclusions were reached by O'Riordan (1988, 8), who described British environmental groups as "politically active but only sporadically influential." Any power they may have had, he argued, was a product of their respectability and campaigning credibility, and, for the most part, they had tended to take on the attire, reasoning, and behavior of the establishment. Truly radical environmental opposition in Britain had steadily declined.

In recent years, however, there have been signs that the new emphasis on the role of the individual in the creation of environmental problems has led to a new level of activism, among both groups and individuals. Citizens have shown themselves more willing not only to change their consumption patterns to make them more environmentally friendly, but also to engage in direct action where needed. This is part of the "dramatic upsurge in single-issue protest activity and unconventional forms of political participation" observed by Evans (1997, 188). Margetts (2000) cites three examples of environmental issues that prompted political activity in the late 1990s: the antiroads movement, direct-action protests in 1998–1999 against the use of genetically modified foods, and concerns about rural issues and the state of the countryside. The latter has been interesting because it has galvanized individuals who have not normally taken part in unconventional forms of political activity. A combination of concerns about the ravages wrought by the BSE (mad cow disease) crisis, the urban bias of the Labour government, and attempts to ban fox hunting prompted an estimated 100,000 people to turn out in a London protest in 1997, and 250,000 to turn out in 1998 (Margetts 2000). A MORI poll found that 80 percent of the participants in the 1998 protest were Conservative voters, a group not usually given to protest activity.

Policy Performance

Most Britons enjoy a high quality of life when measured by indicators such as human health, clean air and water, access to amenity, and controls on urban blight, and certainly the overall quality of life has improved markedly since World War II. Superficially, at least, Britain retains the reputation it has long enjoyed for spectacular natural beauty, and policy trends in recent years have been positive, despite the fact that Britain is a small, crowded, postindustrial society and one of the most urbanized in the EU. Its most pressing environmental concerns revolve around managing the relationship among its people, industry, and what remains of nature in a country with one of the highest population densities in Europe (about 620 people per square mile, compared to 280 per square mile in France and just 80 per square mile in the United States). The most important environmental policy debates have long tended to focus on two key areas: land-use planning and pollution control.

In a country with so many people and so much competition for access to land and amenity, optimizing the use of land for as many general benefits as possible has been a major challenge. Postwar policy was driven by the Town and Country Planning Act of 1947, which provided Britain with one of the most comprehensive land-use planning systems in the world and required that all proposed development be subject to planning permission from local authorities. Unfortunately, farming was largely excluded from these planning requirements, and Britain's drive to greater self-sufficiency in food quickly brought massive, and often damaging, change to the countryside. The use of intensive agricultural techniques and factory farming expanded, chemical fertilizers and pesticides were increasingly used to help improve yields, forests and hedgerows were removed to make way for bigger and more productive fields, and ecologically important wetlands, grasslands, heaths, and marshes were "reclaimed" for agriculture. About 75 percent of Britain is today covered by agricultural land and about 10 percent is covered by urban settlement.

The Agriculture Act of 1947 not only introduced guaranteed prices to farmers for all major agricultural produce, but also provided the National Farmers Union (NFU) with the right to be consulted on all matters relating to agricultural policy (Garner 1996, 158–159). This paved the way for a symbiotic relationship between farmers and government. While farmers were once widely seen as guardians of the interests of the countryside, it had become increasingly clear by the mid-1970s that they had placed the profit motive above concerns for maintaining wildlife and nature. This became more obvious to Britons as increased mobility and a desire to seek recreation and relief from urban life in the countryside gave them direct experience of the kinds of changes that had taken place.

On the positive side of the ledger, about 20 percent of the land area of Britain was given protection as national parks and other protected areas, the area of greenbelts around towns and cities grew substantially in the 1950s and 1960s, and there was a steady decline in the area of derelict land or vacant lots. However, a 1992 government report warned that nearly one-fourth of hedgerows were lost between 1984 and 1990, the area of moorland had decreased by about 20 percent since the 1940s, the area of grassland in England and Wales had fallen by 40 percent since the 1930s, water levels in rivers and fens had fallen as extraction increased, and, although the area of forest in Britain grew by nearly two-thirds between 1947 and 1991, nearly all the increase was in the form of coniferous forest (Department of the Environment, 1992, chaps. 5, 10).

More recent research has shown some positive trends in several critical areas of land use. According to the *Countryside Survey 2000* (Department of the Environment, Transport and the Regions, 2000), the declines in the lengths of hedges and walls have been halted, and even reversed in some cases. Reductions in plant-species diversity have been slowed or even halted in most, but not all, habitats. While the total area of coniferous woodland in Britain remained unchanged between 1990 and 1998, the area of broadleaved woodland grew by 5 percent. The area of grassland and bogs fell over the same period, but the number of lowland ponds increased, and the biological condition of streams and small rivers improved.

The second major issue has been pollution control. As the cradle of the industrial revolution, Britain has longer experience with generalized air and water pollution than any other country, but while it took a number of early initiatives to deal with the problem, they tended to be ad hoc, to be decided through negotiation with affected interests, and to avoid setting targets and standards. The 1956 Clean Air Act was passed largely in response to public and political concern arising out of the 1952 London smog, the most serious in a line of smogs for which the city had become notorious and which was responsible for as many as 4,000 deaths (Ashby and Anderson 1981, 104). The 1956 law resulted in significant decreases in smoke pollution and was strengthened by the Clean Air Act of 1968. Other key pieces of air pollution legislation—most of them passed in response to the requirement of EU law—include the Road Traffic Acts of 1972 and 1974 (controlling emissions from road vehicles), the Control of Pollution Act of 1974 (defining the quality of motor fuels and fuel oil), and the Environmental Protection Act of 1990 (which introduced the concept of integrated pollution control).

The British record on air pollution has generally been positive (Environment Agency, 2000). Sulfur dioxide emissions—which had already fallen during the 1970s—were cut by more than two-thirds between 1980 and 1998. Britain was thus well on its way to meeting the 87 percent reduction target set by EU law and the UN treaty on transboundary air pollution. Carbon dioxide emissions fell by 20 percent between 1970 and 1998, although they are expected to rise after 2010. Lead emissions from road vehicles have fallen by 60 percent since unleaded fuel was introduced in the mid-1980s. Thanks mainly to road-vehicle emission controls, particulate emissions have been halved since

1990 and are expected to fall by two-thirds between 1995 and 2010. After climbing in the 1970s and 1980s, nitrogen oxide emissions have since declined, despite the growth in the number of vehicles on British roads and a doubling since 1970 in the number of miles traveled by road. Critics of government policy on air pollution have taken heart from the steady trend away from ad hoc approaches and toward the setting of ambient air quality standards and specific targets, in line with Britain's EU partners. Policy integration has been further promoted by the launch, in 1996, of the National Air Quality Strategy, a ten-year plan aimed at reductions in eight key pollutants, including sulfur dioxide, nitrogen oxides, and carbon monoxide.

Britain's record with water pollution has been less encouraging, in part because environmental-quality objectives were introduced only with the 1989 Water Act and the 1991 Water Resources Act and in part because of the structural changes that have taken place in water-quality management institutions since 1989. However, most of the trends have, once again, shown some positive shifts. Britain has a per capita water demand that is less than half the average for the European OECD, and 87 percent of Britons are served by wastewater treatment plants (compared to an average of 61 percent in the European OECD) (Organization for Economic Cooperation and Development, 1994, 48, 57), but surface-water quality has deteriorated only slightly since the early 1980s. According to the Environment Agency (2000), organic loads from sewage treatment works fell by one-third between 1990 and 1995 and are expected to continue to fall. Reductions are also expected in nutrient loads, and emissions of hazardous substances into the sea have fallen substantially since the 1980s.

Conclusions: The Changing Place of Environmental Policy

While Britain has a long history of legislative and policy responses to environmental problems, the results of those responses have been mixed, at best. A tradition (at least since World War II) of policy decisions being made largely as a result of consultation between the government and affected interests resulted in an ad hoc approach to problems. Solutions were prompted less by an overall concern with promoting sustainable development than with keeping those interests happy. Among such interests, the industrial and agricultural lobbies were particularly influential and, between them, ensured that approaches to air and water pollution and to land management issues infringed as little as possible on their overriding concerns with efficiency and profits. The prevailing efforts remained far from the kind of integrated and holistic approach to environmental management that would have proved more effective. Meanwhile, the growing environmental lobby largely remained on the margins of the debate.

The last two decades have seen fundamental changes that have considerably reduced the influence of industrial and agricultural interests. First, the body of EU law has grown, and Britain has been obliged to adopt a more structured and integrated approach to the environment, based less on ad hoc consultation between regulators and the regulated and more on the agreement and achievement of standards. The most important environmental initiatives of the past 20 years have come less from domestic pressures than from the requirements of EU law. EU law has not only required the attainment of specific goals on air and water quality, for example, but has also brought structural changes to British approaches to environmental policy (notably environmental impact assessment and the adoption of integrated pollution control). These changes have, in turn, encouraged changes in the structure of domestic policy institutions and the underlying principles of domestic environmental law.

Second, public awareness of environmental problems and demands for changes in policy have grown. Elections are still fought on familiar issues, such as the economy, welfare, education, and public safety. But the environment has become an increasingly common element in the speeches of elected officials and the manifestos of political parties and is increasingly used by voters as a factor in their assessments of the differences among parties. The membership of environmental interest groups has grown, more Britons express themselves willing to accept the economic costs of environmental protection measures, and more are willing to take measures to change their lifestyles so as to place less pressure on the environment.

Finally, and perhaps most ironically, the privatization policies of the antiregulation Thatcher administration resulted in much greater public access to information on the activities of farming, industry, and the electricity- and water-supply companies, which combined with the requirements of EU law to result in new regulatory obligations being placed on each of these sectors. Institutional and legislative reforms may still be confused and confusing, but they have resulted in considerably greater obligations being placed on newly privatized industries to meet stricter environmental standards.

The cumulative results of these changes have been an improvement in the quality of the British environment (although much remains to be done) and significant changes in approaches to dealing with environmental problems. However, domestic actions taken by the British government—and by the governments of other EU member states—are already decreasing in importance as the powers and the reach of EU law expand. The future is likely to see the British government taking fewer independent policy initiatives, becoming increasingly involved in negotiating policies with its EU partners, and being responsible less for policy initiation than for implementation and enforcement.

References

Ashby, E., and Anderson, M. (1981). *The politics of clean air.* Oxford: Clarendon Press.

Carter, N., and Lowe, P. (1998). Britain: Coming to terms with sustainable development? In K. Hanf, and A.-I. Jansen, eds., *Governance and environment in Western Europe: Politics, policy and administration.* Harlow, Essex: Longman.

Department of the Environment. (1992). *The UK environment.* London: Her Majesty's Stationery Office.

Department of the Environment. (1995). *Annual Report 1995.* London: Her Majesty's Stationery Office.

Department of the Environment, Transport and the Regions. (2000). *Countryside Survey 2000.* London: Her Majesty's Stationery Office.

Environment Agency. (2000). *Environment 2000 and Beyond.* London: Her Majesty's Stationery Office.

Environmental Data Services. (1995, August). *ENDS Report 247.* Concord, NH: Environmental Data Services.

Evans, M. (1997). Political participation. In P. Dunleavy, A. Gamble, I. Holliday, and G. Peele, eds., *Developments in British politics 5*. Basingstoke: Macmillan Press.

Garner, R. (1996). *Environmental politics*. Hemel Hempstead: Prentice Hall.

Haigh, N., and Lanigan, C. (1995). Impact of the European Union on UK environmental policy making. In T. S. Gray, ed., *UK environmental policy in the 1990s*. London: Macmillan.

Jordan, A. G. (2000). Environmental policy. In P. Dunleavy, A. Gamble, I. Holliday, and G. Peele, eds., *Developments in British politics 6*. Basingstoke: Macmillan Press.

Jordan, A. G., and Richardson, J. J. (1987). *British politics and the policy process*. London: Allen & Unwin.

Kavanagh, D. (1990). *British politics: Continuities and change*. Oxford: Oxford University Press.

Long, T. (1998). The environmental lobby. In P. Lowe, and S. Ward, eds., *British environmental policy and Europe: Politics and policy in transition*. London: Routledge.

Lowe, P., and Goyder, J. (1983). *Environmental groups in politics*. London: Allen & Unwin.

Lowe, P., and Ward, S. (1998). *Britain in Europe: Themes and issues in national environmental policy*. In P. Lowe, and S. Ward, eds., *British environmental policy and Europe: Politics and policy in transition*. London: Routledge.

Macrory, R. (1997). The United Kingdom. In G. Enyedi, J. Giswijt, and B. Rhode, eds., *Environmental policies in East and West*. London: Taylor and Graham.

Margetts, H. (2000). Political participation and protest. In P. Dunleavy, A. Gamble, I. Holliday, and G. Peele, eds., *Developments in British politics 6*. Basingstoke: Macmillan Press.

McCormick, J. (1991). *British politics and the environment*. London: Earthscan.

Milton, K. (1991). Interpreting environmental policy: A social scientific approach. *Journal of Law and Society*, *18*(1), 4–17.

Ministry of Agriculture, Fisheries and Food. (2001). http://www.maff.gov.uk

Morphet, J. (1998). Local authorities. In P. Lowe, and S. Ward, eds., *British environmental policy and Europe: Politics and policy in transition*. London: Routledge.

National Trust. (2001). http://www.nationaltrust.org.uk

Organization of Economic Cooperation and Development. (1994). *Environmental performance review: United Kingdom*. Paris: Organization for Economic Cooperation and Development.

O'Riordan, T. (1988, October). The politics of environmental regulation in Great Britain. *Environment*, *30*(8), 5–9, 39–44.

Royal Society for the Protection of Birds. (2001). http://www.rspb.org.uk

Smith, R. A. (1872). *Air and rain: The beginnings of a chemical climatology*. London: Longmans, Green.

Wildlife Trusts. (2001). http://www.wildlifetrusts.org

Young, H. (1990). *One of us*. London: Pan Books.

Young, S. C. (1993). *The politics of the environment*. Manchester: Baseline Books.

5

Environmental Policy and Politics in Germany

Helmut Weidner

Legal regulations and government measures to protect nature and human beings against environmentally hazardous activities of commercial and industrial firms have a long tradition in Germany. The Water Rights Act and the Factories Act have a particularly long tradition. The various regulations enacted to prevent noxious and offensive emissions from becoming a nuisance or being prejudicial to health and property were brought together and systematized, for the first time, in the Prussian Industrial Statute of 1845.

In the years following the Second World War, social and economic considerations were in the forefront of political and social debate. The legislative and administrative framework designed to protect the environment was only partially developed in the years that followed, though during this period of the "economic miracle," considerable damage was caused to the environment (Wey 1982).

It was not until the Social Democrat–Liberal coalition came to power in 1969 that environmental policy, responding to powerful currents in the United States, developed into an independent policy area based on a comprehensive concept of environmental protection, in the sense of protecting and conserving the basic natural means of sustaining life (Hartkopf and Bohne 1983). In the following years, environmental policy has been characterized by marked ups and downs. Among the population, however, environmental awareness has risen to a high level, and even in times of economic recession, environmental protection has still been deemed a relatively high priority. At times, environmental issues have triggered major (and partly violent) social and political conflicts; since the late 1980s, a more cooperative policy style has developed among the various actor groups and institutions in the environmental arena. In the end, environmental interests and their proponents have become widely integrated into established institutions and, for some time, Germany—from an international comparative perspective—has belonged to the pioneers of environmental policy. But until recently, hardly anyone would have expected that a radical ecoactivist and system critic like Joschka Fischer would someday even become foreign minister.

Modern Environmental Policy

A Dynamic Beginning

The term of office of the Social Democrat–Liberal coalition government was the starting point for the development of a systematic environmental policy—that is, the emergence of a separate program and its institutionalization as an independent policy arena.[1] This was surprising because there was no noticeable public demand for this, nor pressure exerted by organized interest groups to which government would have had to respond. The main conditions favoring the steep career of environmental policy were the general reform-oriented climate at that time[2] and the specific interest of the then–Minister of the Interior, Hans Dietrich Genscher, who wanted to endow his small Free Democratic Party (FDP, Liberals) with a better reform image. The central actors were a small number of politicians and senior officials who cooperated intensively with representatives of trade unions and economic-interest organizations. This way, consensus on the basic goals of environmental policy could be achieved relatively fast. Trade unions and economic organizations were willing to cooperate because they hoped to have a greater say in the design of environmental policies. They also expected that environmental protection costs could be partly covered by higher prices and partly be compensated by further economic growth (see Weßels 1989). A "regulatory approach," with strong emphasis on technical and economic measures, resulted from this close cooperation, in part because the participating administrators tended to have a legal background.

Basic, though not all, responsibilities relating to pollution control were transferred to the Ministry of the Interior.[3] As early as September 1970, a comprehensive crash program for environmental protection was adopted. This announced measures for clean air, noise abatement, water pollution control, waste disposal, chemicals, and nature and landscape protection. In the Environment Program of 1971, general guidelines for environmental policy were set down, which at the time were among the most progressive in Europe.

From these and other deliberations in the environment program, three principles were deduced that were to act, and still act, as central guidelines for environmental policy: the principles of precaution (*Vorsorgeprinzip*) and cooperation (*Kooperationsprinzip*) and the polluter-pays principle (literally: the principle of causation = *Verursacherprinzip*). The principle of precaution had come to play a dominant role in political statements and is always listed as the first of the three principles.

Several pieces of federal legislation concerned with pollution control were quickly passed between 1971 and 1974 (see Kloepfer 1989),[4] such as the Air Traffic Noise Act (1971), the Leaded Petrol Act (1972), the Waste Disposal Act (1972), the DDT Act (1972), the Federal Pollution Control Act (1974), and the Act on Environmental Statistics (1974), as well as a number of regulations and administrative directives (e.g., the decree of 1971 establishing a Council of Environmental Experts). In 1974, the Federal Environmental Agency was established.

A Period of Stagnation: Growing Conflicts and the Return of Dynamics

Sections of society that demanded more ambitious environmental measures, regardless of the economic recession, continued to expand. Initially, nuclear risks apart, major environmental concerns and associated conflicts related to the dangers arising in the chemical industry. Later, the problem of dying forests, and the related matter of clean air policy, became environmental issue number one (Boehmer-Christiansen and Skea 1991; Weidner 1986).

In parallel with the growing politicization of environmental issues in society, the organizational basis for the protection of environmental interests was improving: the so-called new social movements increasingly turned toward environmental issues, the number of environmentally oriented pressure groups grew steadily, and the first "green parties" were set up. As early as 1977, green groups participated (as "green lists") in elections to the district parliaments under the slogan of environmental protection. In the European elections of 1979, several such groups put up candidates with a "green label," attracting almost a million votes.[5]

Since the established parties were often not trusted to bring sufficient pressure to bear on the politico-administrative system—and, in fact, for a long time, they attached only minor importance to environmental protection issues (Hucke 1990)—the newly founded green parties were increasingly successful electorally at both local and regional levels. These parties failed to enter parliament in 1978 in the states of Lower Saxony and Hamburg, with 3.9 percent and 4.5 percent of votes, respectively. However, after 1979, they were able to demonstrate increasing electoral support. For the first time in German history, a representative of a green party (Bremen Green List) entered the parliament of the City-State of Bremen, in 1979. Baden-Württemberg followed in 1980, Berlin in 1981, and Lower Saxony, Hamburg, and Hesse in 1982. In the federal elections of 1980, however, they received only 1.5 percent of the votes, not enough to enter the national parliament. Before the Social Democrat–Liberal coalition government collapsed toward the end of 1982, the Greens were represented in six state parliaments (Poguntke 1993; Raschke 1993).

The combination of the growing and successful protest of environmental groups against private and public projects, the increasing criticism of an environmental policy assessed as being too lenient by the media, and the provocation the "green" response presented to the established parties themselves, ensured that in the 1980s the latter began to concern themselves much more intensively with this new policy area (Malunat 1987). Business and trade union organizations weakened in their opposition to environmental goals. This happened, not least, because these vested interests changed their own assessment of the tension between ecology and the economy. While they previously argued, almost without reservation, that pollution control measures would have a negative impact on economic growth and employment, they came to recognize that such measures can constitute an important factor in improving both the economic climate and the structure of the economy.

All this merely corresponded to scientific findings that had been ignored for some time (Sprenger 1979; Wicke 1989). And, finally, the new assessment of the situation was also supported by the emergence of a new branch of economic activity, the environmental protection industry, also labeled the ecoindustrial complex (Jänicke 1990). The general change of perception by industrial and trade union (see also Jahn 1993) bodies did not, of course, prevent specific sectors from strongly opposing an environmental policy. This opposition came especially from sectors that were undergoing some form of crisis or were otherwise particularly affected, such as the iron and steel industry, mining, energy utilities, and the automobile industry.

The governmental institutions responsible for environmental protection reacted to this climate of opinion by paying more attention to the environment. This reorientation of environmental policy began to take shape around 1980, without, of course, implying immediate translation of programs and regulations into action.

Some of the final bursts of energy from the Social Democrat–Liberal coalition government were aimed at improving environmental protection. As late as September 1, 1982, shortly before the change of government, it made far-reaching decisions on the future design of environmental policy.[6] This action, however, could not prevent the coalition's internal demise when the Liberals joined the Christian Democrats (CDU), with the subsequent shift of power to a Conservative-Liberal coalition government under Chancellor Helmut Kohl (Süß 1991). As an opposition party, the Social Democratic leaders then became more sensitive to ecological issues than ever before (Malunat 1994, 7; Malunat 1987). They made great programmatic efforts to attract "green voters"—for example, by emphasizing the need for an "ecological modernization of industrial society" in their party program.

Accomplishments of the New Business-Oriented Conservative-Liberal Government in Environmental Policy

The new Conservative-Liberal government that took over in October 1982 was considered by many to be very close to economic interests. Therefore, it was generally expected to give environmental issues low priority. However, Friedrich Zimmermann, the new minister responsible for environmental protection, quickly pushed through some strict environmental regulations, overriding the sometimes-strong resistance of the affected industrial circles. This is especially true of clean air policy, which was extremely controversial at that time because of the rapid increase in forest damage (*Waldsterben* or "dying forests"). Zimmermann enacted the Ordinance on Large Combustion Plants after only about nine months in office. It contained Europe's strictest regulations for limiting the emission of air pollutants from large combustion plants and became a model for other European countries (Mez 1995). The minister's unanticipated attempt, in July 1983, to have the U.S. ceilings for automobile-exhaust emissions adopted as a European Community (EC) directive caused a major political stir and finally led to stringent EC emission standards for passenger cars (see Boehmer-Christiansen and Weidner 1995; Holzinger 1995).

The onset of forward-looking clean air policy at the national level accelerated the West German government's international activities, and it also set the pace, more and more frequently, at the level of the EC. These activities were not only rational from a global perspective, but also coincided with the country's own interest in spurring internationally coordinated measures against acid rain and long-range currents of air pollutants, since German industry could offer the necessary environmental technology.

Minister Zimmermann was able to overcome resistance from industry by convincing them that, in the long run, these measures would strengthen the competitiveness of the German economy. The government also supported its policy financially through tax reductions, R&D funding, and so on. And, finally, business circles anticipated major political turbulence if environmental damage from air pollution increased significantly, because even stricter regulations than those proposed by the minister could result (see Boehmer-Christiansen and Weidner 1995, 48–52, 103–106).

While measures in the sphere of clean air policy led to large emission reductions, improvements in environmental quality in some other areas were minimal. Therefore, attacks on Minister Zimmermann's environmental policy grew steadily.

The Chernobyl Effect: The Establishment of the Federal Ministry for the Environment

The loss of confidence in the Minister of the Interior's competence in matters of environmental policy climaxed shortly after the Chernobyl nuclear catastrophe in April 1986.[7] He hesitated to act in the subsequent period, merely attempting to allay the public's widespread concern about the spread of radioactive fallout. He also attempted to explain away some of the serious deficiencies in the planning and organization of public safety measures designed to handle nuclear disasters and similar catastrophes—deficiencies that became increasingly evident (Peters et al. 1987). This response led to sharp criticism of the Minister of the Interior in particular, and of the handling of environmental protection in general.

In this situation, the federal government made a quick, politically astute decision. On June 5, 1986, the federal chancellor issued an organizational decree establishing the Federal Ministry for the Environment, Nature Conservation, and Nuclear Safety. Walter Wallmann, a politician with little experience in environmental matters, was appointed the first full-time Minister of the Environment in the Federal Republic of Germany. He was soon replaced (in May 1987) by Klaus Töpfer, a professor of regional planning and former Minister of the Environment in Rhineland-Palatinate. His very active environmental policy, especially his international initiatives, rapidly gained him respect among experts and the general public, as well as with environmental organizations.

The concentration of environmental competencies in a special Ministry of the Environment had long been called for by different experts and organizations (Drexler and Czada 1987; Pehle 1998). This integration of responsibilities was expected to increase the viability of environmental concerns in the government's internal decision-making process and to favor cross-sectional policy approaches. After all, some of the authority, especially authority important for designing and establishing preventive environmental policy, remained in other ministries. Moreover, the Ministry of the Environment was one of the small ministries, in terms of its staffing and the size of its budget. In 1993, it still accounted for only about 0.3 percent of the 1993 federal budget. In 1994, the budget was raised by about 7 percent, to a total of 1.33 billion DM. In subsequent years the size of the budget remained by and large the same (1999: 1.30 billion DM).

After the establishment of the Ministry of the Environment, environmental policy activities accelerated. Several strict laws and ordinances were enacted, new environmental institutions were founded, and the producers' liability for the environmental consequences of their activities was increased. In sum, Germany became—with respect to environmental innovations—one of the pioneer countries and an international player in the international environmental policy arena (see Umweltbundesamt, 1995, 2000; Jänicke, Mönch, and Binder 1993; Weidner 1995). The program for reducing CO_2 emissions by 25 percent by the year 2005, in particular, attracted much attention worldwide.

After the Unification of the Two German States: New Challenges to Environmental Policy

In spite of shared boundaries and a common language, the disclosure of information about the catastrophic ecological situation in the German Democratic Republic (GDR) (Franke 1992; Petschow, Meyerhoff, and Thomasberger 1990) surprised even West German experts, who had been critical of the GDR system. They discovered that large parts of a country of 16 million people had to be decontaminated and cleaned up.

With the formal unification of the two German states on October 3, 1990, West German environmental laws came almost completely into operation in the five new states that made up the former GDR. The Unification Treaty of 1990 (Art. 34) provided the essential legal basis for the stated goal: to protect the natural bases of existence and to create uniform, high-quality ecological conditions throughout Germany.

The desolate environmental conditions in many parts of East Germany made it highly improbable that the ambitious official goals for environmental restoration could be achieved by the end of the millennium, as planned. Other developments unfavorable to an effective environmental policy, such as the worsening general economic situation and the dramatic rise in unemployment, also held up progress. However, with massive administrative, technical, and monetary support—there was, and still is, huge transfer of public money to the eastern German states (partly raised by a general "solidarity tax")—East Germany underwent an extraordinarily rapid transformation into a region with almost–West European environmental standards.

This was not only an effect of deindustrialization (above all, the drastic decline of resource-intensive, inefficient industrial branches and the pollution-prone agricultural sector), the great improvements in the energy sector, the relatively rapid exchange of high-pollution cars, and the broad diffusion of modern pollution control equipment. It was also the effect of general industrial modernization that employed the latest technology. Further, a broad state-subsidized job-creation program led to about 100,000 (temporary) new jobs in environment-related areas. This not only accelerated environmental improvement trends, but also contributed to a positive image of environmental policy among large segments of the population.

These positive ecological effects of unification were accompanied by negative developments in some important areas: the breakdown of recycling systems, the side effects of increased consumption (littering, waste, and so on), growing traffic, and the increasing land consumption due to the construction of new housing, large commercial areas, and roads. Especially in the early 1990s, this environmental deterioration led to numerous conflicts.

The growing environmental consciousness in Germany was reflected in citizens' initiatives and in the protests of environmental organizations against large industrial and public development projects (especially landfills, waste incineration plants, airports, and highways). The federal government's main strategy for overcoming this opposition was to pursue a policy of deregulation (privatization of environmental tasks) and of reducing public participation in decision-making processes, with the objective of speeding up planning and permit and licensing procedures. After the federal elections of 1994 led to the continuation of the Conservative-Liberal coalition government under Chancellor Kohl, Angela Merkel, a physicist, was appointed Minister of the Environment. During her term in office, key areas of environmental policy stagnated (Weidner 1999). This was not so much the effect of her (compared to her predecessor Klaus Töpfer's) lower strategic "will and skill," but was primarily a result of the worsening economic situation, which hit the new federal states particularly hard (unemployment rates were, at times, as high as 25 percent). Issues such as mass and long-term unemployment, social welfare, state expenditures, and public debts dominated the governmental agenda, leaving environmental issues far behind. In public opinion, too, environmental protection lost significance, as compared to other social issues, though it still ranked high (Bundesministerium für Umwelt, Naturschutz und Reaktorsicherheit/Umweltbundesamt, 2000). During the 1990s, the debate on the economic challenges of globalization intensified, which also had negative effects on environmental policy. All these developments contributed to an increasing trend toward neoliberal economic approaches, both within the government and among economic interest groups. However, it would be wrong to say that the most severe economic recession in Germany since World War II entailed a massive rollback in environmental policy. By and large, Germany continued to play a progressive role in the global environmental policy arena. This can be explained, among other things, in terms of the strong interest of relevant parts of German industry in an expanding national and global "environmental market," the positive economic net effects (employment, resource and energy efficiency, and so on) of many environmental policy measures, and a stable institutionalization (along with increasing interrelationship) of environmental policy proponents in all spheres of society.

The "Red-Green" Government Since 1998: A Failure to Meet High Expectations

The general elections of fall 1998 terminated the roughly 16-year-long Kohl administration, and a coalition government of the Social Democratic Party (SPD) and Bündnis 90/Die Grünen (in the following: the Green Party) took over. The formation of this coalition was facilitated because the two parties involved already had experience with "red-green" coalitions at the state level. In addition, they both favored, in principle, an ecological tax reform. In simple terms, the objective of this reform was to shift the tax burden from labor to pollution. Jürgen Trittin, of the Green Party, became the new Minister of the Environment. Large parts of the population put high hopes in the new government and the "green" minister, not only with respect to a consolidation of social welfare and reduced unemployment, but also with regard to environmental policy. In particular, they hoped for a more active policy against "global environmental challenges." These expectations were by no means unfounded. Both parties' election manifestos stressed a strong commitment to a strategy of "ecological modernization" of economy and society to promote innovation and employment. The capacity for global competition was slated to be increased, more jobs were to be created, and environmental pollution was to be lowered at the national and international levels, especially through increased resource efficiency.

The high expectations on the part of the public, and especially on the part of environmental organizations, were, in part, bitterly frustrated. From the very beginning, the Ministry of the Environment concentrated on the so-called "nuclear power phaseout" (in 2000, the share of nuclear power in electricity generation was 31.2 percent). This led to an extremely contentious and wearisome process. Most of the influential business organizations, the major power-generating companies, the opposition parties (Conservatives, Liberals), and some state governments strongly opposed the plan and even prepared for court action. The new government faced the risk of having to pay billions in compensation. Consequently, the conflicts impeded the government's capacity for action in environmental policy. After lengthy negotiations with nuclear power plant operators, the so-called "nuclear consensus" was finally achieved. It fixed a timetable and measures for the "nuclear power phaseout": construction of new nuclear power plants (NPPs) is forbidden by law; in 2021, the last NPP will be closed down; and plutonium-generating reprocessing of nuclear waste must be reduced markedly and completely phased out in five years.

As early as April 1999, the first stage of the ecological tax reform, which attempts to reduce fossil-fuel consumption by levying higher taxes on energy (including gasoline), was implemented. Due to an imprudent political strategy pursued by the Ministry of the Environment (and representatives of the Green Party), a broad and polemic press campaign against the tax reform was launched. The tax reform met massive opposition from the business sector, and large parts of the public also protested against it because they feared adverse social and economic consequences, in view of rising energy prices. The "ecotax" is thought to have a number of flaws, including the fact that the tax rates for industries are much lower than those for the residential sector, that some industries are even exempted, and that the tax is not based on the carbon content of fuels. In 1999, a total of DM 19 billion were levied with the ecotax. According to the stepwise increase in tax rates, the estimated amount for 2003 is DM 35 billion. The ecotax revenue is used to cut monthly social security contributions from employers and employees in order to lower labor costs and encourage job creation. It is also meant to act as an incentive to efficiency-related and thus emission-reducing measures. These potential effects are usually called the "double dividend" of the ecotax. As a result of nationwide protests against high fuel prices, in 2000 the government was forced to introduce compensation measures for those hit hardest by the ecotax (especially commuters and low-income households).

Concerning general environmental and nature protection policy, the new government has—with few exceptions (e.g., waste management policy)—neither achieved anything worth mentioning nor created innovative programs. In spite of repeated announcements since 1999, there still is no comprehensive "national environmental plan," as has been set up in many other countries. Also repeatedly announced since 1999 was the establishment of an independent "Council for Sustainable Development," but only in April 2001 was the council finally established.

In terms of energy policy, however, the new government set ambitious goals and began implementing appropriate measures to reach them—for example, by raising the share of renewable energy in power generation to 10 percent (currently it is approximately 5 percent) by 2010. Already, Germany is the world's leading nation (far ahead of the United States) in wind-energy generation. In view of the energy policy goals, it is conceivable that at least the objective of the Kyoto Protocol of 1997 to reduce greenhouse-gas emissions can be achieved. Compared to 1990, Germany succeeded in lowering CO_2 emissions by 15 percent (supported by industrial shutdowns in the former GDR). Among the four Western industrialized countries that have achieved any CO_2 reductions, Germany ranks first. In November 2000, the federal government passed an ambitious climate protection program. In international climate policy, Germany continues to play an active and progressive part. In particular, Germany attempts to stipulate—over the opposition of the United States and some other countries (e.g., Japan, Canada, and Australia)—that the industrialized countries actually reach much of the envisaged CO_2 reduction within the respective country itself, not only by means of "emission trade" (see Bundesministerium für Umwelt, Naturschutz und Reaktorsicherheit, 2000). This commitment was reflected in the official disagreement on global climate policy between Chancellor Schröder and U.S. President Bush in early 2001. German industry has not only renewed its voluntary commitment to reduce CO_2 emissions by 20 percent from 1996 to 2005, but it now pledges an even higher CO_2 reduction rate (28 percent by 2005).

In addition, for all six greenhouse gases listed in the Kyoto Protocol, emissions are to be reduced by a total of 35 percent by 2012, based on 1990 levels. The achievements are monitored by an independent institute. Its first monitoring report indicates that German industry is performing well. There is no worldwide commitment to climate protection comparable to that of German industry. Another outstanding energy policy measure is the highly controversial electricity "feed-in" law adopted in 1999. It is designed to support the development of renewable energy by ensuring guaranteed prices to be paid by power companies to producers of electricity from renewable sources. Further measures include the "100,000 Roofs Solar Power Program" (running from 1999 to 2005), based on government subsidies. There are also several measures to cut energy consumption in private households, including stricter construction regulations and tax breaks for modernization of older buildings; investments in public transportation; and promotion of more efficient cars.

The present German government has now been in office for about two and a half years. This is much too short a time frame for a well-founded evaluation of its environmental performance. Therefore, the critical judgment given here must be read with caution. In addition, it is necessary to consider the particular restrictions with which the environmental policy of the "red-green" government has had to cope. Although public

Table 5.1
Germany: A Statistical Profile

Land	
Total area (1,000 km^2)	349.00
Arable and permanent crop land (% of total) 1997	34.50
Permanent grassland (% of total) 1997	15.10
Forest and Woodland (% of total) 1997	30.10%
Other land (% of total) 1997	20.30%
Major protected areas (% of total) 1997	26.90%
Nitrogenous fertilizer use (t/km^2/arable land) 1997	14.80%
Population	
Total (100,000 inhabitants) 1998	823.00
Growth rate % 1980–1998	5.10%
Population density (inhabitant/km^2)	230.50
Gross domestic product	
Total (billion U.S. dollars) 1998	$1,522.00
Per capita (1,000 U.S. dollars/capita)	$18.50
GDP growth (West Germany only) 1980–1998	44.90%
Current general government	
Revenue (% of GDP) 1997[b]	45.00%
Expenditure (% of GDP) 1997[b]	45.80%
Government employment (%/total) 1997[b]	15.30%
Energy	
Total supply (MTOE) 1997	347.00
% change 1980–1997	−3.70%
Consumption (MTOE) 1998[c]	243.19
% change 1988–1998[c]	−5.10%
Road-vehicle stock	
Total (10,000 vehicles) 1997	4,403.00
% change 1980–1997	60.20%
Per capita (vehicles/100 inhabitants) 1997	54.00
Air profile and greenhouse gases	
CO_2 emissions (tons/capita) 1998[a]	10.68
CO_2 emissions (kg/USD GDP) 1998[a]	0.47
Chlorofluorocarbons (tons/ozone depletion potential) 1986	131,046.00
Chlorofluorocarbons (tons/ozone depletion potential) 1990	78,470.00
Halons (tons/ozone depletion potential) 1986	19,749.00
Halons (tons/ozone depletion potential) 1990	15,910.00
Methane (metric tons) 1990	6,218.00

Table 5.1
(continued)

Nitrogen oxide emissions (kg/capita) 1998[a]	22.00
Nitrogen oxide emissions (kg/USD GDP) 1998	1.20
NO emissions (1,000 metric tons) 1980	3,540.00
NO emissions (1,000 metric tons) 1987	3,570.00
NO emissions (1,000 metric tons) 1990	3,230.00
Nonmethane volatile organic compounds (million tons) 1990	3.15
Sulfur oxide emissions (kg/capita) 1998[a]	18.0
Sulfur oxide emissions (kg/USD GDP) 1998	1.00
SO_2 emissions (1,000 metric tons) 1980	7,520.00
SO_2 emissions (1,000 metric tons) 1987	7,550.00
SO_2 emissions (1,000 metric tons) 1990	5,690.00
Pollution abatement and control	
Total (household excluded) expenditure (%/GDP) 1995	1.50%

Sources: OECD, *OECD Environmental Data, Compendium 1999* (Paris: OECD, 1999), except as noted.
[a] OECD, The OECD Observer, *OECD in Figures* (Paris: OECD, June 2000).
[b] OECD, *National Accounts of OECD Countries* (Paris: OECD, 2000).
[c] IEA/OECD, *Energy Balances of OECD Countries, 1997–1998* (Paris: IEA/OECD, 2000).

opinion polls indicate that Germans still think environmental issues are very important, other issues (e.g., unemployment) have gained higher priority during recent years. Business organizations have used the "globalization debate" in trying to fend off stricter environmental regulations, although so far there have been no valid indications that globalization puts the economy under significant competitive pressure or has triggered a race to pollution havens (relocation of production to countries with lower environmental standards). Chancellor Schröder supports environmental policy only halfheartedly and tends to side with the interests of large industries—in particular, the automobile industry. On his initiative, for instance, the EU "scrap-car directive," requiring manufacturers to take back and recycle cars of their own type registered after July 2002, was changed in favor of the automobile industry. Last but not least, the previous successes of environmental policy seem to have given an "all-clear signal," so that many Germans feel existing regulations are sufficient to further improve environmental quality. The absence of recent environmental catastrophes (forest dieback, smog, and so on) has reinforced this feeling of safety.

Overall, compared to the last term of the preceding government, the environmental policy of the red-green coalition has been somewhat more dynamic. Recently this has also applied to agriculture. For a long time, the risks of mad cow disease were played down, but when the political failure to act appropriately had become public knowledge and consumers stopped buying beef, two ministers of the red-green government (health and agriculture) were forced to resign. A new Ministry for Consumer Protection, Food, and Agriculture was established and soon thereafter programs were passed, which included the strong promotion of eco-farming. In an evaluation of red-green environmental policy, Martin Jänicke, member of the Expert Council for Environmental Issues, stated that in view of the restrictive political-social conditions, the new government's energy policy can be considered a pioneering achievement, by international standards. Concerning goals and measures taken in the other environmental policy areas, the government only ranks average (Jänicke 2000, 53).

The Constitutional Division of Powers and Organizational Structure of Environmental Policy

Legislative Power

Germany is a federal republic. According to the basic Constitutional principle, legislative power lies with the states unless the Constitution explicitly assigns it to the federal government (Art. 30, 70 GG).[8] The present German Constitution divides the legislative authority between the federal government and the 16 states (11 "old," i.e., Western, and 5 "new," i.e., Eastern, states) in three ways: in some (unusual) areas, the federal government has exclusive jurisdiction; in all others it has concurrent or "framework" jurisdiction (see Rose-Ackerman 1995). Since the constitutional amendments of 1971 and 1972, the federal government has the following environment-related powers (only the most important ones are listed here):

- Exclusive jurisdiction (Art. 73 GG)—only for areas relating indirectly to environmental protection: federal railways, air traffic, statistics, international affairs, bilateral and multilateral agreements
- Concurrent (competitive) jurisdiction (Art. 74) for:
  - Some aspects of commercial, civil, and criminal law
  - Noise abatement
  - Nuclear energy (siting, radiation protection)
  - Waste management
  - Air pollution control
  - Poisonous substances
  - Plant protection
  - Animal protection
  - Coast protection
  - Road traffic, highway construction
- Framework jurisdiction (Art. 75)[9] for:
  - Nature protection and hunting
  - Regional (land-use) planning
  - Water supply and protection
  - Coastal preservation

The federal government is also granted jurisdiction for general administrative directives and regulations issued to implement federal statutes, but only if explicit authorization is given in the several environment-related laws. This, however, requires the consent of the Upper House, which is true of all laws that concern genuine state responsibilities, such as implementation and enforcement.

Implementation and Enforcement

Implementation and enforcement of both federal and state environmental laws are the responsibility of administrative agencies at the state level. Only in a few areas—for example, regulation (screening/registration) of chemicals, licensing of nuclear power plants, and highway planning—is a federal agency responsible or does the state agency act as an "agent" of the federal government. In the latter case, the state agency's activities are subject to legal and actual supervision.

The local authorities implement tasks under the self-government principle, explicitly granted by the Constitution (Art. 28, sec. 2 GG), and tasks delegated by national (very rare) and state governments (very often) to them. The first are under only legal supervision by state authorities; the latter are under legal and actual supervision and control. Their legislative powers are confined to issuing bylaws. Environment-related bylaws usually deal with land-use rules, charges for waste handling and sewage, establishment of noise-abatement zones, certain effluent standards, waste management plans, and so on.

Local authorities play an important role in the implementation of environmental policy, both directly, by fulfilling their environment-related responsibilities and duties, and indirectly, by their development and land-use policies, which often interfere with environmental protection goals. Especially with economically strong and politically entrenched local authorities, there have been problems concerning their legal powers, which have become a matter for litigation. The City of Munich, for example, enacted a ban on no-deposit beverage containers and the City of Kassel introduced a local tax on certain foods (see the rule of the Constitutional Court BverfGE 98, 106), both of which have been successfully challenged in court. Although the scope for discretion is restricted by several forms of federal and state control, studies on local power and implementation have shown that local bodies actually still enjoy a high degree of flexibility. Therefore, the quality of local environmental policy is highly dependent on the degree of environmental consciousness of local administrators and citizens, "political will and skill," and the economic structure and situation in the local area, including the financial situation of the local authority. The public transport system and the establishment of special protected areas (e.g., car-free zones) are examples of areas where the scope and the form of environment-related activities depend strongly on the capacities and willingness of local institutions. Other examples include action plans for sustainable development formulated and implemented at the local level—as called for by Agenda 21, a document approved by over 170 nations at the 1992 Earth Summit in Rio de Janeiro. (Further information is available at http://www.un.org./esa/sustdev/agenda21.htm.) Despite these broad local competencies, the general nationwide trends of environmental policy and quality, especially with regard to mass pollutants, such as sulfur dioxide, dust, nitrates, and so on, are almost exclusively the result of federal policies.

Organizational Structure of Environmental Protection

The organizational structure of environmental protection in Germany is complex. This section briefly describes this organizational structure at the federal, state, and local levels. Because of space limitations, the federal-level organization for environmental protection is described only in key words and only the most relevant institutions are considered.

The establishment of the Federal Ministry for the Environment, Nature Conservation, and Nuclear Safety in 1986 (see Pehle 1998) was the last major restructuring of the administrative system dealing with environmental policy. At present, all 16 federal states have established a ministry mainly or exclusively responsible for environmental matters. Some of them also have responsibilities for activities with a close, but indirect, relation to environmental policy, such as regional planning, nature conservation, health, nuclear safety, agriculture, and urban development. The state ministries for the environment are not subordinate to the federal ministry.

National Level The constitutional division of powers means that the states (with rare exceptions) are responsible for the implementation and enforcement of federal laws. This is reflected in the organization of environmental policy at the federal level. With the exception of the regulation of toxic substances under the Federal Chemicals Act and the regulation of nuclear safety, where federal implementation authorities are responsible, there are no central implementation agencies. Even in the area of siting of nuclear plants and radiation protection, where the federal government is granted implementation powers, most of the tasks have been delegated to state authorities.

With regard to powers, responsibilities, and the principles of coordination and conflict resolution, the formal internal structure of the federal government is determined by the constitution. According to Article 65 GG, it is the federal chancellor who sets the general guidelines of governmental policy and bears responsibility for it. Within these guidelines, each federal minister manages his or her department independently. As for environmental policy, this means that all ministers with environment-related tasks have great scope for maneuvering that is restricted only by the general guidelines and, of course, by the law. Thus, there is great potential for interministerial conflict, especially because as many as 16 ministers have had environment-related powers for some time. The competing interests between environmental and other policies (financial, economic, agriculture, energy, transport, and so on) have created further potential for conflict. This situation has improved since the establishment of the Ministry of the Environment and the accompanying concentration of environmental duties and powers. The remaining conflicts are managed by formal and informal conflict-resolution procedures. The formal procedures again are structured by the Constitution, which requires that interministry conflicts be settled by the federal government—that is, by all ministers on the basis of majority rule. In the event of a deadlock, the chancellor decides. Therefore, it is an ongoing task of the Minister of the Environment to seek support for his or her policies through internal and external coalition building, negotiation and bargaining, and support from outside. To avoid and manage conflicts, various coordination institutions have been established at the federal level.

State Level Germany is a federation of 16 states. In general, implementation and enforcement of environment-related laws and regulations lie within the power of the states. They independently organize their organizational and implementation structure, which explains the high degree of variety in this structure. Some states have established special authorities independent of the general administration; in other states, the general administration is the ultimate authority but is advised and supported by separate technical agencies.

Local Level According to Article 28 GG, the local authorities have the right to self-administration. Their environment-related tasks include mandatory and voluntary duties, which they perform "in their own right," and others delegated to them by state governments or the federal government—for example, the establishment of noise-abatement plans under Article 47a of the Federal Emission Control Act.

Laws in other areas also grant the local authorities wide-ranging jurisdiction in the field of environmental policy. For example, urban traffic and transport lies within the planning responsibility of the local authorities. The Road Traffic Act (a federal law) gives them numerous additional powers, such as the power to impose speed limits, introduce environmental traffic management schemes, and establish pedestrian zones.

There is no general rule as to how many environment-related departments should exist under the jurisdiction of local administration and what specific functions they should have. As a rule, several departments and authorities are responsible for various functions within one local body, mostly depending on its size and problem structure. The larger local authorities (e.g., cities), but also associations of several smaller authorities (e.g., districts), have often established special environmental departments.

Supranational Level: The European Union The complex environmental policy regime is further complicated by the fact that Germany is a member of the European Union. Since its inception, this supranational institution has expanded its competencies and activities in environmental policy tremendously. And the rule that "EU law transcends national law" still applies. At the level of the European Community (EC; or European Union (EU), as it was renamed after the Treaty of Maastricht), environmental protection, as a task, was established as late as 1972 on the basis of a (far-reaching) interpretation of the EEC treaty of 1957, the so-called Treaty of Rome (Rehbinder and Stewart 1985). A year later, the first of a total of five EC Environmental Action Programs—policy statements of the Community—was established. The bulk of EC environmental legislation, almost exclusively in the form of directives, followed.[10] These directives have to be implemented and enforced by the member states within a fixed period of time. The general rule is that national legislation is subordinate to EC legislation.

For a long time, the environmental legislation was vaguely, and sometimes confusingly, based on certain articles (especially Article 2) of the Treaty of Rome. This situation was improved by a series of amendments enacted by the Single European Act of July 1, 1987, which granted the EC environmental policy and lawmaking powers. In addition, Article 100 A required a high level of environmental protection for all EC harmonizing regulations relating to the establishment of the internal market.[11]

For various reasons, the outcomes of the EC's environmental policies had been heavily criticized for some time as being inadequate, stressing the lowest common denominator, and often putting brakes on environmental pioneers. Strong criticism was also directed at the "centralist pattern" of the EC decision process and its general bias in favor of economic interests. Members of the European Parliament, too, pointed to the fundamentally undemocratic structure of the EC decision-making process due to the better legal and organizational position of the executive branches (e.g., the EU Commission, the representatives of the national governments of the member states) in influencing policymaking and legislation. As a rule, representatives of the national government exclusively negotiated and participated in basic decision-making processes at the supranational level.

This kind of criticism was mirrored at the state and local levels because the democratically elected bodies at these levels only had a limited opportunity to control and supervise the policymaking activities of the federal government in "Brussels." This applies, in particular, to nongovernmental organizations (see Hey and Brendle 1994). In Germany, these problems are seen as undermining the constitutional principles of a federal state, especially one based on local self-government. As a remedy, some new forms of participation and consultation between the states and the central government have been established (see Rehbinder 1989, 1992). This began with the ratification law for the Single European Act. Later, a new Europe article was added to the German Constitution (Article 23 GG). It stipulates that any transfer of sovereignty to the EU level requires the approval of the Upper House (Bundesrat, where the states are represented). Furthermore, all German states have established offices in Brussels. From there, they not only "observe" EU institutions, but also lobby and build alliances and networks. For example, the states lobbied successfully for the incorporation of a "subsidiary principle" into the Treaty of Maastricht. A Committee of the Regions and Local Authorities (CoR) with advisory status was also established, and a new Article 146 allows for any EU member country to send subnational ministers to act as its delegation to the Council of Ministers, the EU's most powerful legislative body (Bomberg and Peterson 1998). Furthermore, representatives of the single German states often participate in EU committees and working groups as members of central government delegations. This is meant to guarantee the federal principle in the EU (created by this treaty) and provide for more regional and local (and also, to some degree, national) autonomy, with respect to EU policies.

With the Treaty of Maastricht, the environment has achieved full status as a policy area and has become one of the Union's priority objectives. The follow-up Treaty of Amsterdam of 1997 even made the integration of environmental requirements into other policy sectors a central objective. This is meant to stimulate systematic interpolicy coordination. Such coordination is badly needed in Germany but is hard to achieve because of the high degree of sectorialism and fragmentation in the German multilevel environmental policymaking system (including the neocorporatist mode of interest mediation).

Therefore, the new generation of EU environmental policy poses a bigger challenge to Germany than to many other member countries. The EU increasingly favors flexible, incentive-based instruments over regulatory instruments, and multistakeholder involvement and cooperation are now often called for. The focus has definitely shifted from policy formulation to implementation.[12] This change in EU policy style and the rise of a certain regulatory competition among the member countries at the EU level—trying to bring EU legislation as far as possible in line with their own regulatory approaches—have had significant effects on Germany (Héritier, Knill, and Mingers 1996). The need for institutional adaptation to EU policies has not only led to changes in administrative structures and processes, but has also influenced the position of business and nongovernmental actors. Examples of the changes are the Drinking Water Directive, the Access to Information Directive, the Environmental Impact Assessment Directive, the Integrated Pollution Prevention and Control Directive, and the Eco-Audit and Management Scheme (EMAS). By and large, the new EU environmental policy style contributes positively to the modernization of Germany's environmental policy by promoting market-oriented, self-regulatory, flexible, stakeholder-oriented, cooperative, and increasingly transparent instruments (Lahusen 2000). Thus, EU policy is much less a brake on progressive initiatives than it used to be.

There still is a bias in favor of economic interests. However, environmental interest groups have gained a lot more influence over the last decade. Since the Treaty of Maastricht and the subsequent legislation, the position of subnational authorities has been strengthened, too. In spite of the marked contradictions remaining between environmental policy and other policy areas, such as transport, agriculture, and energy, recent EU legislation and programs have strongly supported cross-sector perspectives and activities. The 5th Action Program of the EU, established in 1992, explicitly emphasized sustainable development and the need for an integrative policy (among other things, by specifying five target sectors: industry, energy, transport, agriculture, and tourism). All in all, the formerly rather sharp conflicts between economic and environmental interests within the EU decision-making system have subsided significantly. High standards of environmental protection are now increasingly seen as key to global competitiveness, not only of the economic sector but also of other sectors (see Weale et al. 2000, 75–85).

Coordination of Environmental Protection Policy in Germany

The complexities of a three-tier federal system (Bund-Länder-Gemeinden) with separated and overlapping powers of jurisdiction, as well as implementation and enforcement tasks, require a high degree of vertical and horizontal coordination and cooperation among the various levels. The attempt to achieve a high degree of uniformity in implementation is encouraged by common bureaucratic traditions within state governments and by the predominant legal background of senior officials. Furthermore, permanent employment of senior officials and low turnover rates allow for the accumulation of expert knowledge and favor the establishment and maintenance of informal communication networks. However, there are conflicts that the courts are called on to solve or, alternatively, the higher authorities issue formal "instructions" to the authorities, which have followed a "nonconformist" path.[13]

Overall cooperation in the field of environmental policy bears the mark of the special German brand of federalism. Features of this kind of federalism are the strong interdependence between executive institutions at all levels and the coordination mechanisms that have developed, especially between the central government and the states. Complex procedures of representation and consultation are fundamental to this system.

Much legislation and many other important decisions result from bargaining behind the scenes and in the various bodies between politicians (of the government and the opposition) and administrators. From an international comparative perspective, the high degree of formal and informal cooperation is an outstanding feature of the German federal system (see Norton 1994, 259).

The system is even more complex due to the fact that it is closely meshed with the neocorporatist elements of industry and society:

> Germany possesses a powerful system of interest group organisations, recognised in law and closely linked with the political parties. It includes statutory chambers of trade and industry and trade union and voluntary bodies associated with the churches. They are a recognised part of the institutional system in each Land and larger municipal areas and receive support from public revenue. They work through a complex network of consultative procedures which underlie and integrate policy-making at all levels. The interweaving of interests and influence between sectors and levels of government and society (*Politikverflechtung*) is an outstanding aspect of German political life. (Norton 1994, 243; see also Lehmbruch 1992)

Environmental Policy Performance

Germany's geographic and political situation, as well as its economic structure, bring with them specific environmental problems and, as a rule, require greater efforts in environmental protection than in many other countries. Population density and the degree of industrialization are among the highest in Europe. Germany also has a large proportion of environmentally problematic industrial sectors and highly industrialized agriculture, one of the densest transport networks, and high (and increasing) traffic volume. Germany has a long coastline and—because it shares borders with nine other countries—most of its main rivers have their sources in neighboring countries.

Remarkable progress has been made in several environmental areas. Since 1990, substantial reductions in national emissions of major air pollutants have been achieved (e.g., SO_x by 76 percent, NO_x by 34 percent, volatile organic compounds (VOCs) by 47 percent, and SO_2 by 52 percent). Even the road-transport sector contributed to this positive development (total emissions 1990–1999: –32 percent NO_x, –60 percent CO, –75 percent VOC), despite rising numbers of passenger cars (+15 percent) and trucks (+43 percent) from 1991 to 1999. Total CO_2 emissions were 15 percent lower in 2000 than in 1990. Ambient air quality has generally improved.

In the area of water pollution control, the huge investments in municipal and industrial wastewater treatment plants have led to improvements in water quality, with respect to some parameters and pollutants (e.g., biological oxygen demand (BOD), chemical oxygen demand (COD), suspended matter, and heavy metals). In most aquatic systems, there have been considerable improvements, notably in the Rhine and the Elbe. However, groundwater quality has deteriorated in many areas, due mainly to agricultural activity. The generation of waste has been stabilized or, for some waste categories, even reduced. The recycling rate has been increased.

Today, the most problematic areas include air pollution and the noise and damage to nature caused by private transport (the growing number of passenger cars with increasing mileage, the trend toward cars with more powerful engines, the increase in road transportation of goods, and the high growth rate in air traffic: 75 percent in the period from 1991 to 1999). Also problematic are increased illegal dumping and the growing opposition to conventional waste treatment (landfills, incineration), as well as the tens of thousands of sometimes heavily contaminated sites (as a result of improper waste disposal and production practices in the past). There has been a general increase in developed land (1997: 120 hectares per day; 1999: 129 hectares per day), a high rate of tree damage (21 percent), soil and water pollution by the agricultural sector, and increasing volumes of sewage sludge with a high concentration of toxic substances (due to improved techniques of effluent treatment, making its disposal, especially on agricultural land, a matter of conflict). Deplorable deficits in nature protection policy are evident (loss of biodiversity has not been halted), and global challenges caused by the so-called greenhouse gases are increasingly apparent (e.g., the transport sector's CO_2 emissions have increased in recent years).

The most comprehensive assessment of the present government's environmental policy was published by the Council of Environmental Experts in various reviews (e.g., Rat von Sachverständigen für Umweltfragen, 1987, 1994, 1996, 1998, 2000) and a special report (Rat von Sachverständigen für Umweltfragen, 1989). The measures and outcomes of governmental activities are judged with ambivalence. The SRU was especially positive about the clean air policy measures, the construction of sewage treatment plants, the ban on leaded gasoline, the commitment at the international level, and some of the programs offering economic incentives.

The SRU directed sharp criticism at major shortcomings in the gathering and publication of environmentally relevant information in nature conservation, landscape planning, the protection of the soil and groundwater, noise abatement, and the protection of food from impurities. The Rat von Sachverständigen für Umweltfragen (SRU) called for greater clarity and openness in the decision-making process involving environmental policy, especially in the procedures for standard setting (Rat von Sachverständigen für Umweltfragen, 1987, 61). In its published reports of 1996 and 1998, the SRU intensified its criticism of the policy approach. In addition, the council demanded that authorities grant wider participation to environmental organizations and cooperate more effectively with them (see Rat von Sachverständigen für Umweltfragen, 1996, 230ff.). In its 2000 report, the SRU criticizes the new red-green government for not dealing adequately with general environmental matters and for being too preoccupied with nuclear energy issues.

The first Organization for Economic Cooperation and Development (OECD) report on environmental performance in Germany, published in 1993, reached a positive judgment overall. It especially emphasized that "the de-coupling of economic growth from the flow of several major pollutants over the past two decades is indicative of Germany's remarkable achievement in reconciling economic growth and environmental objectives" (Organization for Economic Cooperation and Development, 1993, 205). However, it also mentioned that great challenges remain—

for example, in the areas of biodiversity, agricultural pollution, and transport.

The second OECD performance report (Organization for Economic Cooperation and Development, 2001) actually produced a slightly more positive evaluation of Germany's environmental achievements. In particular, it highlights the improvements in emission reductions and ambient air quality and the generally impressive environmental progress made in the five new German states. It mentions favorably the continuing efforts to reconcile economic growth and environmental protection, and it considers Germany's progress in decoupling economic growth from emissions and use of resources impressive. Enforcement and compliance are assessed as generally effective and based on good monitoring and institutional capacities, as well as on an increased range of economic instruments. The 1999 ecological tax reform is seen as an important step in the right direction, and the wide-ranging and successful program of international cooperation is acknowledged. Compared to this long (and selective!) list of laudable achievements, the list of shortcomings is brief, and few major areas of deficits are highlighted. Bigger problems and the need for accelerated efforts are seen in the following areas: nature conservation, biodiversity (although some remarkable successes are mentioned), pollution of aquatic systems (especially by agricultural effluents), subsidy policy, public participation in evironmental policy, the legal power of environmental organizations, and the transport sector. According to the OECD report, progress in developing a national sustainable development strategy has been slow, and the new states have not yet entered a sustainable development path.

Comparing the state of the art of environmental policy in Germany with that in other industrialized countries, one finds that one or another of the three basic principles (prevention, polluter pays, and cooperation) are effectively pursued in some countries, but that no country is the leader regarding all three principles. In this context, Germany is, at least, above average (see Rehbinder 1991).

If the trends in environmental quality, emissions, institutional capacities, and use of technologies to cut emissions are compared internationally, Germany's environmental achievements rank among the world's best, comparable to those of Sweden, Switzerland, Denmark, Austria, and the Netherlands.

Analysis of Environmental Policy

The effects of environmental policy, described in the preceding section, are the result of an interplay between political, administrative, legal, economic, social, and ideological factors and structures shaping the mode of policymaking and implementation—that is, the policy style—in the environmental policy area. A systematic theoretical treatment of this complex of topics is a formidable task that would require a separate study. Therefore, only some of the categories usually mentioned in political science studies (e.g., Jänicke 1990; Weale 1992; Jänicke and Weidner 1995, 1997) as being fundamental to explaining a country's environmental performance are considered here.

1. *Problem pressure* Due to the highly industrialized economy, the large sector of pollution-intensive industries, high population density, and the degree of urbanization, motorization, and traffic in Germany, the "pressure of problems" (pollution levels, polluting accidents, and so on) was extremely high in earlier years and is still relatively high. Although pollution levels were even higher in the 1960s (high air and water pollution levels in various industrial regions), they were not capable of creating corresponding "political pressure" at that time. It took deeper public awareness to transform the pressure of environmental problems into political pressure intense enough to provoke systematic government action. This happened in the course of the 1970s, having been encouraged by favorable socioeconomic conditions: the existence of a flourishing economy and a high level of material wealth, the growth of the service-oriented sector of the economy, and, combined with these, the rise of postmaterial values. But despite broad legal and organizational measures and remarkable achievements in pollution control, the overall pollution load continued to increase during the 1970s and 1980s. Several serious pollution incidents and the growing recognition that pollution threatened human health and, particularly conspicuously, forests (*Waldsterben*), led to heavy pressure on government and industry to adopt effect-oriented policies. Overall, it can be said that a medium to

high level of pollution was a fact, not fiction. As an explanatory factor for the fundamental environmental policy changes, however, it was only relevant after it had become a public issue due to frequent media coverage, increasing research activities, and mobilization of the public by environmental organizations. With the creation of green parties, environmental issues also became a matter of party political competition. Thus, the emergence of a broad environmental movement, made up of groups that represented a cross-section of society and that were able to increase public awareness of environmental issues, led to a transformation of environmental problems into a widespread environmental consciousness and, finally, to pressure on government institutions to take effective action.

2. *Economic capacity* Advanced economies exhibit not only a high level of strain on the environment, but also a better capacity to combat it. The type of economic structure a country has also helps determine its capacity for innovation. If, for example, the economic structure is shaped by "selective monopolies" (strong power position but low flexibility), prospects for an ecological modernization of the economy are poor (Jänicke 1992).

Germany's economic capacity is comparatively high. There are, of course, old-fashioned and highly subsidized branches of industry, such as iron and steel or coal mining; considerable ups and downs in the business cycle; and, since the 1990s, the country has been facing a deep economic recession. Yet, compared to many other OECD countries, Germany was—and still is—a basically rich country with a competitive economy and a flourishing "ecoindustrial" complex, which also means that there is considerable expertise in pollution-abatement technology. This ecocommercial sector includes a broad variety of "green businesses" (consultancies, ecological research institutes, ecofarming, producers of ecologically sound products, and so on) and manufacturers of pollution-abatement equipment, monitoring devices, waste incineration plants, and the like. It is estimated that at present, about 1 million jobs (nearly 3 percent of the total workforce) are created directly or indirectly by this sector (Umweltbundesamt, 1995, 5). In the 1990s, Germany's annual expenditure on pollution abatement and control (about 1.5 percent of GDP) remained one of the highest among OECD countries. Germany also belongs, together with the United States and Japan, to the leading export nations of environmental technology.

Nevertheless, the environmentally problematic industries as well as fluctuations in the economic cycle have had a strong influence on environmental policy decision making. In the 1970s, the economic slump after the first oil-price crisis led to an "antienvironmental coalition" of high-ranking government officials, business federations, and trade unions whose objective was to stop or at least slow down progress in environmental decision making. It was, however, not very successful because, even under worsening economic conditions, most Germans still gave (and give) environmental protection high priority. This forced the federal government to launch an ambitious cleanup program directed, in particular, at powerful branches of industry, such as public utilities and the automobile industry, at a time (early 1980s) of rising unemployment. In the 1990s, under the Conservative-Liberal government, influential business federations and government bodies again tried to downplay environmental policy endeavors by putting more emphasis on economic concerns, at the expense of environmental protection. They were relatively successful in reducing the dynamics of environmental policymaking, but a dramatic backlash did not happen (Rat von Sachverständigen für Umweltfragen, 1996).

The 1994 national elections demonstrated that public interest in environmental matters has not waned fundamentally: the Green Party (renamed Bündnis 90/Die Grünen) won 7.3 percent of the vote, and in some later state elections they also performed well. In the 1998 national elections, a red-green coalition government came into power. In the beginning, this created considerable fear in business circles that they would go ahead with the most ambitious environmental regulations without attending to the changing global economic situation ("a race to the top") and possible competitive disadvantages to German industry. Some influential business federations threatened to use the "exit option" (i.e., relocation of production to other countries) if environmental regulations became too demanding, but no significant transfer of firms actually occurred. In spite of initially deep conflicts between the government and industry on the nuclear phaseout and ecotax programs, a modus vivendi could be reached and a compromise-based cooperation style developed. This is due to the fact that the Social Democratic coalition partner (and especially Chancellor Schröder) demonstrated an increasingly pro-business stance in environmental conflict matters. But the Greens were also more and more prepared to compromise, accepting deviations from their election platform. Of greater importance, however, was the general strategic environmental concept of the government, setting ecological modernization, job creation, economic competitiveness, and innovation as priorities. Striving for win-win strategies became a cornerstone of government policy. The trade unions, in sharp contrast to their attitude in the 1970s, now generally support the policy of ecological modernization, expecting positive synergies of environmental protection and job creation. The establishment of the Council for Sustainable Development, in April 2001, might be seen as a symbol of a reinforced cooperative style of environmental policymaking in Germany. This council provides a forum for exploring common ground, stimulating consensus, and mediating conflicts. Its 16 members represent all segments of society (business, labor, environmental organizations, science, and so on).

3. *Integration capability of the political system* In the 1970s, German society became deeply divided over environmental issues. Soaring environmental conflicts, especially conflicts regarding nuclear power plants—resulting in mass rallies and a violent and increasingly militant protest movement—posed a significant challenge to established political institutions and their ability to achieve consensus on fundamental issues.

Several unsuccessful attempts by government and economic institutions to marginalize and discriminate against the environmental and antinuclear movements had negative consequences. The traditional coalition of interest groups, including government branches, business federations, trade unions, and established political parties, with its strong commitment to economic growth and social welfare, began to lose the ability to create and maintain a nationwide consensus on the fundamental goals of society. As a result of the increasing political costs accruing from the exclusion of environmental interests from decision making (see Delwaide 1993) and in the face of challenges from newly founded green parties, the neocorporatist network[14] opened some channels. It reluctantly gave in to new ideas on the relationship of the economy and the environment and allowed environmentally minded persons access to established institutions.

The system of proportional representation for election of members of parliament[15] makes it difficult for any single party to gain enough seats in the parliament to form a government by itself. Thus, in most cases, there is a need to form coalition governments. This system encourages political bargaining and consensus politics at, and between, all levels of government because it also applies to all federal states and, by and large, all local bodies.

This system of proportional representation, together with a relatively generous funding system for political parties, also made it possible for the environmental movement to enter parliaments (and governments). Once the so-called 5 percent hurdle is overcome (to enter parliament, a political party needs at least 5 percent of the votes), a party is entitled to official funding for its parliamentary and other work. Further, all political parties campaigning in elections and attaining more than 0.5 percent of the votes get financial compensation for their activities in proportion to the number of votes they get, irrespective of their actual expenditure.[16]

The system of federalism, in turn, provides many opportunities for new parties to "learn politics." Most of the green political activists learned their political skills at the local and state levels before they participated in federal and European elections (Raschke 1993).

The party-political system has been as significantly altered by green politics as the German political culture has been by environmentalism (Frankland and Schoonmaker 1992). The system has become much more open to new political values and minority issues. Starting in the 1980s, formerly radical "system critics" began to occupy jobs as ministers or agency directors, mainly in the field of the environment. The same applies to leading members of environmental organizations.

The relationship between representatives of environmental organizations, public administrations, and the business sector has generally improved. It is less conflicted and more cooperative (Roth and Rucht 1991). This could be explained, on the one hand, by the growing pragmatism of their ideology as well as the degree of professionalization in their active membership, and on the other, by a "greening" within certain units of public administration and some business sectors. More and more often, environmental organizations receive paid work assignments from public administrations or sponsoring contracts from business enterprises. Cooperation among these groups of actors also was smoothed fundamentally by a general consensus on the need for an ecological modernization of society and the economy, if the goal of sustainable development is to be reached. Further, there has been a notable deradicalization among environmental organizations. In contrast to their previous view, almost all of them, and especially their leading members, now consider it possible to improve environmental conditions within the existing political and economic system (see Hajer 1995; Koopmans 1995). Similar reorientations took place within the Green Party. After many battles between the two major factions—in which the so-called "Realos" (the realistic-pragmatic faction) defeated the "Fundis" (the uncompromisingly fundamentalist faction)—the Greens no longer see themselves as the extraparliamentary opposition within parliament (Raschke 1993). At present, the Green Party is represented in almost all state parliaments. In many state governments, they have been, or still are, coalition partners of the Social Democratic Party, and in some (albeit rare) cases, they have even formed local-level coalition governments with the Conservatives (CDU). Since 1998, they have formed the national government together with the Social Democrats, and the formerly radical political activist Joschka Fischer is now in office as Federal Foreign Minister, functioning, at the same time, as Vice-Chancellor of Germany. Even in the 1980s, such a development would have been inconceivable.

Other established institutions also modified their paradigms and organizational structures to include environmental protection goals. By now, almost all traditional institutions—trade unions, political parties, business federations and individual enterprises, churches, professional and scientific organizations, private and public interest groups, public administration, the educational sector, and so on—have instigated organizational changes to adapt to the environmental challenge. Thus, it now seems possible to speak of an "ecologicalization" of established institutions, though of course in varying degrees and sometimes only symbolically.

The growing impact of environmentalism on established political institutions, public agencies, and other societal institutions has caused a general shift of priorities. By and large, a new "contrat social" has emerged in which environmental protection (especially the concept of "ecological modernization," and also increasingly since the Rio Summit, "sustainable development") is a constituent. However, this does not mean the conflict over whether environmental protection should be a top priority of government has been settled once and for all, as the recent challenges from industry and influential politicians clearly demonstrate.

4. *Does federalism matter?* The German federal states (Bundesländer) are highly autonomous bodies with constitutionally guaranteed legislative and executive competencies. The Constitution (Basic Law) also provides the states with powerful means and institutions to participate in the creation of national legislation and to influence national administrative procedures. With the constitutional amendments—their enactment needed the consent of the states—in the 1970s, the federal level increased its environmental lawmaking power in order to streamline and stimulate environmental policy. Aside from some rare exceptions, the states have remained responsible for the implementation and enforcement of federal laws, and they independently create the administrative structure. The municipalities also have relevant environment-related tasks and competencies.

The three-tier federal system and EU membership create a complex multilevel decision-making system, with numerous vertical and horizontal (formal and informal) coordination and cooperation institutions for streamlining and integrating environmental policy (see Zimmermann and Kahlenborn 1994). Formally, this kind of close and permanent cooperation is required by the principle of "cooperative federalism" contained in the Constitution.

This specific type of federalism, which differs greatly from that in the United States or Canada, has advantages and disadvantages. For instance, it makes it almost impossible for the central government to rapidly and fundamentally change the main content and direction of environmental policy—either for better or for worse. It constrains the discretion of the central government in international and supranational policymaking, because the government must always consider whether it will get the support of the states for the position it takes, especially with respect to implementation. The EU legislation of the 1990s has even strengthened the position of the states by granting them more opportunities to influence the federal government's environmental policy deliberations at the EU level. These new opportunities have led to a more active role on the part of the states in EU policy, as well as to increased cooperation with the federal government, among themselves, and with subnational institutions of other countries. New forms of "transnational governance" have come into being that can act as a shelter against environmentally unfavorable globalization dynamics. The federal states and municipalities increasingly have become "environmental innovators" (especially in terms of social and institutional innovation) in trying to solve complex regional or local problems.

On the other hand, the large room for maneuvering provided to subnational authorities might lead to increased competition among them to attract industry—for example, by lowering certain environmental requirements. However, this has rarely happened so far, mainly due to the detailed national environmental standards for the siting of plants, the watchdog role played by NGOs and competing state authorities, and especially the risk that court proceedings might be initiated by various proponents of environmental policy. A new challenge to cooperative environmental federalism arose after German unification: the new states obviously put more emphasis on economic factors when balancing economic against environmental interests.

However, whereas formerly the federal states and local authorities frequently used to water down or obstruct the federal government's environmental protection projects, in the 1990s, environmental policy at the state and local levels became more progressive in certain areas than the central government's policy. This has been combined with an increasing responsiveness, especially of local authorities, to the environmental demands of the public. The "greening" of the federal states and many local bodies—in combination with their increasing self-confidence during the evolution of competitive federalism—has created effective barriers against attempts of the federal government to weaken environmental policy through federal regulations.

Therefore, the general question, "Does federalism matter?"—that is, "Has the federal system had more positive or negative impacts on environmental policy achievements?"—is by no means easy to answer. A systematic answer would go beyond the scope of this chapter; only a few points can be touched on here. There surely cannot be a simple yes or no answer to this question because the influences differ not only according to circumstances that have changed over time, but also depending on the analytic perspective taken. The matter is further complicated by the fact that the kind of federalism can change tremendously in substance over time, while keeping its formal structure. This is what has happened in Germany during the period of environmental policy investigated here. Considerable evidence shows an increase of competitive elements within "cooperative federalism" without making it a purely "competitive federalism."

The term *cooperative federalism* is used to characterize a federal system with a close-knit formal and informal communication network between the federal and state levels based on consensual negotiation and bargaining among the policy elites ("consensus politics"). In environmental policy, this feature is reflected in the cooperation principle (one of the guiding principles of German environmental policy). This principle calls for close cooperation between regulators at all levels of government, regulatees, policymakers, and the affected public (via its interest organizations), in both policymaking and implementation. In reality, however, cooperative federalism has been strongly biased, in the sense that participation in the cooperation network was almost exclusively a matter for policy elites from federal and state executive bodies, keeping parliaments and local bodies largely outside.

It was mainly due to the challenges environmentalism posed and the success of the green parties, combined with the rise of new values and ideas, such as regionalism, decentralization, "small is beautiful," comprehensibility, substantial participation, and so on, that these arrangements of executive-based, closed-shop consensus politics changed. The political decision-making process became more open and tolerant of conflict. The success of green parties in public elections, first at local and state levels and later at the federal level, brought new players into the game of politics who were not bound to the traditional basic "area of consensus" set up over decades by the established political parties and administrations. In addition, when the established parties started to compete in environmental protection policy, the traditional "area of consensus" lost its appeal and cooperative federalism was transformed into moderately competitive federalism. Concerning environmental policy, this means that a dynamic trend toward regulatory and organizational reform, as well as experimentation with new instruments and concepts, took place at state and, especially, the local level. Now, there is much more flexibility, variety, and pluralism in the German environmental policy arena. In this context, the constitutionally fixed power of the states for implementation and enforcement has also played a decisive role.

In summary, the impact of federalism on the "quality" of environmental policy in Germany is decisively determined by situational, structural, institutional, and political factors and their changing interrelationships over time. Federalism provided a political training field for the new green party organizations, as well as channels for getting access to the politico-administrative system (remember that the Greens first entered parliaments at state and local levels and, then, in 1983, at the national level).

And, last but not least, federalism provides the public not only with a political structure allowing for direct participation in concrete politics—that is, in decision making with direct visible effects "in their own backyards," but also, and relatively frequently, with the opportunity to campaign or run for political office in one of the numerous elections that take place.

5. *The Courts* The forms of judicial control of governmental actions and environmental conflict settlement are laid down in public (administrative), civil, and criminal law, for each of which different sets of courts are responsible, both at the federal and at the state level. The courts have had, at times, especially in the 1970s and 1980s, a significant influence on environmental policy implementation. However, unlike in other key political areas of government policy (e.g., abortion), the Federal Constitutional Court has not intervened on a massive scale with adjudication on environmental matters, although some of its decisions have helped to make the concept of precaution more viable. Up to now, criminal law and the civil courts have played only a minor role in environmental

policy. A few spectacular decisions by the Federal Supreme Court—in particular, on questions of liability—have helped to increase industry's risk awareness with respect to both products and activities.

Nevertheless, administrative courts play an important role in environmental policy and conflicts. They exercise comprehensive judicial control over administrative actions, whereby the rights of individuals have been violated (see Article 19 GG), and they are often mobilized by third parties in the course of licensing or planning procedures, with the aim of achieving tighter environmental standards or stopping projects or operating plants. But the regulatees also frequently initiate court action against the administration. The lower courts, in particular, have often ruled in favor of environmental concerns. These courts also started quite early with a thorough review of the interpretation and application of broad statutory terms by the public administration, as well as the adequacy of standards set in administrative directives. They have often challenged the administration's decisions and proceeded to develop standards on their own initiative. This has resulted in a heated debate on the administrative courts' increasing intervention in the government's areas of competence. The higher courts and the Federal Administrative court have overruled a number of these decisions and generally emphasized government's scope for policy design.

Taking legal action is only possible if the injured party's individual statutory rights have been unlawfully harmed by the act or omission of the administration. This concept of the judicial review of administrative action is reflected in the restrictive criteria for access to administrative courts, which, for example, generally do not allow class action or "altruistic" action by an individual. Only the nature conservation laws of (most) states give recognized NGOs access to the courts. In stark contrast to the situation in the United States, further obstacles, not only to standing-to-sue, but also to substantive participation in environmental decision making, are presented by limited public access to administrative and private information relevant to the case. Implementation of the EU Directive on Access to Environmental Information has improved the situation, however. A Federal Supreme Court decision in 1984 and the Environmental Liability Act of 1990 also relaxed certain restrictions concerning burden of proof.

The European Court of Justice increasingly comes into play. It is responsible for clarifying and interpreting European environmental law and is activated through appeals from various actors, including environmental organizations and referrals from member state courts. Since the Treaty of Maastricht (1992), it is allowed to fine national governments for failing to comply with its judgments. As a rule, the decisions of the European Court have supported environmental measures.

All in all, increasing court involvement—virtually all major environmental disputes have ended up in court—and successful obstructions of private and public projects have encouraged experimentation with new approaches to conflict resolution, especially in the form of mediation procedures. The basic impetus came from the United States, where so-called environmental dispute resolution techniques have been applied for a longer period. The first major mediation procedures in Germany were introduced in 1990. Meanwhile, there have been promising experiences with about 200 cases. The use and acceptance of these negotiation-based instruments is a further indication of the growing popularity of cooperative instruments among all relevant environmental actor groups (see Weidner 1998).

6. *International Influences* Germany's environmental policy is, in many respects, highly sensitive to developments in foreign countries and to global environmental policy. As noted, it shares borders with nine other countries; all of its major rivers have their sources abroad; it is dependent on the import of nonrenewable natural resources; it is one of the world's leading export countries; and it is a member of several international organizations, most importantly, the EU. The major stimuli for the development of modern environmental policy in the 1970s came from abroad.

Whereas in the 1970s the federal government frequently played a restrictive role in the international environmental policy arena, it became a more proactive and progressive player in the 1980s, because such a changed attitude seemed to serve the German (especially economic) self-interest better. There are also important areas in which German environmental policymaking has been positively influenced by international developments—for example, by the laws on environmental impact assessment and access to environmental information and regulations on toxic waste export, as well as policies against ozone depletion and global warming. Further, there is plenty of evidence that the influence of globalization on national environmental policy is not as negative as often claimed. This is not only due to the business sector's considerable interest in well-functioning and rising global environmental standards—for example, because of their interest in expanding export opportunities for pollution control technology or limiting economic competition by "ecodumping." There are also increasingly clear signs that ecologically unsound strategies are becoming more and more vulnerable. Facilitated by new communication technologies, NGOs' networking, and so on, a kind of globalization of environmental policy is advancing rapidly. Networked NGOs and international institutions (such as UNEP, OECD, the World Bank, and the EU Environment Agency) are monitoring environmental activities more and more closely, and the dissemination of information on deficits, as well as innovation and progress, is rapid and worldwide. Local Agenda 21 processes are just one of many examples of such rapid diffusion of innovative practices in Germany because of the efforts of international institutions. Although globalization and the influence of international institutions are often used as an excuse for environmental deficiencies in Germany, the arguments have a weak empirical basis. Shortcomings in national environmental policy are often the main cause (Jänicke and Weidner 1997). On balance, international institutions and regimes have played a stimulating role in the development of Germany's environmental policy.

Conclusion

For over a quarter of a century now, Germany has had a systematic modern environmental policy, which is both institutionally and legally well grounded within the general political system. The foundations for this policy were laid within the space of a few years by the Social Democrat–Liberal coalition government, which came into power in 1969. The coalition between the Conservatives and Liberals, which took shape in late 1982, did not confirm the initial fears of many people who thought that they might implement a weak environmental policy due to their strong ties to economic interest groups. A variety of factors led to progressive developments. These factors included a heightened environmental consciousness in society, some spectacular and widespread pollution events (especially "forest dieback"), and growing interest within the business sector in the environmental markets (rapidly expanding in the EU and some other regions). Other factors included the challenges in the political arena coming from the ecological movement, green parties, and the "greenified" Social Democratic Party (Padgett 1994), which had made the concept of "ecological modernization" one of its battle cries. Thus the new government passed pioneering legislation, brought about massive drops in pollution levels in specific problem areas, and stimulated the development of environmental policy, both at EC and international levels. The measures taken and their outcome have assured Germany high standing in the field of progressive environmental policy.

The main winner in this relatively new political field is the so-called ecoindustry. Its rate of growth is above average, and it has reached a leading position in the world market. Since the 1980s, "green" enterprises and environmentally committed business associations—for example, B.A.U.M., Future, and Unternehmensgrün—have become the motor of ecological modernization by way of "integrated" cleaner technologies and as a generator of "environmentally friendly" demand. Also, some indications of a transition from mainly reactive, curative measures and instruments to a more precautionary approach should be noted. The reform of environmental liability legislation that was finally passed by the Bundestag in September 1990, after years of discussion, was an important step in this direction. The introduction of liability for risks irrespective of fault related to industrial plants, the easing of proof of causality, and the extension of the definition of liability to also apply to the normal operation of plants is more than a simple cosmetic improvement to environmental law along conventional lines. These regulations have put pressure on the owners of plants that are potentially damaging to the environment to take greater precautions to avoid environmental hazards than in the past. The concept of climate protection, especially the ambitious goal of reducing CO_2 emissions by 25 percent of their 1990 level by 2005, has also supported a new ecological orientation—in this case, in the field of energy policy. The Packaging Ordinance of 1991[17] and the comprehensive Waste Management Act of 1994[18] opened up new paths in waste management policy toward recycling of waste or avoidance of its production wherever possible. In November 1994, an amendment to the Constitution (Art. 20a GG) made environmental protection a "state goal" (*Staatsziel*).[19]

The new challenges to environmental policy posed by the economic recession, which began after the unification of the two German states, put a brake on the more progressive developments just mentioned. However, in contrast to the great challenge to environmental policy posed by economic interest groups after the first oil crisis in the 1970s, the government units responsible for environmental protection have defended their territory more offensively this time, (Umweltbundesamt, 1993). A massive rollback in environmental policy would also be hard to achieve because a substantial ecological consciousness has become a permant part of the German value system (Jänicke and Weidner 1997; Weidner 1997).

The search for outstanding differences in the strategic approach and range of instruments used in the environmental policy of the Conservative-Liberal government, as compared to their Social Democrat–Liberal predecessors, leads to the conclusion that, for a long time, the strategies and approaches have remained fundamentally the same. This is surprising in view of their very different ideologies and party programs.

This could largely be explained by the peculiarities of the German political system, which makes it difficult to bring about swift, radical change in environmental policy. This is so even though conditions favorable to a modernization of environmental policy have already existed within society and its institutions, as well as in the political and administrative system and in important economic sectors. These favorable conditions include a broad distribution of environmental expertise, many varied forms of cooperation between environmental policy innovators from all areas of society, extensive technical and informational resources, well-organized networks of environmental interests, and, not least, a relatively broad and stable consensus as to the necessity of an environmental modernization of industrial society.

The slow pace of change in environmental politics is consistent with the general hypothesis that the institutional conditions in German politics make a fundamental change of strategy a long drawn-out process.

As a rule, major changes require broad political and social consensus, typically gained through a complicated and time-consuming process. The roots of this approach are in the neocorporatist pattern for solving problems and in the associated institutional fabric. This fabric is made up, for example, of the specific form of German federalism and of the requirement for the rule of law and the constitutional opportunities to scrutinize fundamental political decisions that this entails, as well as of the proportional voting system, which seldom leads to clear-cut political majorities.

The old neocorporatist mode of cooperation of the postwar period, between state, business associations, and unions (the "iron triangle"), is gradually developing into a network (the "green triangle") in which organizations specifically representing environmental concerns are now also able to participate in political decisions. More and more parts of the business sector have developed a keen interest in stringent environmental regulations, creating new markets and expanding those already established.

Further stimuli for going ahead with environmental policy have come from abroad. The German environmental movement has been strengthened not only by relatively progressive EU directives and programs (e.g., the 5th Environmental Action Program), but also by initiatives from international organizations, such as UNEP or OECD, and from the new environmental "pioneer countries" (e.g., Sweden or the Netherlands). All in all, the influence of international organizations on German environmental policy has been salutary.

Further, an expanding, independent, highly specialized "ecobureaucracy" has arisen and is developing its own esprit de corps. It has opened the door to a new kind of employee: an expert with a specialized environmental education gained both before and during employment, younger and often highly motivated, and brought up in a social atmosphere of increasing "greenness." Membership in environmental organizations is no longer a rarity. Cooperation with such organizations is no longer a problem for these administrators.

The victory of the Social Democrats and the Greens in the national elections of 1998–both parties had emphasized "ecological modernization" of German society and industry during their election campaigns—

demonstrates that a majority of the German population supports a progressive environmental policy, even in times of globalization. However, the economic recession and the challenges of globalization, along with essential changes in the attitudes and interests of the key environmental actor groups supported by the upcoming paradigm of "sustainable development," have created an increasing preference for flexible and cooperative instruments and procedures, including voluntary agreements and economic instruments. Therefore, the policy style has changed significantly from a command-and-control, reactive, and curative type to a flexible, cooperative, and effect-oriented one. This is the result of a complex process of institutional change, public pressure, influences from abroad, new cognitive "framing" of environmental issues among the central actor groups, growing environmental capacities of all kinds, and expectations that a stringent environmental policy will strengthen the country's position in global competition—just to name the most relevant factors.

Nevertheless, it is doubtful whether the concept of ecological modernization is sufficient to pave the way for sustainable development. The concept basically rests on expectations that conflicting goals can be resolved by consensus and by developing win-win strategies. However, it is already clear that transforming the most problematic areas—agriculture, land use, stabilization of resource use, and transport—in the direction of sustainable development will not only create "losers," but will also come up against structurally vested, powerful interests. As yet, the new red-green government has not systematically focused its environmental policy on those most restrictive areas and powerful target groups—with the exception of the agricultural sector, where the mad cow disease opened a window of opportunity. But the short period in office (about two and a half years) is admittedly, not enough to make a fair prediction about whether the government will make progress in the problem areas mentioned, similar to the progress achieved in energy policy.

Notes

1. See the fundamental study by Müller (1986) on the environmental policy of the Social Democrat–Liberal government.

2. Optimism concerning the government's ability to move society toward greater social equity and participatory democracy was so strong that this period has been described as one of planning and reform euphoria. See Glaeßner, Holz, and Schlüter 1984.

3. On the reasons for the choice of the Ministry of the Interior as the body with competencies for the environment, see Müller 1986, 55–60.

4. In 1972, an amendment to the Federal Constitution (Article 74) increased the legislative power for environmental regulations of the national parliament.

5. For a comprehensive description of the environmental movement, see Brand, Büsser, and Rucht 1986; Brun 1978; Ellwein, Leonhard, and Schmidt 1983; Frankland and Schoonmaker 1992; Hrbek 1988; Leonhard 1986; Linse 1986; Poguntke 1992, 1993; Raschke 1993; Roth and Rucht 1991; Rucht and Roose 1997; Rucht 1980.

6. The environmental decisions of the Social Democrat–Liberal government are reproduced and commented on in *Umwelt* (BMI) *91*, September 14, 1982.

7. The Chernobyl catastrophe not only led to increased antinuclear sentiments among Germans (see Koopmans 1995, 203ff.), but also to a revision of basic policy on nuclear power plants within the Social Democratic Party (see Rüdig 1990).

8. GG = *Grundgesetz* (the German Constitution).

9. This allows the federal government to enact general (i.e., framework) laws, which then are "filled in" by the state (Länder) legislatures. Several initiatives to establish concurrent jurisdiction for the water and nature protection law have failed because of opposition from the Upper House. According to Article 79, section 2 GG, an amendment changing the Constitution requires a two-thirds majority in both houses.

10. Directives are addressed to the governments of the member states. With respect to their implementation, in most cases, it is the *goal* of the direction that counts. The *way* this goal is to be achieved is not fixed and can be decided by the national governments.

11. These provisions were modified and amended by the Treaty of Maastricht. Among other things, majority voting was introduced for most matters of environmental policy.

12. For an excellent and comprehensive assessment of the development of EU environmental policy, see Weale et al. 2000.

13. For example, in 1993 the Federal Minister of the Environment issued instructions as prescribed under Article 85, paragraph 3 66, in the conflict over the storage of radioactive waste to the Minister of the Environment for the state of Hesse. This was possible because the federal minister in this case has jurisdiction over the federal states in matters of "nuclear law."

14. The German system of political interest mediation is—compared to competitive-pluralistic systems—often characterized as a neocorporatist system in

which consensus on basic economic and social issues is pursued by the large business federations, trade unions, and the government (see Schmitter and Lehmbruch 1979; Schmidt 1990, 1992).

15. Furthermore, an important feature of Germany's national electoral politics is the lack of connection between the representative and his or her constituency. Thus, the population has little opportunity to reward or punish politicians directly by voting.

16. In December 1993, the Parties Act of 1967/1988 was amended. The new act came into force in January 1994. This created a completely new (and more favorable) system of state financing for political parties. The parties now receive state funds, the level of which depends (as under the old system) on their election success (votes received) and on the level of membership dues and donations received. For each vote received in elections for state parliaments, the Bundestag, and the European Parliament, the party receives 1 deutschmark (DM) and DM 1.30 for each of the first 5 million votes. In addition to this, for every deutschmark received in membership dues and donations, the state coffers add DM 0.50.

17. The Packaging Ordinance obliges industry and suppliers to take back virtually all packaging. It is combined with the "Green Dot" program, named after the recycling logo printed on packaging and containers. Under the program, manufacturers of consumer products pay a licensing fee for the "Green Dot," which is supposed to cover the costs of privately organized collection and recycling of packaging materials. The system is managed by *Duales System Deutschland*, a private enterprise founded to coordinate collection and recycling. Since the introduction of this system, total waste volume has decreased considerably. As a result, fewer waste incineration plants will have to be built than public authorities had anticipated.

18. The *Kreislaufwirtschaftsgesetz*, literally translated: "Law for a Circulation-Based Economy."

19. Since November 1994, environmental protection is—as a state goal—part of the Federal Constitution (Art. 20a GG). A state goal, in contrast to a basic individual right (*Umweltgrundrecht*), primarily functions as a normative guideline for the legislature. A basic right, however, would grant rights to the citizen that could be directly claimed by recourse to the courts.

References

Boehmer-Christiansen, S., and Skea, J. (1991). *Acid politics*. London: Belhaven Press.

Boehmer-Christiansen, S., and Weidner, H. (1995). *The politics of reducing vehicle emissions in Britain and Germany*. London: Pinter.

Bomberg, E., and Peterson, J. (1998). European Union decision making: The role of sub-national authorities. *Political Studies*, 46, 219–235.

Brand, K. W., Büsser, D., and Rucht, D. (1986). *Aufbruch in eine andere Gesellschaft*. Frankfurt am Main: Campus.

Brun, R., ed. (1978). *Der grüne Protest. Herausforderung durch die Umweltparteien* (Green protest: Challenges from environmentalist political parties). Frankfurt am Main: Fischer.

Bundesministerium für Umwelt, Naturschutz und Reaktorsicherheit. (2000). Halbzeit! Zwischenbilanz der Umweltpolitik 1998–2000 (Halftime! Interim balance for environmental policy 1998–2000). Berlin: Bundesministerium für Umwelt.

Bundesministerium für Umwelt, Naturschutz und Reaktorsicherheit/Umweltbundesamt. (2000). Umweltbewußtsein in Deutschland 2000 (Environmental consciousness in Germany). Berlin: Bundesministerium für Umwelt.

Delwaide, J. (1993). Postmaterialism and politics: The "Schmidt SPD" and the greening of Germany. *German Politics*, 2(2), 243–269.

Drexler, A., and Czada, R. (1987). Bürokratie und Politik im Ausnahmefall. Untere Vewaltungsbehörden nach "Tschernobyl" (Bureaucracy and politics in exceptional situations: The lower-tier administration after "Chernobyl"). In A. Windhoff-Héritier, ed., *Verwaltung und ihre Umwelt. Festschrift für Thomas Ellwein* (The public administration and its environment: Publication in honor of Thomas Ellwein) (pp. 66–90). Opladen: Westdeutscher Verlag.

Ellwein, T., Leonhard, M., and Schmidt, P. M. (1983). *Umweltschutzverbände in der Bundesrepublik Deutschland. Research report for the Umweltbundesamt.* Berlin: Umweltbundesamt.

Franke, N. (1992). Zwischen politischer Brisanz und ökologischer Irrelevanz—Die Rolle der Umweltberichterstattung in der ehemaligen DDR. In H. Weidner, R. Zieschank, and P. Knoepfel, eds., *Umwelt-Information. Berichterstattung und Informationssysteme in zwölf Ländern* (pp. 411–447). Berlin: edition sigma.

Frankland, E. G., and Schoonmaker, D. (1992). *Between protest and power: The Green Party in Germany*. Boulder: Westview Press.

Glaeßner, G. J., Holz, J., and Schlüter, T., eds. (1984). *Die Bundesrepublik in den 70er Jahren*. Opladen: Leske + Budrich.

Hajer, M. A. (1995). *The politics of environmental discourse: Ecological modernization and the policy process*. Oxford: Clarendon Press.

Hartkopf, G., and Bohne, E. (1983). *Umweltpolitik*, Vol. 1: *Grundlagen, Analysen und Perspektiven* (Environmental policy, Vol. 1: Foundations, analyses, and perspectives). Opladen: Westdeutscher Verlag.

Héritier, A., Knill, C., and Mingers, S. (1996). *Ringing the changes in Europe: Regulatory competition and the transformation of the state*. Berlin: de Gruyter.

Hey, C., and Brendle, U. (1994). *Umweltverbände und EG. Strategien, politische Kulturen und Organisationsformen*. Opladen: Westdeutscher Verlag.

Holzinger, K. (1995). A surprising success in EC environmental policy: The Small Car Exhaust Emission Directive of 1989. In M. Jänicke, and H. Weidner, eds.,

Successful environmental policy: A critical evaluation of 24 cases (pp. 187–202). Berlin: edition sigma.

Hrbek, R. (1988). Umweltparteien. In O. Kimminich, H. Freiherr von Lersner, and P.-Chr. Storm, eds., *Handwörterbuch des Umweltrechts* (vol. 2, pp. 646–652). Berlin: Erich Schmidt.

Hucke, J. (1990). Die Entwicklung eines neuen Politikfeldes (The development of a new policy area). In K. v. Beyme, and M. G. Schmidt, eds., *Politik in der Bundesrepublik Deutschland* (Policies and politics in the Federal Republic of Germany) (pp. 382–398). Opladen: Westdeutscher Verlag.

Jahn, D. (1993). *New politics in trade unions*. Aldershot: Dartmouth.

Jänicke, M. (1990). *State failure*. Cambridge: Polity Press.

Jänicke, M. (1992). Conditions for environmental policy success: An international comparison. *The Environmenalist*, *12*(1), 47–58.

Jänicke, M. (2000). Ökologische Modernisierung als Innovation und Diffusion in Politik und Technik (Ecological modernization as innovation and diffusion in politics and technology). FFU Report 00-01. Berlin: Freie Universität Berlin.

Jänicke, M., Mönch, H., and Binder, M., eds. (1993). *Umweltentlastung durch industriellen Strukturwandel? Eine explorative Studie über 32 Industrieländer (1970 bis 1990)*. 2nd ed. Berlin: edition sigma.

Jänicke, M., and Weidner, H., eds. (1995). *Successful environmental policy: A critical evaluation of 24 cases*. Berlin: edition sigma.

Jänicke, M., and Weidner, H., eds. (1997). *National environmental policies: A comparative study of capacity building*. Berlin: Springer.

Kitschelt, H. (1985). Political opportunity structures and political protest: Anti-nuclear movements in four democracies. *British Journal of Political Science*, *16*, 57–85.

Kloepfer, M. (1989). *Umweltrecht*. Munich: Beck.

Koopmans, R. (1995). *Democracy from below: New social movements and the political system in West Germany*. Boulder: Westview Press.

Lahusen, C., ed. (2000). The role of cooperation in European environmental policy. *Environment—The Journal of European Environmental Policy*, *10*(6) (Special Issue), 251–308.

Leonhard, M. (1986). *Umweltverbände*. Opladen: Westdeutscher Verlag.

Linse, U. (1986). Ökopax und Anarchie. Eine Geschichte der ökologischen Bewegungen in Deutschland. Munich: Deutscher Taschenbuch Verlag.

Malunat, B. M. (1987). Umweltpolitik im Spiegel der Parteiprogramme. *Aus Politik und Zeitgeschichte*, *B 29/87*, 29–41.

Malunat, B. M. (1994). Die Umweltpolitik der Bundesrepublik Deutschland. *Aus Politik und Zeitgeschichte*, *B 49/94*, 3–12.

Mez, L. (1995). Reduction of exhaust gases at large combustion plants in the Federal Republic of Germany. In M. Jänicke, and H. Weidner, eds., *Successful environmental policy: A critical evaluation of 24 cases* (pp. 173–186). Berlin: edition sigma.

Müller, E. (1986). *Innenwelt der Umweltpolitik. Sozial-liberale Umweltpolitik—(Ohn)macht durch Organisation*? (The interior world of environmental policy: Social-liberal environmental policy—(dis)empowerment through organization?). Opladen: Westdeutscher Verlag.

Müller, E. (1989). Sozial-liberale Umweltpolitik. Von der Karriere eines neuen Politikbereichs. *Aus Politik und Zeitgeschichte, B 47–48*, 3–15.

Norton, A. (1994). *International handbook of local and regional government: A comparative analysis of advanced democracies*. Aldershot: Edward Elgar.

Organization for Economic Cooperation and Development. (1993). *OECD Environmental Performance Reviews: Germany*. Paris: Organization for Economic Cooperation and Development.

Organization for Economic Cooperation and Development. (1994). *OECD Environmental Performance Reviews: Germany*. Paris: Organization for Economic Cooperation and Development.

Organization for Economic Cooperation and Development. (2001). *OECD Environmental Performace Reviews: Germany*. Paris: Organization for Economic Cooperation and Development.

Padgett, S. (1994). The German Social Democratic Party: Between old and new Left. In D. S. Bell, and E. Shaw, eds., *Conflict and cohesion in Western European Social Democratic parties* (pp. 10–30). London: Pinter.

Pehle, H. (1989). *Das Bundesministerium für Umwelt, Naturschutz und Reaktorsicherheit: Ausgegrenzt statt integriert*? (The Federal Ministry for Environment, Nature Protection and Reactor Safety: Separated instead of integrated?) Wiesbaden: Deutscher Universitätsverlag.

Peters, H. P., Albrecht, G., Hennen, L., and Stegelmann, H. U. (1987). Die Reaktionen der Bevölkerung auf die Ereignisse in Tschernobyl. Ergebnisse einer Befragung (The public's response to the events in Chernobyl: Results of a survey). *Kölner Zeitschrift für Soziologie und Sozialpsychologie, 39*(4), 764–782.

Petschow, U., Meyerhoff, J., and Thomasberger, C. (1990). *Umweltreport DDR. Bilanz der Zerstörung, Kosten der Sanierung, Strategien für den ökologischen Umbau*. Frankfurt am Main: Fischer.

Poguntke, T. (1992). Between ideology and empirical research: The literature on the German Green Party. *European Journal of Political Research, 21(4)*, 337–356.

Poguntke, T. (1993). *Alternative politics: The German Green Party*. Edinburgh: Edinburgh University Press.

Raschke, J. (1993). *Die Grünen. Was sie wurden, was sie sind*. Cologne: Bund-Verlag.

Rat von Sachverständigen für Umweltfragen (SRU). (1987). *Umweltgutachten 1987*. Stuttgart: Kohlhammer.

Rat von Sachverständigen für Umweltfragen (SRU). (1989). *Kurzfassung des Sondergutachtens Altlasten*. Bonn: Bundesministerium für Umwelt.

Rat von Sachverständigen für Umweltfragen (SRU). (1994). *Umweltgutachten 1994. Für eine dauerhaft-umweltgerechte Entwicklung*. Stuttgart: Metzler-Poeschel.

Rat von Sachverständigen für Umweltfragen (SRU). (1996). *Umweltgutachten 1996. Zur Umsetzung einer dauerhaft-umweltgerechten Entwicklung*. Stuttgart: Metzler-Poeschel.

Rat von Sachverständigen für Umweltfragen (SRU). (1998). *Umweltgutachten 1998. Umweltschutz. Erreichtes sichern—neue Wege gehen*. Stuttgart: Metzler-Poeschel.

Rat von Sachverständigen für Umweltfragen (SRU). (2000). *Umweltgutachten 2000. Schritte ins nächste Jahrtausend*. Stuttgart: Metzler-Poeschel.

Rehbinder, E. (1989). The Federal Republic of Germany. In T. T. Smith Jr., and P. Kromarek, eds., *Understanding U.S. and European environmental law: A practitioner's guide* (pp. 8–20). London: Graham & Trotman/Martinus Nijhoff.

Rehbinder, E. (1991). *Das Vorsorgeprinzip im internationalen Vergleich*. Düsseldorf: Werner-Verlag.

Rehbinder, E. (1992). Rethinking environmental policy. In G. Smith, W. Paterson, and S. Padjet, eds., *Development of German politics* (pp. 227–246). London: Macmillan.

Rehbinder, E., and Stewart, R. (1985). *Environmental protection policy*. Berlin de Gruyter.

Rose-Ackerman, S. (1995). *Controlling environmental policy: The limits of public law in Germany and the United States*. New Haven: Yale University Press.

Roth, R., and Rucht, D., eds. (1991). *Neue soziale Bewegungen in der Bundesrepublik Deutschland* 2nd ed. Bonn: Bundeszentrale für politische Bildung.

Rucht, D. (1980). *Von Wyhl nach Gorleben. Bürger gegen Atomprogramm und nukleare Entsorgung*. Munich: Beck.

Rucht, D., and Roose, J. (1997). The German environmental movement at a crossroads? *Environmental Politics*, *8*(1), 59–80.

Rüdig, W. (1990). *Anti-nuclear movements: A world survey of opposition to nuclear energy*. Harlow, Essex: Longman.

Schmidt, M. G. (1990). Die Politik des mittleren Weges. Besonderheiten der Staatstätigkeit in der Bundesrepublik Deutschland. *Aus Politik und Zeitgeschichte*, *B9–10*, 23–31.

Schmidt, M. G. (1992). *Regieren in der Bundesrepublik Deutschland*. Opladen: Leske + Budrich.

Schmitter, P., and Lehmbruch, G., eds. (1979). *Trends toward corporatist intermediation*. London: Sage.

Sprenger, R.-U. (1979). *Beschäftigungseffekte der Umweltpolitik*. Berlin: Erich Schmidt.

Süß, W., ed. (1991). *Die Bundesrepublik in den achtziger Jahren*. Opladen: Leske + Budrich.

Umweltbundesamt. (1993). *Umweltschutz—ein Wirtschaftsfaktor*. Berlin: Umweltbundesamt.

Umweltbundesamt. (1995). *Umweltdaten Deutschland 1995*. Berlin: Umweltbundesamt.

Umweltbundesamt. (2000). *Daten zur Umwelt. Der Zustand der Umwelt in Deutschland 2000* (Environmental data: The condition of the environment in Germany 2000). Berlin: Erich Schmidt Verlag.

Weale, A. (1992). *The new politics of pollution*. Manchester: Manchester University Press.

Weale, A., Pridham, G., Cini, M., Konstadakopulos, D., Porter, M., and Flynn, B. (2000). *Environmental governance in Europe: An ever closer ecological union?* Oxford: Oxford University Press.

Weidner, H. (1986). *Air pollution control strategies and policies in the Federal Republic of Germany*. Berlin: edition sigma.

Weidner, H. (1995). *25 years of modern environmental policy in Germany: Treading a well-worn path to the top of the international field*. WZB-paper FS II 95–301. Berlin: Wissenschaftszentrum Berlin für Sozialforschung.

Weidner, H. (1998). *Alternative dispute resolution in environmental conflicts: Experiences in 12 countries*. Berlin: edition sigma.

Weidner, H., ed. (1997). *Performance and characteristics of German environmental policy: Overview and expert commentaries from 14 countries*. WZB-paper FS II 97–301. Berlin: Wissenschaftszentrum Berlin für Sozialforschung.

Weidner, H. (1999). Umweltpolitik: Entwicklungslinien, Kapazitäten und Effekte (Environmental policy: Lines of development, capacities and effects). In M. Kaase, and G. Schmid, *Eine lernende Demokratie—50 Jahre Bundesrepublik Deutschland* (A learning democracy—50 Years Federal Republic of Germany) (pp. 425–460). WZB-Jahrbuch 1999. Berlin: edition sigma.

Weßels, B. (1989). Politik, Industrie und Umweltschutz in der Bundesrepublik: Konsens und Konflikt in einem Politikfeld 1960–1986 (Politics, industry and environmental protection in the Federal Republic of Germany). In D. Herzog, and B. Weßels, *Konfliktpotentiale und Konsensstrategien. Beiträge zur politischen Soziologie der Bundesrepublik* (Potentials for conflict and consensual strategies: Contributions to a political sociology of the Federal Republic) (pp. 267–306). Opladen: Westdeutscher Verlag.

Wey, K.-G. (1982). *Umweltpolitik in Deutschland. Kurze Geschichte des Umweltschutzes in Deutschland seit 1900*. Opladen: Westdeutscher Verlag.

Wicke, L. (1989). *Umweltökonomie. Eine praxisorientierte Einführung*. 2nd ed. Munich: Franz Vahlen.

Winkelmann, Chr. (1992). *Untersuchung der Verbandsklage im Umweltrecht im internationalen Vergleich. UBA-Forschungsbericht Nr. 90104/36*. Berlin: Umweltbundesamt.

Zimmermann, K. W., and Kahlenborn, W. (1994). *Umweltföderalismus. Einheit und Einheitlichkeit in Deutschland und Europa*. Berlin: edition sigma.

6

Environmental Protection in Italy: Analyzing the Local, National, and European-Community Levels of Policymaking

Rudolf Lewanski and Angela Liberatore

Environmental degradation has transformed both the landscape and the quality of life in Italy in many ways. Air pollution in cities and contamination of soil, groundwater, and seas—with their impacts on ecosystems, human health, and Italy's cultural heritage—are some cases in point.

From the mid-1960s, Italy began to institute some policy measures to deal with environmental problems. Initially ad hoc and piecemeal, environmental policy in Italy developed in legislative and institutional terms and acquired higher visibility and political legitimacy, though many problems still hamper its implementation and effectiveness.

Though a number of approaches based on different factors could be usefully adopted to shed light on the features and outcomes (some successes and many failures) of environmental policy in Italy, this chapter focuses on the interactions between the relevant institutional actors at the local, national, and supranational (especially at the European Community or EC) levels. Center-periphery cleavages have characterized Italian environmental policy since its beginning and are likely to continue in the future. At the same time, with the emergence and consolidation of EC[1] environmental policy, the interplay between the EC and member states appears to be of special relevance in shaping the development of Italian environmental policy.

This chapter first describes the main environmental problems that exist in Italy and then discusses Italian environmental policy's primary stages of development. Many of the processes and problems that will be analyzed are hardly typical of Italy alone. A feature somewhat peculiar to the Italian case, however, is the mix of rapid social and economic modernization and the consequent development of phenomena that negatively affect environmental resources, on the one hand, and the delay, at least compared to other similar countries, in adopting responses to the problems, on the other. Some possible explanations for this situation are offered in the following paragraphs. In particular, the local, national, and supranational actors and their interests and interactions will be analyzed to account for the policy's "style" and outcomes, including implementation deficits.

Environmental Problems in Italy: An Overview

Like many other European countries, Italy entered a phase of accelerated economic growth and industrialization in the 1950s. While this allowed Italy to become one of the "Group of 7" most developed Western countries, it brought about unequal development within Italy (the northern part of the peninsula being economically more advanced than the south), and serious environmental disruption occurred throughout the country.

Italy, still renowned as the "Garden of Europe" in the early 1900s, has suffered a sharp decline in environmental quality due to economic activities, urban sprawl, and traffic. Many of the large industries that grew rapidly in the aftermath of the Second World War—such as chemicals, steel, cement, refineries, and paper—are highly polluting. In 1999, there were 1,194 (179 less than in 1996) industrial plants presenting serious environmental risks, as defined by the so-called Seveso Directives issued by the EC in 1982 and 1996 (Legambiente and Istituto di ricerche Ambiente Italia, 2001, 162). Also, specialized industrial districts (perhaps the real secret of the Italian "economic miracle"), spread across the northern and central regions, are responsible for serious pollution due to the concentration of a great number of small- and medium-sized factories of the same type in relatively limited territories. For example, 250 ceramic tile factories, producing 30 percent of the world's total sales, are located in an area of 50 square kilometers near Modena.[2]

With a population of 57 million living in an area of 300,000 square kilometers, Italy's average population density is 191 inhabitants per square kilometer. In urban areas, especially in the Po Valley and along the coasts, the figure soars to 2,000, making them among the most densely congested areas of Europe. Built-up land has increased from 120,000 square kilometers in 1961 to 260,000 in 1986, occupying approximately 90 percent of national territory and destroying both rich farmland and areas of natural value. In the Po Valley, for example, land "consumption" for construction purposes proceeds at a rate of 0.6 percent per year (Lega per l'Ambiente, 1989, 227). An indirect indicator of this process is the per capita consumption of cement, which is three times higher than that of the United States, Germany, or Great Britain. Public land-use policies have often been unable to manage urban expansion effectively. Construction without the required permits became widespread during the 1960s and 1970s, especially in southern Italy. On average, some 15 percent of all existing civilian buildings are constructed without the required permits, but in certain areas (Sicily, Campania), they represent more than one-sixth of the total (Legambiente and Istituto di ricerche Ambiente Italia, 2001, 237).

Lack of attention to environmental quality has had, and continues to have, negative effects on traditionally important economic sectors, such as tourism, which is a major source of income due to Italy's favorable climate, its world-renowned artistic heritage, and its 7,500 kilometers of coastline. Large industrial plants and infrastructure projects—often promoted by public authorities—have been sited in areas of considerable scenic and natural value, particularly along the coasts. Emissions and vibrations produced by traffic in ancient cities,[3] among other things, are responsible for the decay of the cultural heritage represented by historic monuments and buildings.

Although seemingly improved in quality over recent years, portions[4] of Italy's coastal waters often have had to be declared off limits for swimming due to pollution exceeding limits established by EC directives. Both organic and inorganic substances (toxic metals and agrochemicals, in particular), from household, agricultural, and industrial sources, cause high levels of pollution in surface waters. Italy's water pollution treatment capacity is estimated at about one-half to one-third of total discharges, considerably lower than in other comparable European countries, such as France at 60 percent, Great Britain at 80 percent, and

Germany at 90 percent (Ministero dell'Ambiente, 2001, 279; Conte and Melandri 1994, 301). The lack of adequate treatment of urban discharges causes health hazards involving, for example, the consumption of contaminated seafood caught in polluted waters or of vegetables irrigated with polluted water.[5]

The major rivers, Po, Adige, and Tevere, appear to have seriously deteriorated, and their condition has actually worsened over the last several years (Conte and Melandri 1994, 286, 298–300). Forty-one percent of all lakes have high levels of eutrophication; the same phenomenon, in recent years, has caused algae blooms and consequent fish deaths in the Adriatic Sea, with heavy losses in the tourism sector. Oil pollution is also common along the peninsula's beaches. Twenty-five percent of world oil production goes through the Mediterranean Sea (which represents only 0.8 percent of the globe's total waters), and approximately 44 percent of all petroleum and its products transported by ship in the Mediterranean basin are directed to, or leave from, Italian ports. A number of ship accidents causing major oil spills and serious environmental damage have occurred in the last decade.

On the other hand, tourism itself is a cause of environmental problems. Excessive concentrations of seasonal visitors have serious negative impacts on fragile ancient monuments and city centers (Venice being an extreme example) and on natural ecosystems, such as the coasts and seawaters. For example, 45 percent of the peninsula's beaches are subject to erosion (Ministero dell'Ambiente, 2001, 156), which modifies sea currents. This is often due to human interventions (wave barriers, boat marinas, and so on) connected to tourism.

The above-mentioned phenomena, coupled with the absence of effective preventive measures, are also responsible for numerous environmental emergencies throughout the country. Thirteen percent of Italian municipalities (for a total of 1,037) have been officially recognized as presenting a highly deteriorated environmental situation; more than 18,000 square kilometers—6.2 percent of the country's land area, inhabited by more than 11 million people—have been declared hazardous due to high levels of environmental pollution (Ministero dell'Ambiente, 2001, 36). Intensive stock-breeding and cultivation methods that rely on agrochemicals are heavily concentrated in the four regions of the Po

Valley (Lombardy, Piedmont, Emilia-Romagna, and Veneto) and are causing pollution of soil and underground water reserves by nitrates and pesticides. As a consequence, water-supply systems serving 2 million people in more than 300 municipalities in northern Italy had to be closed in past years (88 percent of their drinking water comes from underground reserves).

Solid refuse is another serious problem because Italy lacks adequate disposal facilities for urban and industrial refuse, including toxic waste. Ninety-seven million tons of waste are produced yearly, including 39.2 million tons of industrial waste, 10 percent of which is classified as toxic. Treatment plants, 90 percent of which are landfills often not in compliance with regulations, have a capacity of about two-thirds of the total amount of urban waste and 40 percent of the industrial waste produced yearly. In 1998, 77.4 percent of urban waste went to landfills (much of which did not comply with legal requirements), 7.3 percent was incinerated (in part, in plants that produce electricity with the heat deriving from refuse combustion), and the rest was treated in other ways and, to some extent, recycled (11.2 percent was collected through the separation of different types of refuse) (Ministero dell'Ambiente, 2001, 197).

Still another relevant issue is soil erosion and landslides (estimated at about 3,000 per year), caused by building and road construction, excessive mining (7,000 sites were counted in 1980), timber cutting without replanting, mechanization of agriculture in hilly areas, and lack of maintenance of water-drainage systems in geologically fragile land (approximately one-sixth of Italy's total territory has been classified as highly unstable), periodically causing disruptive floods.

Emissions of several air pollutants, such as sulfur dioxide, suspended particulates, and nonmethane volatile organic compounds (VOCs), have decreased or remained stable over the last two decades or so, whereas greenhouse gases have continued to increase (Ministero dell'Ambiente, 2001). Thanks to its relatively isolated geographic location—separated from the rest of Europe by the Alps (Europe's highest mountain chain) to the north, and surrounded by water on the remaining sides—and thanks to the prevailing easterly winds, Italy is somewhat affected by pollution produced in other countries, but, on the whole, is a net "exporter" of airborne pollutants.[6]

The effects of acid rain are starting to be recognized in Italy, as elsewhere in Europe. About one-third of the sulfur deposits and one-half of the oxidized nitrogen come from abroad (United Nations Environment Program, 1991, 43), causing forest decline (along with more traditional forms of damage, such as fires that destroy about 0.6 percent of the forests every year). In a 1989 survey, 15 to 17 percent of the forests were classified as damaged by acid rain, an increase from the 4 to 6 percent in 1985. This is certainly less serious than in parts of Central Europe where a 50 percent rate of *Waldsterben*[7] is registered, but a source of concern, nevertheless. The total area of regional and state parks has doubled since the mid-1980s, now comprising 10 percent of the country (though still short of the official policy target of 10 percent). Also, four marine parks have been established. Thirteen of 97 mammal species are on the verge of extinction, whereas 14 flora species are extinct. The survival of dozens of others is seriously threatened.

The Evolution of Italian Environmental Policy: Phases and Actors

For purely heuristic purposes, the development of environmental policy in Italy can be divided into four distinct phases based on several major turning points in its brief history. These turning points are represented by the passage of relevant legislation and the entrance into the policy arena of new institutional actors.

Phase 1: The Birth of National Environmental Policy

The birth of environmental policy in Italy dates back to the mid-1960s. The first explicit piece of legislation that attempted to tackle a problem, namely air pollution, specifically in environmental terms (rather than for its impact on human health or for its repercussions for economic activity) was Act 615, passed in 1966. The aim of this law, as well as several minor laws of the same period, was to "patch up" specific, geographically limited problematic situations. "Antismog" Act 615 dealt mainly with air pollution affecting large urban areas in northern Italy, where frequent thermal inversions occurring during the cold seasons are coupled with mountainous conditions that hinder dispersion.[8] The core of the "Antismog" Law's strategy was to reduce the sulfur content of fuels used for domestic heating; it only enabled local authorities to deal with emissions from industrial sources to a limited extent. Actual implementation and enforcement of air pollution control policies essentially began in a few areas of northern Italy in the early 1970s, when regulations were issued by the central government (i.e., the Health Ministry), with considerable delay with respect to deadlines prescribed by the act, itself.[9]

In this phase, environmental issues had very low political visibility, and for several years, the "Antismog" Law remained an isolated example of remedial action taken against pollution by the central government. In the absence of adequate legislation, responses to increasing pollution in the 1970s came from local authorities and individual judges (at the time called *pretori d'assalto* or "assault judges"), both of whom tried to tackle specific problems by means of the available—antiquated—laws, such as the 1934 "Health Laws" or others actually meant for the protection of particular economic activities, such as fishing or navigation.[10]

Politicians in this period were hardly motivated to get involved with an issue that had no relevance for interparty or intraparty dynamics, and that attracted little attention from the media and public opinion. It was not remunerative in electoral terms, especially in a period in which, after the oil shocks, the economy was a high priority. Ecologists exerted little influence: the only well-established ecological organization, Italia Nostra, was mainly concerned with the protection of the treasures of Italy's cultural heritage, such as monuments. It should be noted that, in the Italian context, the natural and the built/artistic environment are often tightly interwoven, as the later creation of a Ministry for Environmental and Cultural Goods indicates. Other associations, such as the Italian branch of the World Wildlife Fund (WWF), founded in 1966, were still in their infancy and had small memberships.

Phase 2: Supplying the Legislative Toolbox

A second major turning point in environmental policy occurred in 1976. Three significant events took place: (1) the passage of a second major piece of legislation—this time in the field of water pollution; (2) the upgrading of the role of the relevant institutional actors—that is, the regions; and (3) the dioxin pollution accident in Seveso, which greatly increased public awareness of environmental problems.

From the mid-1970s to the mid-1980s, efforts to create sound environmental policy were accelerated as a number of previously missing legislative instruments were supplied to the "toolbox." Act 319, aimed at reducing water pollution, was finally issued in 1976 after about ten years of debate. Several other relevant acts dealing with industrial and household waste disposal, sea protection, and detergent biodegradability were also issued in this period.

The Water Pollution Control Act is relevant, not only because it finally tackles a severe environmental problem, but also because of the approach it adopts to reach its goals. Although mainly based on typical regulatory means (emission standards), for the first time it introduced economic tools in the form of positive incentives—state allocations granted to municipalities and private firms for the construction of discharge treatment facilities—and negative incentives—a tax on water effluents to be levied by municipalities (though, in fact, it has hardly been applied) (Lewanski 1986, 171).

The policy arena during this period was still limited to a small number of actors, although new ones—both institutional and social—began to appear. As far as institutional actors are concerned, a first move toward establishing a central authority in charge of this specific policy area was the creation of an Interministerial Committee for Environmental Protection, which had the task of coordinating competencies on environmental matters dispersed among 16 different ministries. The major institutional actors, regions, and local health units (about 630 decentralized units—USL—of the National Health System instituted in 1978[11]) were established at the peripheral level. The regions were originally created in 1970, but were actually awarded substantial powers only in 1976.[12] Many of the regional responsibilities, such as nature and natural beauty protection, parks, hunting and fishing, pollution, health, land use, city planning, agriculture, and mining, are directly or indirectly related to environmental policy.

With respect to social actors, a strong antinuclear movement, together with local authorities who opposed the siting of nuclear plants, contributed to limiting nuclear power programs. (Other factors were also involved, such as the greater strength of the oil lobby than the nuclear lobby.) The movement, because of its leftist roots, also characterized the environmental issue in anticapitalist terms.[13] Together with the antinuclear movement, the existing environmental organizations began to grow and others were established. Lega per l'Ambiente, presently the largest Italian environmental association, was founded in 1980. Several local "Green Lists," which eventually formed the Green Party at the national level, were also founded during this period.

The visibility and political relevance of environmental problems increased considerably in this period, as a result of popular reaction to the central government's attempt to launch a large nuclear power plant construction program (as in many other European countries). Other catalysts included serious industrial accidents—especially in the chemical sector—that attracted media attention and caused a profound shift in public opinion on environmental issues. Particularly important in this respect was the July 1976 accident at the Swiss-owned ICMESA chemical plant in Seveso, a little town in the Lombardy region. As a consequence of dioxin emissions, thousands of residents and workers were evacuated from the area and 220,000 people were placed under medical and epidemiologic surveillance. Many suffered health problems from which they have not recovered, even today.

Phase 3: Approaching Maturity

A third turning point in the development of environmental policy in Italy was the creation, in 1986—after a number of unsuccessful attempts—of the Ministry of the Environment, an authority specifically in charge of environmental policy at the central level. The new ministry was empowered with responsibilities in the fields of air, water and soil, noise pollution, solid waste, mining, parks, and sea and coastal protection. However, the preexisting ministries were successful in maintaining control of other fields related to the management of natural resources. For instance, the Ministry of Agriculture was able to retain jurisdiction over hunting, an activity obviously relevant for the protection of wildlife, a major natural resource. Hunting is a hotly debated issue in Italy and traditionally of great political interest, since 2 million hunters/voters represent a substantial pressure group. Ecologists have promoted several national and regional referenda against hunting, as discussed later in the chapter. The Ministry of Public Works maintained its jurisdiction over water management and land-use planning. In other areas of critical importance for the environment as well, such as energy, industrial, and agricultural policies, the Environmental Ministry has only limited power to impose ecologically oriented measures, as opposed to more traditional technical and economic ones. This, together with other problems, such as its relatively limited staff, budget, and technical expertise, has weakened the effectiveness of the Environmental Ministry. Nevertheless, the establishment of the ministry officially marked the upgrading of the status of environmental policy vis-à-vis other sectoral policies.

The new ministry's first priority was to remedy previously nonexistent or inadequate legislation pertaining to hazardous waste, industrial air emissions, landscape, soil protection, use of water resources,[14] noise pollution, and so on. Solid and, especially, hazardous waste represented an issue to which the ministry was forced to direct considerable attention because of having to enact EC directives. Action was also required because of a number of crises that broke out in that period, both at home and abroad, such as the drinking-water pollution problem caused by illegal dumping and the discovery of illegal toxic waste dumps that Italy was responsible for in developing countries.

Public expenditures by central, regional, and local authorities with environmental mandates increased substantially in the late 1980s. In 1988, the environmental expenditures reached 1 percent of the gross national product (GNP) (Ministero dell'Ambiente, 1989, 332–337), considered internationally to be the minimum necessary threshold for a developed nation. However, in the last few years, it has slipped back to 0.8 percent—in comparison to an EC-country average of 1.2 percent—as a result of the government's priority to reduce public indebtedness.[15] It should be pointed out that the introduction of substantial distributive elements in environmental policy represents an important source of consensus, both for the Environmental Ministry and for the policy itself. On the other hand, some of these resources are actually used on a regular basis to deal with environmental disasters (floods, landslides, and so on) that are often caused by environmentally unsound policies. These policies allow, for example, for destructive building projects and deforestation, or they fail to require preventive action, as in the case of cleanups of heavily polluted areas declared to be of "high environmental risk," like the Lambro and Bormida rivers and the Venice and Naples areas. Furthermore, until 1998, the ministry's actual ability to spend allocated funds proved rather low, with only about a third of the allocated sums spent. This was the lowest spending rate among all central ministries; the average spending capacity of central ministries is 50 percent.[16]

Although allocations for scientific research in the environmental field have certainly increased over the years, from 0.6 percent of total expenditures for research in 1975 to 2.8 percent in 1991, public expenditures in this area remain low compared to those of many other industrialized nations. In 1990, public research and development expenditures in the environmental field amounted to 216,196 million liras.

The lack of funding, however, does not appear to be the only problem, since Italian research institutions typically do not take full advantage of funds made available at the international level—for example, from the EC or the European Science Foundation. A further weakness of research carried out in this area is the lack of coordination of activities promoted by a variety of public authorities, such as the National Research Council (CNR), the Ministry for University and Scientific Research, the Environment and Energy National Agency (ENEA), the Ministry of Industry, and the Environmental Ministry itself.[17] Finally, and perhaps most important, there is little evidence of the actual relevance of scientific research to, or spillover effects of this research into, the policymaking process.

During this period, the visibility of environmental problems increased, due to accidents like those at Bhopal and Chernobyl, which were widely reported in the media. A growing demand for environmental protection by the public, partly related to the shock caused by the disasters, caused voters to reject the Italian nuclear power program through a referendum held in 1987.[18]

An indicator of the demand for environmental quality, represented by data provided by polls and nationwide surveys, shows that the concern of Italians for local, national, and international environmental problems is at least as strong as that of people in other European countries, if not stronger (Organization of Economic Cooperation and Development, 1993, 288–290). Such concern, however, does not necessarily translate into willingness to take action, as the limited electoral successes of the Greens in local and national elections seem to indicate. After participating in local and regional elections since the early 1980s, the Greens obtained seats in the national Parliament for the first time in 1987, with 2.5 percent of the votes. In the elections for the European Parliament of 1989, the Greens received 6.2 percent of the votes. In the political elections of 1994, they received 2.7 percent, and in the European ones of the same year, 3.2 percent. With the exception of 1989, when two different lists including other smaller parties were competing, the Greens' share of the votes appears to have stabilized at around 2.5 to 3 percent. Significant regional variations exist, however. The Greens generally do better in the northern areas, and receive lower-than-average percentages of the votes in the south and on the major islands Sicily and Sardinia.

The main environmental organizations are the Italian branches of the WWF, Greenpeace, the Friends of the Earth, as well as a national association, Lega per l'Ambiente. Ecological organizations had some 1.5 million members in all in 1993, a figure that makes this the largest sector among all voluntary organizations. However, this total includes the 600,000 members of the Touring Club and the Alpine Club, which are not really—or at least not primarily—environmental groups.

During this period, the attitudes of the business world toward environmental issues also began to change. Although vigorously opposing extreme policies, many industries began to realize the inevitability, if not the necessity, of natural resource protection.[19] As a result of public policies that imposed antipollution measures, some sectors of industry discovered that environmental protection could represent a profitable opportunity. The annual proceeds from green business, at present, are in the range of approximately 6,500 billion liras, more than 4 billion U.S. dollars.[20] The emergence of green business can be explained in light of three interwoven factors: (1) increasing government regulation, which has forced polluting industries to modify technologies and install emission-abatement equipment; (2) the growing availability of public financial resources allocated for environmental purposes, which have created substantial business opportunities; and (3) growing environmental sensitivity among consumers in favor of greener products. Also, concerning the spread of the cleaner technologies and products just mentioned, national firms have had an incentive to modify attitudes and strategies because of the realization that manufacturers in other EC countries were enjoying a competitive advantage due to the fact that they had started moving in this direction years earlier. This might help explain the adoption of environmental protection measures by several large firms, such as the FARE (Fiat Auto REcycling) project set up by the major national automobile manufacturer and aimed at solving the waste disposal problem represented by discarded vehicles.

Finally, criminal organizations, such as the Mafia and Camorra, play a special role with respect to environmental policy. They are often involved both in large public works—for example, highways, dams, and so on—with significant negative impacts on the environment, and in activities directly connected with environmental policy, such as refuse collection and disposal.[21]

1996–2001: The Emergence of the Environmentalists

Until 1992, the development of environmental policy took place in a relatively stable political context in which governments were typically formed by a coalition including the Catholic and Socialist parties, jointly with other smaller moderate political partners. With the disclosure of extensive corruption within the governing parties and the administration (*Tangentopoli*, meaning the "city of kickbacks"), the Italian "First Republic," born after World War II, came to an end. The two main parties disappeared from the political scene and new parties (such as Forza Italia) emerged. Other parties to the right and left that had previously been kept out of the governing coalitions—such as Alleanza Nazionale and the Democratici di Sinistra respectively—started to play an important role in the political game. The crisis opened a period of great uncertainty, from which a new political order has yet to emerge.

Under such circumstances, environmental issues were hardly a priority during the transitional phase following the "Tangentopoli" crisis. In the Berlusconi government (1994–95), the Environmental Ministry was headed by a representative of the extreme right who showed no interest in environmental protection and openly supported nuclear energy, highways, and hunting in national parks. In the following "technical" Dini government (1995–96), environmental issues received greater attention, though the fact that the Minister of Public Works headed the Environmental Ministry met heavy criticism from the environmental movement.

The national elections in April 1996 were won by the "Olive" (center-left) coalition, headed by Romano Prodi. Somewhat paradoxically, the Greens, while having one of their worst showings yet (2.5 percent—i.e., 937,000 votes), were rewarded with the largest number of representatives ever in the lower chamber (a gain of 10 seats, from 18 to 28), thanks to preelectoral agreements among the parties of the coalition. They thus acquired considerable bargaining power within the coalition and became decisive for the survival of the center-left government. The Green Party, which entered a government for the first time in Italy, was assigned the posts of the Minister of the Environment (Edo Ronchi) and that of undersecretary within the Ministry of Public Works. However, it should be noted that the "green" component of the coalition was larger than the Green Party, since representatives of various environmental organizations were also given a number of relevant posts within public agencies, such as the National Electricity Board (ENEL) and the state-owned railway company (FS). More important, the environment also represents an important concern on the agenda of the major party in the coalition, the Democratici di Sinistra. In October 1998, a reshuffling of the government took place, as the coalition's center of gravity shifted rightward and Massimo D'Alema replaced Prodi as prime minister. Nevertheless, there was considerable continuity between the two governments. The coalition supporting the new government remained, by and large, the same (though the extreme left exited the coalition and was replaced by a small center party). Many ministers also kept their posts, including the Minister of the Environment, Ronchi. The D'Alema government ended on April 19, 2000, after poor results in the election of the new regional councils and presidents. A new cabinet, headed by Giuliano Amato (previously a member of the Socialist Party), was formed with the votes of the same coalition as the former government, which remained in power until May 2001.[22]

Despite the new political situation represented by the first Italian government to include a relevant green component, environmental issues were not initially among the government's priorities. Economic issues (inflation, employment, public debt, and especially the process of entering the European Monetary Union) together with institutional matters (federalism, electoral rules) were at the forefront of the government's concerns. To boost employment and the economy and to gain consensus, the government was often tempted to bring back traditional "recipes," such as public works policies (especially new roads and highways, on the grounds that poor transportation negatively affects Italian competitiveness in international markets). Environmental issues, however, were soon recalled, especially after the disaster in the Campania region in May 1998. A huge landslide of some 3 million tons of mud overran several small towns in the provinces of Salerno, Avellino, and Caserta, killing 250 people and severely damaging houses, farms, and industries. In the context of this disaster, environmental policy acquired both legitimacy, in policy terms, and a considerable public profile. Further, such "natural" disasters convinced the government—and, to some extent, public opinion—that stronger environmental management policies were required. For example, during recent floods that hit the Po Valley, Prime Minister Giuliano Amato implicated the greenhouse effect as the cause.

In any case, during this five-year period, the center-left-green governments have, in fact, passed a number of significant measures, providing for the following:

- Creation of many new land and marine parks,[23] along with an increase in resources allocated for their management (art. 2 of Act 426, 1998); also, to cope with serious threats to natural areas, special legislation was passed to prevent forest fires (Act 353 of 2000).[24]
- Promotion of forms of transportation having a reduced impact on the environment, especially for freight (see the General Transportation Plan passed by the government, in the year 2000, covering the next ten years[25]) and in urban areas.
- Promotion of recycling and reuse of urban and industrial solid waste (Decrees 22/97 and 389/97; Ministero dell'Ambiente, 2000); national legislation stipulated that 30 percent of municipal waste should be recycled by the year 2002, though municipalities have complained that they are unable to comply.

- Protection of the stratospheric ozone layer (Act 179/97 has strengthened previous measures; Act 409 of 2000 authorized Italy's financial contribution to the Multilateral Fund for the Montreal Protocol).
- Prevention of relevant industrial accidents (provisions 137 of 1997 and 334 of 1999), enacting the Seveso II directive (no. 96/82).
- Imposition of a 1 percent tax on a number of pesticides and types of livestock feed; the income obtained from this tax will go into a fund for promoting organic agriculture[26] and enhancing food quality.
- Standards limiting exposure to electromagnetic emissions (from radio, television, and cellular phone antennas, as well as from high-voltage transmission lines) on the basis of a national plan to reduce the number of sites with TV and radio antennas (from 700 to 487) and the transmission power of such antennas by a factor of 30 (Act 381 of 1998). Specific legislation (Act 55 of 2001) was passed with the objective of protecting the population and workers from electromagnetic radiation. All authorizations that have been issued in recent years to telecommunications companies have included specifications concerning environmental and health protection. The financial law of the year 2000 (art. 112 of Act 388, 2000) allocates funds to prevent and reduce such emissions.
- Promotion of sustainable development, especially in urban areas.
- Reduction of noise emissions, especially from transportation facilities such as railways and airports (e.g., the decree of November 29, 2000).
- Use of environmental projects (water management, renewable energy sources, national parks) to create new job opportunities.
- Cleanup of highly polluted and hazardous land and water sites (such as industrial areas; funds have been allocated by Act 426 of 1998).
- Demolition of illegal structures built inside protected natural areas; the Environmental Ministry can take action in case municipalities fail to do so (Act 426 of 1998).
- Reduction of CO_2 emissions, enacting the Kyoto agreement (Italy has committed itself to reducing emissions by 6.5 percent by 2008–2012, relative to 1990 levels; at present, emissions have actually increased by 4 percent; Ministero dell'Ambiente, 2001, 119).

• Promotion of energy saving, energy efficiency, and renewable-energy sources:[27] since the liberalization of the electricity-generation market—previously a public monopoly—each firm is required to produce a minimum of 2 percent by means of renewable-energy sources (art. 11 of Act 79, 1999).

• Increased use of nonregulatory policy instruments, such as taxes (e.g., on carbon emissions, airplane noise, waste, and so on),[28] and increased use of voluntary tools and agreements, discussed below.

• A substantial increase in financial resources allocated (from 747 to 1,164 billion liras in 1998), as well as in the actual spending capacity of the Environmental Ministry (approximately 40 percent in the same year; LegAmbiente and Istituto di ricerche Ambiente Italia, 2001, 201).

Intergovernmental Relationships

As mentioned, in its initial phase in the mid-1960s, environmental policy was by and large local policy in which local authorities, municipalities,[29] and provinces,[30] joined by regions after their creation in the 1970s, played an important role, not only in implementing national laws but in actually initiating policies. Such activism of the "periphery" continued into the 1980s and, in some respects, continues today, despite a number of difficulties and limitations that will be described below. To local authorities attempting to cope with spreading pollution, the national laws at times represented more of a hindrance to their initiatives than a useful resource. On the other hand, in recent years, the Environmental Ministry appears to have used the state-controlled legal and financial resources in an attempt to shift the tide from the periphery to the center. The present section examines the conflicting trends that characterize center-periphery relations in this policy area.

Regional and local authorities have played an important role in stimulating the development of environmental policy. Municipalities, being the institutions nearest to citizens, are often the first to grasp emerging social demands. In attempting to respond, they either become agents of political and administrative innovation by directly experimenting with solutions that are subsequently adopted and extended by the central state, or they act as pressure groups, requesting state legislation and intervention in new areas, such as the environment. Recycling policies aimed at reducing the quantities of refuse to be disposed of and policies prohibiting the use of certain products, such as plastic shopping bags (to limit the danger of chlorine compounds produced by incinerators), are examples of administrative innovation. An example of the second type of process would be pressure exerted by health officials and mayors of major northern cities, such as Milan, Turin, and Genoa, who asked the central government to help resolve the frequent smog problems, thus bringing about the passage of the "Antismog" Law (no. 615/66).

Regions have fulfilled similar roles. The Water Pollution Act of 1976 is to a large extent based on legislation passed by the Lombardy region two years earlier. The Tuscany region and the province of Trient[31] had issued discharge limits of their own. In this case, the central ministries apparently sensed that they were losing an opportunity to establish their authority over a new policy area. They were also caught in the middle between regional initiatives, on the one hand, and the first EC directives—issued in 1975–76—on water quality, on the other. Furthermore, economic interests were concerned about a possible fragmentation of the national market, with different rules applying to various areas.

Still another example is offered by policies aimed at containing eutrophication of the Adriatic Sea in the late 1980s and early 1990s. One of the simplest measures to take was to reduce the phosphate content of detergents. By the late 1970s, the Emilia Romagna region, which was particularly motivated to take action because of the importance to its economy of Adriatic coast tourism, had persuaded manufacturers to sign a voluntary agreement stipulating that detergents sold in the four provinces nearest to the coast would have a reduced phosphate content. The solution was not satisfactory, however, since huge amounts of pollutants were carried to the Adriatic from the rest of northern Italy by major rivers—the Po in particular. And, during the following decade, eutrophication and algae blooms became more and more frequent. When the algae problem became a serious threat to tourism in the summer of 1988, numerous municipalities along the coast—Venice and Ravenna among others—issued ordinances reducing the allowable phosphate content of detergents to 1 percent. This was the lowest possible limit foreseen by the national law in 1986. Although the administrative courts (T.A.R.), pressed by detergent manufacturers, overturned these municipal provisions, the Ministry of Health was compelled to revise its previous 2 percent limit, bringing it down to 1 percent. Starting in 1996, local and regional authorities were proactive in introducing processes that Agenda 21 (adopted at the 1992 Rio Earth Summit) recommended for implemention by local authorities. At present, some 200 administrations have signed the Aalborg and Göteborg charters. But the central government responded belatedly, eventually offering support to local authorities in connection with policies concerning urban areas, such as the Environmental Ministry decree of October 8, 1998, or of July 19, 2000.

Another dimension of local activism is represented by bargaining relationships that local authorities, at times, are able to establish with the central state in order to obtain the additional resources required to resolve local environmental problems. Emergency situations can offer the opportunity to strengthen their power vis-à-vis their national counterpart. A particularly interesting case of this type was the discovery of illegal toxic waste from Italian companies that was sent to Nigeria and Lebanon in 1988. The exposé of this problem nearly caused a diplomatic crisis. After the waste was shipped back at the expense of the Italian government, the problem became how to dispose of it, given the lack of hazardous waste treatment facilities and the opposition of local communities in Italy. Ultimately the regions of Tuscany and Emilia Romagna—ruled by the major opposition force, the Communist Party—stepped in. The authorities there enjoyed sufficient credibility to guarantee acceptance of the waste by local residents. The areas governed by the parties in power in Rome, in contrast, proved totally hostile. But an important provision was added to the agreement. A portion of the Emilia Romagna territory was determined to contain extremely hazardous environmental conditions, which increased its regulatory powers and brought in additional national financial resources. Also, the provision allowed for the building of treatment plants and the upgrading of professional knowhow in the local public agencies at the expense of the central government. These resources were subsequently used in the cleanup of illegal dumps

Table 6.1
Italy: A Statistical Profile

Land	
Total area (1,000 km^2)	294.00
Arable and permanent crop land (% of total) 1997	37.20%
Permanent grassland (% of total) 1997	15.50%
Forest and Woodland (% of total) 1997	23.30%
Other land (% of total) 1997	24.10%
Major protected areas (% of total) 1997	7.30%
Nitrogenous fertilizer use (t/km^2/arable land) 1997	8.40%
Population	
1998 (100,000 inhabitants)	577.00
Growth rate % (1980–1998)	2.20%
Population density (inhabitants/km^2)	191.50
Gross domestic product	
Total GDP (billion U.S. dollars) 1998	$1,050.00
Per capita (1,000 U.S. dollars/capita)	$18.20
GDP growth 1980–1998	36.80%
Current general government	
Revenue (% of GDP) 1996[b]	45.60%
Expenditure (% of GDP) 1996[b]	49.40%
Government employment (%/total) 1997[b]	15.80%
Energy	
Total supply (MTOE) 1997	163.00
% change 1980–1997	17.80%
Consumption (MTOE) 1998[c]	128.89
% change 1988–1998[c]	14.60%
Road-vehicle stock	
Total (10,000 vehicles) 1997	3,389.00
% change 1980–1997	74.90%
Per capita (vehicles/100 inhabitants) 1997	59.00
Air profile and greenhouse gases	
CO_2 emissions (tons/capita) 1998[a]	7.40
CO_2 emissions (kg/USD GDP) 1998[a]	0.41
Methane 1990	3,901,300.00
Nitrogen oxide emissions (kg/capita) 1998[a]	31.0
Nitrogen oxide emissions (kg/USD GDP) 1998	1.70
Nonmethane volatile organic compounds 1990	2,400,600.00
Sulfur oxide emissions (kg/capita) 1998[a]	23.10
Sulfur oxide emissions (kg/USD GDP) 1998	1.30

Table 6.1
(continued)

Pollution abatement and control	
Total expenditure (%/GDP) 1997	0.90%
Household expenditures excluded	

Sources: OECD, *OECD Environmental Data, Compendium 1999* (Paris: OECD, 1999), except as noted.
[a] OECD, The OECD Observer, *OECD in Figures* (Paris: OECD, June 2000).
[b] OECD, *National Accounts of OECD Countries* (Paris: OECD, 2000).
[c] IEA/OECD, *Energy Balances of OECD Countries, 1997–1998* (Paris: IEA/OECD, 2000).

found in the region itself and in the treatment of locally produced toxic waste.

The third type of "input" provided by local authorities has been more negative than the first two. In Italy, as elsewhere, the siting of projects with highly negative environmental impacts, or at least perceived as having such impacts, led to increased environmental awareness and generated harsh center-periphery conflicts (Knoepfel 1984, 2298), as shown by attempts to build nuclear power plants in the latter half of the 1970s. In such cases, municipalities and regions became the institutional expression of the local communities by taking an antagonistic position toward the central authorities, even before the rise of the antinuclear movement at the national level. In the early 1970s, conflicts arose between local and central authorities—in particular, the Ministry of Industry and the Electric Energy National Body (ENEL) in the Regions of Molise, Sardinia, Latium (in relation to the Montalto di Castro plant), Piedmont (Trino Vercellese), and Emilia Romagna (Caorso). An agreement was finally reached only in the last two cases, due especially to substantial economic compensation provided by ENEL.[32] As far as the other cases are concerned, Parliament attempted to overcome the local authorities' opposition by introducing two laws, Acts 880/1973 and 393/1975, aimed at curtailing the powers of regional and municipal authorities in the siting procedures (Spaziante 1980). But opposition was so strong that, in the end, the central government proved unable to actually go ahead with its projects. Similar dynamics have occurred in relation to numerous large infrastructure and industrial projects, such as chemical plants, waste disposal facilities, and, more recently, the construction of high-speed train lines.

The central government and the Environmental Ministry, in particular, have resorted to several instruments to increase their control over sectoral policy. Policy tools introduced by the ministry have included the so-called program agreements. According to Act 305/89, "Triennial Plans" should coordinate all activities and expenditures carried out by public authorities in the environmental field. The operational instruments of such plans are represented by program agreements that indicate priority actions and the allocations to be contributed by each level of government (the central state, regions, and local authorities). It is hard to disagree with the philosophy underlying such an approach, since the idea is to coordinate the efforts and resources of various authorities in a situation where considerable institutional fragmentation exists. In practice, however, the use the ministry makes of such decision-making procedures indicates that they are a powerful means of shaping regional and local choices. An analysis of the ministry's action, in this respect, reveals its intention to determine the allocation of financial resources by spelling out, in a detailed manner, the projects to be financed, thus substantially limiting regional and local autonomy.

Another channel through which central authorities have attempted to exert influence over environmental policy can be found in an important piece of legislation, Act 183, passed in 1989 after about 20 years of discussion. Although the act deals with soil protection, it is also concerned with water management, since water causes erosion and floods (more than 45 percent of Italian municipalities face a high or very high level of hydrogeological risk; Ministero dell'Ambiente, 2001, 144). When the 1976 Act on water pollution was under consideration, the newly formed regions strongly opposed the proposal to create ad hoc water authorities based on successful French and British models, as an attempt by the central state to rescind powers that had just been assigned to them. The opposition of the regions, backed by pro-regional political parties such as the Communists, was successful in the end, and the idea of creating water basin authorities was set aside at that time. The 1989 Soil Protection Act, however, was more successful. According to the size of the water basin, three different types of authorities—national, interregional, and regional—were to be created, for a total of 39. By 1998, all national authorities, but only some regional and interregional authorities, had actually been set up; recent legislation (Act 180 of 1998) has accelerated the process. National basin authorities in reality cover most of the territory and are overseen by the central Ministries of Public Works and the Environment, leaving only minor areas under the jurisdiction of the regional authorities.

All in all, the Environmental Ministry has definitely played an active role in recent years. It has promoted policies even on typically minor matters, such as urban waste practices (by promoting recycling, for example) or urban air pollution (by promoting production of less polluting vehicles), as previously mentioned.

The EC Context and Its Influence on Environmental Policymaking in Italy

Importance of the EC Context

The analysis of environmental policy in Italy, as in any other member state, cannot be separated from the context of the European Community, for three main reasons: (1) the priority of EC law, including environmental legislation, over national law; (2) the process of economic integration, mainly through trade liberalization, that increases the interdependence of EC countries with respect to environmental regulation (different regulations could, in fact, represent "nontariff barriers" to free trade); and (3) the reciprocal influences between member states and the EC institutions in policy formulation.

The principle of the supremacy of EC law—already stated, though not in clear terms, in the 1957 Treaty of Rome establishing the European Community—has been developed through jurisprudence in the European Court of Justice. On several occasions, the European Court affirmed that national courts can invalidate national legislation if it contradicts EC law and that the Community's rights can be invoked before national judges.[33] The Italian Constitutional Court did not accept the full supremacy of EC law over national legislation until 1984.

As far as the second factor is concerned, EC legislative activism on environmental matters started in the early 1970s. A main reason for establishing a specifically EC environmental policy was economic—that is, the need to avoid excessive competition within the Community and to gain or maintain access to non-EC markets, especially those of the United States and Japan. An example of this is Directive 85/210 on the coordination of member states' legislation concerning the lead content of gasoline, following the decision by Germany and subsequently other countries to introduce lead-free gasoline. Different standards in EC member states not only hampered the EC gasoline trade, but this lack of coordination also represented an especially important economic and environmental issue in Italy. The tourist sector had an objective interest in standardization, given the significant number of German tourists who come to Italy by car. The oil-refinery sector and the car manufacturers also could not avoid meeting the challenge of adapting to a new product being launched in the EC market.

Finally, on the reciprocal influences of member states and EC institutions, it must be stressed that the EC Commission—particularly the Directorate General for the Environment, Civil Protection, and Nuclear Safety (DG XI)—is the institution that initiates, mainly in form of communications to the EC Council of Ministers, the process of formulating and drafting proposals for new EC directives and other kinds of instruments aimed at protecting the environment. The commission, however, acts on the basis of pressure from and in consultation with member countries. In this respect, it is commonly acknowledged that some member countries, like Denmark, Germany, and the Netherlands, are more active than others, including Italy, in pushing for the adoption of EC environmental measures. The approval of directives and other actions is determined by negotiations between the EC institutions (mainly the commission) and the representatives of member countries and interest groups,[34] on the one hand, and negotiations among member countries, especially within the Council of Ministers, on the other. It is not unusual for the negotiations leading to the adoption of a directive to take several years.[35]

Two aspects of the reciprocal interaction between EC institutions and member states in the development of the Community's environmental policy can, therefore, be identified. First, member states' initiatives and intergovernmental negotiations are crucial "motors" of EC policymaking. Second, EC legislation and policy shape member countries' legislation and policies. The latter aspect is especially important for traditionally "laggard" countries, such as Italy.

Italy in the EC

A large number of environmental regulations have been introduced through Italian legislation only because of pressure to comply with Community directives. This has been the case with legislation related to industrial air emissions, ambient-air-quality standards, environmental impact assessment, toxic waste, and other issues. Italy's reluctance to get involved in this field has been so pervasive that EC directives have even preceded Italian legislation with respect to problems of specific, while not exclusive, concern in Italy. A case in point is the EC directive on major accident hazards, also called the "Seveso Directive" after the name of the little Lombardy town where the chemical-plant accident took place in 1976. The Seveso Directive was issued in 1982 after long negotiations within the Community and after accidents had occurred in other European countries. The directive addressed a field not yet regulated in Italy despite the Seveso accident, and it was incorporated into Italian legislation only in 1988 (DPR 175 of May 1988).

Such a delay is not unique. Until recently, Italy held the far-from-enviable record of EC environmental law infringements (Capria 1988). To find a way out of this situation, a first enabling act was passed in 1987 (Act 183/1987) that gave the government power to enact EC legislation and overcome parliamentary delays. Two years later, another law, known as "La Pergola" Law (no. 86 of March 1989), was adopted to complement Act 183/1987. La Pergola Law ensures the acceptance of the EC Court of Justice judgments, provides for regular communications to the Italian Parliament on the implementation of EC legislation, and defines the role of the regions in the implementation of EC law (regions, with the exception of special regions, could not apply EC law directly; a national law has to be enacted first). The enactment of these laws, as mentioned, represented an opportunity for the Environmental Ministry to introduce previously missing environmental legislation in the latter half of the 1980s (Alberti and Parker 1991; Liberatore 1991).

In the early 1990s, besides improving its "laggard" position with respect to direct regulation, Italy took a proactive stance in introducing economic and other non–"command-and-control" instruments designed for environmental protection. It provided input to EC policy-making in two ways: by introducing (albeit with considerable difficulty, as discussed below) the use of economic instruments within the country, and by supporting the adoption of such instruments at the Community level.

The introduction of economic and fiscal instruments in the field of environmental policy, together with the adoption of procedures aimed at improving cooperation between Environment ministers and ministers responsible for Energy, Finance, and other policies, was one of the priorities of the Italian presidency of the EC in 1990 (CEEP/Ambiente, 1990). Joint EC Council meetings of ministers, particularly of Environment and Energy ministers and Environment and Financial ministers, are now common, as a result of the Italian presidency's effort, and represent important areas of intersectoral policy cooperation. Moreover, the former EC Commissioner for the Environment and subsequent Environment minister in Italy, Carlo Ripa di Meana, used his influence to overcome Italian resistance to the adoption of certain directives—for example, the resistance of the Italian auto industry to the introduction of EC regulations on automobile emissions. He also strongly supported the introduction of economic instruments, such as the adoption of an EC energy/CO_2 tax aimed at achieving the stabilization of CO_2 emissions within the Community and at providing an incentive for energy efficiency.[36]

To conclude, while usually an environmental laggard, Italy, in recent years, has started playing a more active role at the EC level.

National Policy Style

The political culture of a nation, though a dimension hard to define, may be considered the result of the historical sedimentation of collective experiences. The features often referred to as peculiar to Italian political culture include individualism, familialism, localism, clientelism, fatalism, a fragmented political culture, and a lack of trust in others and in public institutions—in particular, absence of a "sense of the state" and of public interest, absence of a "public spirit," and frailty of public ethics (Cavalli 1992, 393). These features are the result of political fragmentation of the peninsula until a century ago, government in the past century by despotic and often-corrupt rulers, the lack of influence by the Protestant Reformation, the long-standing secular power of the Vatican, the absence of a hegemonic class and culture when unification finally occurred, and the weakness of the national bourgeoisie. Obviously, such features are not shared to the same extent by all individual and collective actors, nor are they homogeneously distributed across various parts of the country (cities vs. countryside, north vs. south, Catholic vs. more secular areas, and so on).[37]

For the purposes of this discussion, one relevant consequence is that public goods, including environmental quality, enjoy low priority in the prevailing political culture. There are no collective goods as such, but only goods of the state of which each private faction tries to get its share (Galli 1992). As far as the political system is concerned, such a trait constitutes the terrain on which patronage—a direct exchange relation of favors versus support, votes, and so on—flourishes as a typical mode of consensus generation and relationship between the political parties and citizens.

Particularistic interests of specific groups and clans, if not individuals, rather than wider perspectives of collective interests, dominate public policies. Thus, a considerable amount of Italian legislation entails the granting of immediate and direct gains to specific groups or corporations (Di Palma 1977). Several analyses of the Italian political system in the 1970s (Di Palma 1977; LaPalombara 1988) have pointed out its limited capacity to shape the actions taking place within the social system as a whole. Political parties, especially those in power during the last 50 years or so, such as the Christian Democrats and the Socialists, represent the channel through which such particularistic interests have coalesced. Italian policymaking is often described, by both laypeople and political analysts, as being hegemonized by parties to an extent that goes beyond "physiological" party government. Party representatives occupy positions and make decisions that in other countries are assigned to experts, bureaucrats, or actors representing societal interests. Thus, although many parties have paid lip service to environmental issues in recent years due to the growing public concern and electoral successes of the Greens and the activism of environmental organization, the actual interest in such collective issues remains low, and governing parties seldom actively promote environmental measures. In some cases, the particularistic nature of policymaking has assumed a more pathological character. Investigations carried out by the judiciary in 1992–93 revealed that decisions on public works, whether involving infrastructures having a highly negative impact on the environment or supposed environmental protection measures, such as treatment facilities, were actually made merely on the basis of their capacity to provide kickbacks to political parties and individuals, both politicians[38] and bureaucrats.

While such allegations are obviously difficult to substantiate, they appear valid at least in general terms. From a somewhat different perspective, analysis shows that Italian environmental policy typically, at least until recently, could be characterized as (1) highly reactive, (2) conflictual, and (3) largely based on a regulatory approach.

To begin, policymakers in Western democracies tend to be *highly reactive*, responding to issues as they arise rather than anticipating them before they become policy problems. Examples of truly anticipatory policies are the exception rather than the rule. Yet the reactive nature of environmental policy seems much more extreme in Italy than in other Western countries. Italy is often slow to resort to collective solutions, typically doing so only under duress. Thus, it is not surprising that policy measures are usually triggered by external stimuli, involving either international obligations (especially EC directives) or environmental emergencies, such as those mentioned previously.

An example of the reactive nature of the Italian policy style involves air pollution caused by vehicle emissions. In 1991, the Environmental Ministry decided that something had to be done to reduce traffic-induced air pollution, at least in the major urban areas. It issued a decree ordering the 11 largest municipalities to adopt emergency measures, such as a partial or complete ban on private traffic when specified pollution levels were exceeded. Though such measures were ineffective in actually reducing air pollution levels, their psychological impact has caused a considerable shift toward cars with catalytic converters in urban areas in recent years.

A previous attempt to obtain the same results by ordinary means had failed. The ministry had signed an agreement with Fiat whereby the manufacturer would install catalysts in all new vehicles—anticipating EC requirements—in exchange for tax measures that would provide incentives to vehicle buyers. The government, however, did not follow through in the appropriate time frame.[39]

The reactive character of such an approach becomes even more evident when it is compared to the long-standing policies of Central and Northern European countries, such as the Netherlands or Germany, which adopted tax-rebate measures to encourage consumers to buy vehicles with catalytic converters in the mid-1980s (Boehmer-Christiansen and Weidner 1992). In a sense, the frequent outbreak of national environmental emergencies occurring now might be interpreted as the upgrading of policy for dealing with environmental problems that have been neglected for many years and that are now being brought into the open, thanks to more controls and greater awareness. On the other hand, these emergencies are also symptomatic of the considerable delay in the development of Italian environmental policy, running some 5 to 10 years behind other comparable countries (using as an indicator—however approximate—the years in which basic legislative measures were passed).

Turning to the second of the three main characteristics of Italian environmental policy identified earlier, the relations between institutional actors have been extremely *conflictual* so far. An increasing number of actors are staking out their areas of influence in this new field (with inevitable conflicts among central ministries or between the state and local authorities, as noted), rather than coping with the substantive problems that need to be resolved. Despite the growing need for innovative solutions to environmental problems, many of these actors appear to perceive the distribution of power as a zero-sum game, perhaps also because of the prevailing juridical culture in the public sector.

Finally, Italian environmental policy depends mainly on a *regulatory approach* based on emission and product standards (ambient-air-quality standards have been introduced more recently) that are applied uniformly throughout the country. This is true regardless of the characteristics and uses of the specific areas of the environment to be protected.

In the case of water pollution, for example, all industries can discharge pollutants, within prescribed concentration limits, into a river, regardless the treatment capacity of that specific waterway, of the number of industries discharging into it, or of its various possible uses (recreation, fishing, drinking water, and so on). Such an approach is explainable on the basis of the prevailing legalistic culture of the political-administrative system, which is suspicious of any behavior not directly regulated by law. But it allows only limited flexibility, particularly to the local authorities usually responsible for implementation, to tackle issues according to actual demand for environmental quality or to deal with levels of pollution existing in their specific areas. There is, however, some evidence that bargaining does, in fact, emerge in some local situations.[40]

More recently, however, the Environmental Ministry has attempted to introduce a more diversified set of policy strategies, such as positive and negative incentives. Besides the above-mentioned water taxes introduced in the 1970s, one of the first examples of an ecological product tax was instituted on plastic shopping bags in 1989 to encourage consumers to cut down on their use (and thus on the release of plastics into the environment). A proposal in the early 1990s to impose such taxes on certain activities (pig breeding, pesticide use, and airplane noise, among others) and products (pesticides like atrazine as well as several types of plastic) met with strong opposition from the Ministries of Agriculture and Industry—on behalf of their respective constituencies—forcing the Environmental Ministry to accept a compromise. The 1992 Financial Act was the beginning of ecotaxes and tariffs on refuse collection, water supply, and excavations. Since then, ecological taxes have grown in number, the most relevant example being the CO_2 tax.

Another component of the Environmental Ministry's strategy is the creation of compulsory consortia set up to promote recycling of selected kinds of materials, such as glass, packaging, plastic,[41] aluminum cans, mineral oil, and car batteries. The consortia are nonprofit organizations whose members are producers, importers, and user firms. They receive financial resources through levies established on certain products; minimum collection targets were set for each type of substance,[42] and deposit-refund systems and other product taxes were planned in the event that these targets were not met (Malaman and Ranci 1991).

The ministry also pursues approaches based on voluntary agreements with major industrial groups, such as petroleum companies,[43] Fiat, ENICHEM, and Montedison, in which firms guarantee they will adopt cleanup measures in exchange for specific benefits to be granted by the government.[44] Also, voluntary policy tools, such as ecolabels and EMAS-ISO,[45] are increasingly diffused among Italian firms.

Policy Effectiveness: The Implementation Deficit

Effectiveness can be defined as the ability to obtain compliance with policy programs, as well as to actually improve environmental quality. With respect to the second part of the definition, the initial section of this chapter on the state of the environment shows that the environmental quality in Italy is quite poor. Some local successes have admittedly been achieved—for example, the reduction in SO_2 concentrations in the air and the partial rehabilitation of certain highly polluted areas, like the ceramic district in Emilia-Romagna. But overall environmental quality in Italy, especially in the cities, leaves much to be desired.

As far as compliance with policy programs and goals is concerned, one conclusion that the above description of the evolution of environmental policy in Italy allows is that it has gradually—although with considerable delay—been catching up with other more advanced nations. Adequate legislation now covers most ecological issues, substantial financial resources have been allocated, and ad hoc public institutions have been set up at various governmental levels.

If one considers the actual effectiveness of such policy measures, however, the "missing link" is the implementation phase of environmental policy, where deficits appear somewhat more severe than in other comparable Western countries.

These implementation failures can be attributed primarily to peculiar features of the political and bureaucratic systems, though a clear-cut distinction between the two systems is difficult.

The political system is responsible for the implementation failures with regard to environmental policy in a number of respects. In the first place, the considerable "permeability" of parties and institutions to particular clienteles, rather than to some broader interest, however this might be conceived of, allows pressure groups to hinder implementation of unwelcome measures. Further, in an attempt to respond to the growing demand for environmental quality, the political system is motivated to look for short-term answers, mainly with symbolic or placebo effects and with little thought to actual implementation. This obviously occurs in other countries as well, but since the Italian system has been slower to respond to environmental problems, the phenomenon has had more acute consequences in Italy than elsewhere.

Implementation problems can arise from inadequate or unclear legislation. The long-awaited legislation on national parks (Act 394/91) has not provided a precise definition of protected areas, nor does it deal with the administrative structures that are supposed to manage these areas. Also, the lack of clear technical norms (passed by Parliament or the government) accompanying environmental legislation often causes confusion and hampers implementation. For example, delays in specifying the scientific methodology to be used in determining the biodegradability of plastic bags (a condition for exemption from the tax on each bag) allowed producers to obtain certificates from laboratories and to continue production, as before. Thus, consumption levels, after an initial downward trend (a 9 percent decline in 1989, as compared to the previous year), spiraled again, with an increase of 16 percent in 1990.[46] The fault does not lie with central policymaking alone: many regions have proved unable to pass complementary legislation or general plans that national "frame" laws require in fields such as water and air pollution and parks.

Moreover, public policies are often characterized by a tendency toward "regulation without rules." In other words, formal overregulation is typically matched with substantive underregulation. Even in sectors regulated by stringent norms, these norms, in fact, are not enforced (Giuliani 1991, 89). Factual noncompliance is so widespread that three institutional mechanisms aimed at dealing with it are frequently used: prorogation or deferment of deadlines; "*condoni*" (remissions or pardons), used especially in fiscal policy because of the state's inability to collect revenues; and a relaxation of standards—for instance, when levels of pesticides above EC limits were found in drinking water, a decree simply raised the limits.

Inability to attain policy goals, at least within declared time frames, causes loss of faith in the authorities and loss of credibility in the eyes of the target groups. These groups end up with the impression that deadlines will not actually be respected or even that some sort of prorogation or postponement will eventually be passed. (Culturally, respecting deadlines is considered a bit foolish in Italy!) The constant modification of national legislation concerning deadlines and other parameters has serious consequences for policy legitimacy and, therefore, for its effectiveness. These modifications create a vicious circle by which the less policies are implemented, the less credible and effective they are. Legislative delays and cancelations therefore become the basis for further policy failures and create difficulties for officials responsible for environmental policy implementation. This was exactly what occurred with water pollution control policies in the late 1970s.

Finally, implementation authorities—typically regional and local governments—are considerably constrained in their action because strategic, financial, human, and organizational resources are controlled and allocated directly by the central government.

With respect to financial resources, until recently Italy was the Western nation with the highest degree of fiscal centralization; only 0.7 percent of taxes were collected by local authorities (Dente 1985, 117). Almost all the resources needed to finance the public system were collected by the central state, which in turn, often with considerable delay, redistributed the appropriate percentage to local authorities based on historical expenditures. The local government's ability to raise its own taxes has been extremely constrained. In addition, the resources made available by the state have tended to be bound to specific targets and uses, which has substantially limited the local governments' autonomy in making policy choices. Sixty percent of regional budgets, for example, have gone to the health sector, where "macro" decisions, such as personnel salaries, are primarily made at the national level in negotiations with trade unions. Finally, high levels of annual inflation have, for many years, considerably reduced local-expenditure capabilities in real terms. The consequences of this situation have been political as well, since it has interrupted the "responsibility circuit" between local administrators and citizens/voters. It has been hard to hold local elected representatives responsible for policy results when they do not control basic resources. Political parties have also had difficulty promising specific tax rates for services delivered.[47] This situation has changed in recent years, because the administrative system has undergone an in-depth reform in the direction of greater powers given to local and regional governments, if not of outright "federalization." As a result, the local and regional authorities today have broader fiscal autonomy as well.

The second set of causes that account for low levels of performance in the implementation phase of public policy are to be found in the specific characteristics of the Italian bureaucracy. In quantitative terms, the size of the Italian public administration system is comparable to that of other similar European countries.[48] But the qualitative aspects of the system make the Italian case distinctive. Italy's bureaucratic culture appears dominated by legalistic rather than problem-solving attitudes and values (Aberbach, Putnam, and Rockman 1981, 52). The guiding criteria are the respect for, and execution of, norms typical of a nineteenth-century administrative model (Freddi 1989). The actual attainment of results is of secondary importance and often remains completely out of the picture. Furthermore, technical competence, particularly relevant for environmental policy, is rare and enjoys low status in the eyes of the political and administrative actors. The low status and scarcity of technical compentence are reinforced by national legislation—based as it is on rigid standards, formalized procedures, the absence of flexibility and administrative discretion,[49] and excessive complexity of procedures (the construction of an average-sized chemical plant requires 14 different authorizations, granted by 9 different agencies, instituted by 20 distinct pieces of legislation!).

Bureaucrats, who largely come from southern, less developed areas of Italy, which offer few employment opportunities outside the public sector, often place a high value on job security and tend to share rigid, conformist attitudes. Also, there is little incentive for civil servants to actually pursue policy goals, since career advancement is based on seniority. The professional weakness of bureaucracies characterizes, with some exceptions, all levels of administration and is one of the factors accounting for the disproportionate role political appointees often play in policymaking.

Conclusions and Future Perspectives

An overview of the points made in this chapter leads to the conclusion that environmental policy in Italy has certainly come a long way since its initial steps in the mid-1960s, though it has achieved less and advanced more slowly than in other industrialized nations, at least until the mid-1990s. The main reasons for this state of affairs lie in two sets of factors that can be roughly grouped under the headings of "demand" and "supply," with respect to the formulation of national environmental policy.

Demand, as discussed above, has come mainly from local and regional governments[50] and the environmental movement. However, these forces have often been outweighed by the opposition of potentially affected interests and their political representatives, because the influence of economic interests that favor environmental policy, such as ecobusiness, still remains low. Further, Italian culture tends to attribute less importance to nature and environmental quality than do other national cultures, such as those of Central and Northern Europe.

A more significant role has been played by supply factors, mainly represented by international inputs, especially from the European Community, and by the Environmental Ministry, since its creation in 1986. In spite of a political system that prefers the "particularistic" modes of policymaking described above, demand-and-supply pressures favoring the upgrading of national environmental policies have succeeded in forcing institutions to respond, at least to some degree, to maintain their legitimacy. Results have included general policy decisions expressed through or reflected in legislation, plans, and the creation of institutions, as well as the allocation of financial resources. Both of these stimuli, up to now, have not been able to completely offset the compliance and implementation failures, the cause of which is to be found in the weakness of the administrative system.

If this is the situation with respect to environmental policy at present, what are the prospects for the future? The future of Italian environmental policy will be influenced by four main factors: (1) the future of domestic politics, including the related impact on sectoral policies, including environmental policy; (2) the relations between environmental protection and economic cycles; (3) the completion of the EC internal market and its impact in Italy; and (4) the internationalization of environmental issues.

Concerning the first factor, only radical changes in the Italian administrative system can bring about an improvement in the implementation of environmental policy. The political-administrative system is encountering severe criticism from society because of its inefficiency, its excessive costs, and occasionally its outright corruption. Moreover, conflicting views on the center-periphery system are under discussion, partly due to the demands of the Lega Nord (Northern League), a party whose anticentrist platform has had considerable electoral successes on several occasions. Lega Nord represents one of the three major parties in the center-right coalition that originally supported the government elected in March 1994, and it is part of the center-right coalition that won the May 2001 elections. At the moment, the final outcome of this situation and its effects on environmental quality are not clear. One possible outcome could be a scenario in which

- Political institutions at the central and peripheral levels improve their decision-making capacity—for example, through the modification of the electoral system and a greater stability of representative bodies.
- The administration, due to the need to cut costs because of public indebtedness and the need to become more productive in response to societal demand, gradually becomes more efficient through enhanced problem solving and technical capabilities.
- A redistribution of power and resources in favor of local and regional authorities occurs.

This trend toward greater fiscal autonomy of local governments, as mentioned, is underway. Local authorities, as well as environmental policy, at least in some areas, might benefit from this trend, and more decentralized management of national and EC environmental policy (perhaps including direct implementation of EC directives at the regional level) could be expected. The role of the Environmental Ministry in a situation characterized by stronger peripheral governments, on the one hand, and growing powers of the EC, on the other, remains open to debate.

The potential of public authorities to exercise power in the environmental field in the future, however, also depends on the economic trends. If the problem of the public debt is not solved, there will obviously be diminishing financial resources available for environmental policy unless the environment itself is seen as an opportunity for development. For the time being, the danger is that economic difficulties (such as high unemployment) may result in actions with negative impacts on the environment. Examples could include a large public works program to ensure jobs, or the sale—without constraints on their use—of areas of natural value previously belonging to the state, resulting in short-term revenues but long-term damage. The center-left-green Olive government, in fact, attempted to reconcile both targets by means of integration of environmental goals within sectoral development policies. It remains to be seen whether the center-right coalition that took office in 2001 is likely to put less of a priority on environmental issues.

With respect to the third element, the EC context represents a strong incentive, in economic and political terms, for Italy to upgrade its national environmental policy. On the other hand, the completion of the European internal market is likely to do further damage to the environment, through an increase in the transportation of goods and people and other negative development, if no countermeasures are implemented (Task Force, 1989). In this respect, preventive measures are being taken with the adoption of EC directives, with the formulation of new policy instruments, and with the adoption since 1992 of environmental action programs explicitly aimed at achieving sustainability.

It should be noted, however, that Italian environmental policy has just started to pay attention to sustainability. Moreover, standardization of environmental regulations is generally viewed as a basis for trade liberalization and is thus likely to accompany it. Finally, and related to this last point, more ecofriendly products, production processes, and technologies will increase competitiveness. In this context, Italian industry and other economic sectors will have an incentive to take environmental matters more seriously than in the past, though it remains to be seen how much of this process will be automatic—that is, a result of market mechanisms—and how much will have to be "guided"—in other words, how much the market will have to be regulated—in order to become sustainable.

Turning finally to the fourth point, events such as the United Nations Conference on the Environment and Development as well as some 150 international environmental agreements reached so far indicate a trend toward an increased internationalization of environmental problems. This implies that policymakers at the local, national, and EC levels must take this broader context into account, since the adoption of international conventions, along with the establishment of international monitoring networks and institutions, are politically relevant, albeit not legally binding. Somewhat paradoxically, Italian policymakers tend to be more eager to present a good external image (as indicated by the large sums Italy has contributed to several international organizations) than to respond to their internal constituencies. The increasingly high international visibility and political status of environmental problems could, therefore, represent a powerful stimulus to improvement at the domestic level as well.

Notes

1. With the advent of the Maastricht Treaty in 1992, the European Union (EU) was established. In this chapter, we use the term *European Community* (EC), since the EU did not replace, but includes, the EC (together with new powers) and because we mainly discuss developments prior to Maastricht.

2. On the ceramic industry, see Lewanski 1992a.

3. Also, because of the lack of effective public transportation systems, urban mobility by and large depends on private vehicles with unacceptable levels of congestion and air and noise pollution. In urban areas, 150 billion passengers per kilometer are transported daily by private vehicles, whereas only 20 billion passengers per kilometer use public transportation. Moreover, only 6 percent of goods travel by rail (even though the Italian rail network is among the denseest in Europe), compared to 13 percent in France, 14 percent in Germany, or 36 percent in Austria (Legambiente and Istituto di ricerche Ambiente Italia, 2001, 192). There are 106 vehicles for each kilometer of road (as compared to 32 in the United States); all in all, there are more than 43 million vehicles circulating on Italian highways. Due to heavy traffic, pollution levels are constantly above legal limits, so that authorities have been forced to take emergency measures to limit traffic in many medium-sized and large cities, especially during the winter. Nitrogen oxides and particulates have increased by 20 percent during the 1980s. Only 15 percent of circulating vehicles had catalytic converters in 1994.

4. It covered 5.6% of coastal waters 1999 (Ministero dell'Ambiente, 2001, 239). However, data collected by environmental organizations indicate that the situation might actually be worse than official sources show. Also, the situation varies considerably from region to region.

5. An example is the cholera epidemic that broke out in fall 1994 in the Puglia region.

6. For example, in 1991, Italy imported 216,100 tons of sulfur, while it exported 399,700 tons; in the same year, it imported 97,600 tons of nitrogen, while exporting 160,200 tons (EMEP, 1994).

7. *Waldsterben* is a German term meaning "the death of forests caused by acid rain."

8. The densely inhabited and highly industrialized Po Valley is, in fact, surrounded by mountains on three sides.

9. For a more detailed analysis of the "Antismog" Law and its implementation, see Dente et al. 1984.

10. On the role of the judiciary, including that of the *pretori*—magistrates whose jurisdiction includes all crimes with a maximum penalty of less than 3 years' imprisonment—see Del Duca 1989.

11. In general, USLs (now ASLs) are responsible for managing hospitals and delivering health services. They also guarantee the quality of the environment in which people "work and live" (Act 833/78). USLs include technical apparatus (*Servizi d'Igiene e Prevenzione* and *Presidi Multizonali di Prevenzione*) responsible for controlling and monitoring activities related to environmental conditions in general.

12. Incorporated in the 1946 postwar Constitution as one of the safeguards against future authoritarian backlashes, regions were actually set up (with the exception of five special regions) only in 1970 due to resistance from conservative parties worried about the Communist Party gaining control over certain areas of the country and over the state bureaucracy. The resistance was such that regions were given restricted powers and resources and had to fight to obtain more substantive powers.

13. On the antinuclear and environmental movements in Italy, see Biorcio and Lodi 1988 as well as Diani 1988.

14. See Acts 36 and 37 of January 5, 1994.

15. For purposes of comparison, in that same year, public-sector investments in pollution control and abatement, expressed as percent of gross fixed capital formation, was 1.2 percent in the United States, 3 percent in Japan, 0.9 percent in France, and 2 percent in Germany (Organization of Economic Cooperation and Development, 1993, 295).

16. More precisely, at the end of 1992, only 37.9 percent of the funds allocated for the 1988 annual environmental protection plan had been spent, and only 29.1 percent of the funds for the 1989–1991, three-year plan had been spent

(Istituto Nazionale di Statistica, 1993, 252). After the discovery of "Tangentopoli" the ministry's low-expenditure capability might be seen in a different light. As a matter of fact, none of the environmental ministers were involved in cases of corruption. Perhaps low spending levels, at least to some degree, reflect their lack of interest in spending public money in order just to obtain kickbacks.

17. In the early 1990s, the number of public personnel engaged in research projects concerning the environment has been determined to be only 1,835 individuals (Istituto Nazionale di Statistica, 1993, 254).

18. On the policy and social responses to Chernobyl in Italy, see Liberatore 1993.

19. See Lewanski 1992b, 51–64.

20. The association of ecoindustries—UIDA—has some 150 members. On green business in Italy, see Gerelli 1990. On the broader relations between industry and environment in Italy, see Dente and Ranci 1992.

21. Eurispes and LegAmbiente, 1994; the profits of such organizations in 1998 have been estimated at 11,850 billion liras (LegAmbiente, Fontana, and Miracle 1998, 27).

22. It is interesting to note that on this occasion, the Greens were given the Ministry of Agriculture, rather than that of the Environment. This could be seen as an attempt by the Green Party to integrate environmental concerns into this production sector.

23. By 2000, national parks represented 10 percent of the national land area (as compared to 7.5 percent in 1996) and marine parks included 166,349 hectares (Ministero dell'Ambiente, 2001, 180, 245).

24. The number of forest fires (–27 percent), as well as the amount of surface area destroyed (–54.3 percent), decreased considerably in 1999 as compared to the two previous years (Ministero dell'Ambiente, 2001, 177–178).

25. Ministero dei Trasporti, Piano Generale dei Trasporti e della Logistica, Rome, 2001.

26. Organic agriculture has increased considerably since 1995, when there were only some 10,000 farms (with about 200,008 hectares) using such cultivation methods; in 1999, there were 49,188 firms (with 953,057 hectares) (Di Caro 2001).

27. In 1999, renewable energy sources represented 13 million tons oil equivalent (MTOE) of the total energy supply, out of a total of 183.1 (Ministero dell'Ambiente, 2001, 46).

28. Despite the introduction of a number of new ecotaxes, the total amount of revenue generated by such taxes in 1999 has actually diminished as a percentage of total fiscal revenue relative to 1995 levels (Legambiente and Istituto di ricerche Ambiente Italia, 2001, 235).

29. About 8,100 *comuni* represent the basic elements of the Italian local government system, often dating back to the medieval period; they carry out prac-

tically every type of activity required to satisfy the needs of the local population. Municipalities are responsible for activities relevant to environmental protection, such as public health, traffic, public parks, refuse collection and disposal, sewage treatment, public transportation, and occasionally energy production. Certain activities are carried out through approximately 500 municipal agencies (*aziende municipalizzate*), concentrated mainly in northern and central Italy.

30. The 103 provinces are a vestige of the Napoleonic era. After considerable indecision and confusion in the 1970s on whether certain environmental responsibilities should be concentrated at the municipal or the provincial level, with competencies (for example in the field of water pollution) being passed back and forth, the provinces are now emerging as key actors in this policy arena.

31. The province of Trient, due to the presence of a strong German-speaking minority in the Süd Tirol–Alto Adige region to which it belongs, has been assigned a special status that gives it powers equal to, if not broader than, those of the regions.

32. The construction of a nuclear power plant in Montalto di Castro (Latium) was also begun but never completed. Following the referenda against nuclear power in 1987, the decision was made to complete the plant as a coal and gas plant.

33. On the jurisprudence of direct effect, see Pescatore 1983; for its relevance to the Italian case, see Del Duca 1989.

34. Negotiations with the European Chemical Industry Council (CEFIC) were instrumental, for example, in the adoption of the Seveso Directive of 1982 and other directives regulating chemical substances.

35. Haigh (1989) carefully examines the negotiations leading to the adoption of several EC environmental directives, with special attention to the role of the United Kingdom.

36. On the debate regarding the introduction of this tax, see Liberatore 1995.

37. See Putnam 1993; Cartocci 1994.

38. Even the highest political officials—for example, the ex-prime minister, Bettino Craxi—have been charged with receiving illegal kickbacks in exchange for the approval of public works projects (such as for the installation of desulfurization equipment in publicly owned power plants).

39. It is worth noting that the major Italian car manufacturer, Fiat, started offering cars equipped with three-way catalytic converters in the Netherlands in 1988, showing that it no longer had significant problems meeting stringent standards. In 1990, 68 percent of total gasoline sales were represented by unleaded gasoline in Germany; the figure for Italy was 5 percent. Already in the early 1990s, 80 to 90 percent of new cars sold in the Netherlands and Germany were equipped with catalysts.

40. See Dente and Lewanski 1982, 122–123. On EC law on car emissions, see Arp 1991.

41. "Replastic," the consortium responsible for recycling plastic bottles and similar items, collected 2,000 tons in 1991 and 5,000 tons in 1992 in 600 different municipalities.

42. Glass collection, for example, increased from 377,000 tons in 1990 to 665,000 in 1998; in the same period, plastic collection increased from 3,000 to 151,000 tons, paper from 120,000 to 1 million tons, and organic refuse from 50,000 to 891,000 tons (Legambiente and Istituto di ricerche Ambiente Italia, 2001, 201).

43. Companies in this case voluntarily agreed to distribute cleaner vehicle fuels in the larger and most polluted cities.

44. For a detailed list of such agreements, see Ministero dell'Ambiente, 2001, 458.

45. The number of firms certified according to International Standards Organization (ISO) 14001 in June 2000 was 410; 26 complied with the ECO-Management and Audit Scheme (EMAS) at the same time (Legambiente and Istituto di ricerche Ambiente Italia, 2001, 167).

46. On this topic, see Mariani 1991.

47. The following data provide an idea of the distribution of resources among the various levels of government before the decentralization trend began. In 1987, the regions spent 82,090 billion lire, the municipalities 66,011, and the provinces only 6,588—for a total of 154,689 billion lire, compared to 465,395 billion lire spent by the state. During the second half of the 1980s, municipal and provincial yearly budgets were equivalent to 5.2–5.4 percent of the GNP.

48. Italy's public personnel consist of some 2.2 million employees at the central level and 1.4 million at the regional and local levels. At the regional and local levels, there are 600,000 employees in the health service and 800,000 in local and regional administration. Eighty percent of the latter group are in the municipalities, including 160,000 of the *aziende municipalizzate.*

49. Though, in fact, local authorities responsible for implementation of national environmental policies occasionally seem to resort to informal procedures and bargaining relationships with regulated polluters (Dente et al. 1984, 141).

50. We prefer to consider "local and regional governments" here as an expression of the demand by the local populations, although it would also be possible to include them among the supply actors.

References

Aberbach, J., Putnam, R., and Rockman, B. (1981). *Bureaucrats and politicians in Western democracies.* Cambridge: Harvard University Press.

Alberti, M., and Parker, J. (1991). The impact of EC legislation on environmental policy in Italy. In *European University Institute, European Trends,* 2.

Arp, H. (1991). Interest groups in EC legislation: The case of car emission standards. Paper presented at the Joint Sessions of Workshops of the European Consortium of Political Research, Essex, UK.

Biorcio, R., and Lodi, G. (1988). *La sfida verde: Il movimento ecologista in Italia.* Padua: Liviana editrice.

Boehmer-Christiansen, S., and Weidner, H. (1992). *Catalyst versus lean burn.* WZB papers no. FS II: 92-304. Berlin.

Capria, A. (1988). Direttive ambientali CEE: Stato di attuazione in Italia. In *Quaderni della Rivista Giuridica dell'Ambiente.* Milan: Giuffré.

Cartocci, R. (1994). *Fra Chiesa e Lega.* Bologna: Il Mulino.

Cavalli, A. (1992). Un curioso tipo di italiano tra provincia ed Europa. *Il Mulino, 3,* 391–400.

CEEP/Ambiente. (1990). La priorità dell'ambiente nel semestre Italiano. CEEP Forum. *Ambiente, 10.*

Centro Interregionale Studi E Documentazione (CINSEDO). (1989). *Rapporto sulle Regioni.* Milan: F. Angeli.

Conte, G., and Melandri, G. (1993). *Ambiente Italia* '93. Rome: Koiné.

Conte, G., and Melandri, G. (1994). *Ambiente Italia* '94. Rome: Koiné.

Del Duca, P. (1989). *Italian judicial activism in the light of French and American doctrines in judicial review and administrative decision making* (Working paper). Florence: European University Institute.

Dente, B. (1985). *Governare la frammentazione.* Bologna: Il Mulino.

Dente, B., Knoepfel P., Lewanski, R., Mannozzi, S., and Tozzi, S. (1984). *Il controllo dell'inquinamento atmosferico in Italia: Analisi di una politica regolativa.* Rome: Officina Edizioni.

Dente, B., and Lewanski, R. (1982). Administrative networks and implementation effectiveness: Industrial air pollution policy in Italy. *Policy Studies Journal, 11*(1), 116–129.

Dente, B., and Ranci P., eds. (1992). *L'industria e l'ambiente.* Bologna: Il Mulino.

Diani, M. (1988). *Isole nell'arcipelago: Il movimento ecologista in Italia.* Bologna: Il Mulino.

Di Caro, R. (2001, February 1). Che Business! *L'Espresso,* p. 163.

Di Palma, G. (1977). *Surviving without governing: The Italian parties in Parliament.* Berkeley: University of California Press.

EMAP. (1994). European Monitoring and Evaluation Programme.

Eurispes and LegAmbiente. (1994). *Le ecomafie.* Rome: Eurispes and LegAmbiente.

Franchi Scarselli, G. (1992). *Il Governo regionale dell'ambiente. Analisi giuridica della legislazione a matrice ambientale* (Working paper). Milan: Istituto per la Ricerca Sociale.

Freddi, G. (1989). Burocrazia, democrazia e governabilità. In G. Freddi, ed., *Scienza dell'amministrazione e politiche pubbliche* (pp. 19–65). Rome: Nuova Italia Scientifica.

Galli, C. (1992). La cultura politica degli italiani. *Il Mulino*, *3*, 401–406.

Gerelli, E., ed. (1990). *Ascesa e declino del business ambientale*. Bologna: Il Mulino.

Giuliani, M. (1991). *Giochi regolativi: La politica di protezione del paesaggio*. Unpublished doctoral dissertation, University of Florence.

Haas, E. (1958). *The uniting of Europe: Political, social, and economic forces, 1950–1957*. London: Stevens.

Haigh, N. (1989). *EEC environmental policy and Britain*. 2nd ed. Harlow, Essex: Longman.

Herman, A., and Lewanski, R. (1992). Environmental monitoring and reporting in Italy. In H. R. Weidner, R. Zieschank, and P. Knoepfel, eds., *Umwelt-Information*. Berlin: WZB–edition sigma.

Hoffman, S. (1966). Obstinate or obsolete? The fate of the nation-state and the case of Western Europe. *Daedalus*, *95*, 862–915.

Istituto Nazionale di Statistica (ISTAT). (1993). *Statistiche ambientali*. Rome: Istituto Nazionale Di Statistica.

Knoepfel, P. (1984). La tutela dell'ambiente nell'Europa occidentale. In *Archivio ISAP n.2, Le relazioni centro-periferia* (pp. 2289–2317). Milan: Giuffré.

LaPalombara, J. (1988). *Democrazia all'italiana*. Milan: Mondadori. English edition: *Democracy, Italian style*. Trans. Laura Noulian. New Haven: Yale University Press.

LegAmbiente, Fontana, E., and Miracle, L., eds. (1998). *Rapporto ecomafia*. Milan: Edizioni Ambiente.

LegAmbiente and Istituto di ricerche Ambiente Italia. (2001). *Ambiente Italia 2001*. Milan: Edizioni Ambiente.

Lega per l'Ambiente, and Melandri, G. (1989). *Ambiente Italia, Rapporto 1989: Dati, tendenze, proposte*. Turin: ISEDI Petrini editore.

Lewanski, R. (1986). *Il controllo degli inquinamenti delle acque: L'attuazione di una politica pubblica*. Milan: Giuffré.

Lewanski, R. (1992a). Environmental policy in Italy: The case of the ceramic tile district. *European Environment*, 2(1), 5–7.

Lewanski, R. (1992b). Il difficile avvio di una politica ambientale in Italia. In B. Dente and P. Ranci, eds., *L'industria e l'ambiente* (pp. 27–82). Bologna: Il Mulino.

Liberatore, A. (1991). National environmental policies and the European Community: The case of Italy. *European Environment*, 2(4), 5–8.

Liberatore, A. (1993). Chernobyl comes to Italy: The reciprocal relations between experts, policymakers and the mass media. In A. Barker and G. Peters, eds., *The politics of expert advice in Western Europe: Creating, using and manipulating scientific knowledge for public policy* (pp. 3–47). Edinburgh: Edinburgh University Press.

Liberatore, A. (1995). Arguments, assumptions and the choice of policy instruments: The case of the debate on the CO_2/energy tax in the EC. In B. Dente, ed., *Environmental policy in search of new instruments*. Dordrecht: Kluwer.

Malaman, R., and Ranci, P. (1991, August). Italian environmental policy. Paper presented at the International Conference on Economy and Environment in the '90s, Neuchâtel.

Mariani, S. (1991). *I. processi regolativi in Italia: il caso di una politica regolativa*. Unpublished doctoral dissertation, University of Florence.

Ministero dell'Ambiente. (1989). *Relazione sullo stato dell'ambiente*. Rome: Istituto Poligrafico e Zecca dello Stato. Available at www.minambiente.it

Ministero dell'Ambiente. (1991). *Relazione sullo stato dell'ambiente*. Rome: Istituto Poligrafico e Zecca dello Stato. Available at www.minambiente.it

Ministero dell'Ambiente. (2000). *L'ambiente informa*, *12*, 21.

Ministero dell'Ambiente. (2001). *Relazione sullo stato dell'ambiente*. Rome: Istituto Poligrafico e Zecca dello Stato. Available at www.minambiente.it

Organization for Economic Cooperation and Development. (1993). *Environmental data, compendium*. Paris: Organization for Economic Cooperation and Development.

Pescatore, P. (1983). The doctrine of "direct effect": An infant disease of community law. *European Law Review*, 155–177.

Putnam, R. (1993). *Making democracy work*. Princeton: Princeton University Press.

Rehbinder, E., and Stewart, R. (1985). *Integration through law: Environmental protection policy*. Berlin: de Gruyter.

Spaziante, V. (1980). *Questione nucleare e politica legislativa*. Rome: Officina Edizioni.

Task Force. (1989). *Report on the environment and the internal market*. Brussels: EC Commission.

Time. November 21, 1994, p. 59.

United Nations Environment Program. (1991). *Environmental data report*. 3rd ed. Oxford: Blackwell.

7

Environmental Policy in Australia

K. J. Walker

Ecology and Politics on the Periphery

Understanding environmental policy in Australia requires some appreciation of the internal political processes of "settler capitalist" countries. As in Canada, New Zealand, and some other European colonies in the temperate zone, settlers of European extraction displaced Australia's indigenous population. The resulting colonial economy was part of the grander imperial system, dedicated very largely to the extraction of natural resources and their transfer to Britain. European settlement created serious stresses in a particularly fragile, heretofore isolated ecosystem.

Because ecological systems are fundamental to all life, the scope of environmental policy is necessarily embracive, covering short to very long time frames, concerning itself with processes from microscopic to global in extent. Its challenge extends well beyond political routine. Environment canvasses fundamental issues of political philosophy, such as equality and distributive justice, the rights of humans and nature, and duties owed to future generations. It throws into sharp relief some of the thornier unresolved issues in economics, such as the treatment of stock resources subject to depletion and intergenerational equity. Ecological processes are frequently irreversible: extinct species cannot be restored and disrupted ecosystems generally cannot be reinstated. Threshold effects often conceal changes until dramatic effects appear. Globally, the rate of extinctions is now greater than at any previous time; irreversible damage to ecosystems is universal. The urgency this imparts to environmental issues is ill-suited to incrementalist compromise (Walker 1994; Dovers 1995, 1997).

Ecological sustainability, the goal of environmental policy, cannot be reduced to a cut-and-dried formula with known answers. Scientific uncertainty and outright ignorance are unavoidable, and conceptions of the sustainable reflect changing knowledge. Ecological systems themselves are dynamic and may have multiple points of equilibrium. There is no such thing as "pure," unspoiled nature; human impacts are ubiquitous and must be managed. Human dependence on productive processes in the ecosphere entails concern for sustainability.

The ecological sustainability criterion can define both "negative" and "positive" policy space. Some "solutions" to environmental problems merely displace them in time or space, for example, aggravating the problem in the long term. Nuclear waste storage, neglect of pollution, and refusal to plan for the future are examples of such displacement (Dryzek 1987). "Benign neglect," in particular, may significantly worsen environmental problems. Unprecedentedly severe human impacts threaten sustainability (McMichael 1995). The industrial revolution was based on accelerated exploitation of mineral resources, itself the cause of many contemporary environmental problems (Boyden 1987; Ponting 1992). Stepping up the exploitation of the natural environment—inherent in industrial development—means that, inevitably, limitations on the productive system arise as the ecological consequences reduce resource availability, increase environmental hazards, and create a "funnel" of decreasing choice. But a window of opportunity exists before the adverse consequences so restrict choice that no escape from ecological disaster is possible. Failure to develop radically less damaging productive systems is likely to damage ecological sustainability, most seriously in highly populous human societies enjoying elevated standards of living.

Sustainability requires the preservation and provision of numerous public goods, many (though not all) ecologically based. This poses some serious problems of cooperation and coordination. It is at the heart of some of the thorniest problems of political and economic theory, placing exceptional strain on ad hoc, uncoordinated, "business-as-usual" political systems. Frequently, the only available authority to enforce comprehensive policies is government (Walker 1994). However, environmental management is not traditionally a major role of government and can conflict directly with its "defining" functions (Walker 1989; Rose 1976).

These problems are particularly acute in Australia. The smallest and most arid of the continents, its unique fauna and flora are of exceptional scientific interest and embody important aesthetic values (Smith 1990; Boyden, Dovers, and Shirlow 1989). European settlement accelerated rates both of exploitation and of ecological damage. There are direct, immediate, large-scale threats to flora and fauna; "Australia's record for mammal species extinctions is the worst for any country" (State of the Environment Advisory Council, 1996, 4–39). But Australia is also one of the most urbanized societies in the world, with serious urban pollution problems, including air, water, and toxic wastes.

Australia's settlement involved the transfer of a complex of European crops, livestock, and techniques (Crosby 1993). Sparsely distributed and overmatched, much of the indigenous population succumbed to genocide, losing nearly all their land in the more hospitable, temperate regions. But in the arid center and the tropics, where European agricultural and pastoral technology was less effective, the indigenous population survives. Along with pockets of remnant forest, these are also the areas of greatest conflict over conservation, since native fauna and flora have already been devastated in the more settled areas (Bolton 1992; Lines 1991; Marshall 1966).

Because most of the continent is marginal for agriculture and pastoralism, "development" along European lines always lagged. Its promotion, a central issue in politics, is intimately tied to exploitation of natural resources.

Thus, environmental policy issues in Australia differ significantly from those of the developed industrial nations. Wilderness, conservation of threatened species, both floral and faunal, and issues affecting rural industries gain more emphasis, while "urban" and "industrial" issues, notably pollution and waste management, though important, are less stressed. Acute awareness of important, extensive, relatively undisturbed wilderness, and unique flora and fauna, has engendered conflict between "developers" and conservationists. Such unnecessary polarization is largely due to deeply ingrained, highly persistent patterns of thought and behavior.

Australian Environmental History

Though none of the colonies created by the great European expansion succumbed without human resistance and ecological damage, some adapted relatively easily to imposition of European technology and culture. Australia, remote and forbidding, was far less tractable.

Australia's geographical isolation, lasting nearly 45 million years, drastically limited migration of species; its flora and fauna were largely cut off from the rest of the world. The many unusual ecosystems that survive shed much light on evolutionary processes; lineages known only from fossils elsewhere in the world are present. Australian soils, ancient and severely weathered, tend to be deficient in trace elements essential for plant growth and livestock. Relatively fertile soils are limited to the coast and eastern highlands. Rainfall is seasonal and unreliable; drought is a common occurrence, and water scarcity constrains European-style "development" (Carden 1999). In the tropics, extreme temperatures and high evaporation rates can directly impede plant growth and exacerbate water shortages (Walker 1992b, 1994). The native flora and fauna are well adapted to these conditions, but European crops and livestock suffer, both from water stress and from local diseases and pests.

As modern archaeological discoveries enlarge the time frame of aboriginal settlement to 60,000 years and perhaps considerably more, speculation about its environmental impact has intensified. Modern, highly conservationist cultural practices are now interpreted by some as a reaction to ancient, more profligate land use (Davidson 1992). The dominance in modern forests of fire-resistant and fire-dependent flora, such as *Eucalyptus*, *Acacia*, and *Xanthorrea* (grass trees), has been attributed to extensive Aboriginal use of fire (Pyne 1992; Smith 1992). Aboriginal adaptations were subtle, informed by substantial and intricate knowledge of the environment; a dawning recognition of its importance is now leading to attempts at reconstruction.

Australian ecology seemed to contradict the European rationalist conviction that nature could be "made over." Australia's fauna and flora were rejected as inferior, intractable, and unsuitable as raw material for transformation into a new Europe. So, although naturalists were fascinated by Australia's strange species, "practical" men saw it as infertile bush, to be ringbarked, cleared, and supplanted. But development of techniques for local conditions was slow and painful. Understanding of soil deficiencies was slow to develop, and drought, despite its frequency, is still treated as an exceptional "natural disaster."

Australia's early settlements were far from the main shipping routes; transport costs were high and transit times ran into months, sometimes years. This "tyranny of distance" retarded economic development and reduced the range of goods that could be sold overseas (Blainey 1983). It also led to pervasive insecurity, summed up by the poet A. D. Hope a century later as

> . . . a vast parasite robber-state
> Where second-hand Europeans pullulate
> Timidly on the edge of alien shores.
>
> (Brook 2000, 54)

reflecting the lack of fertile land in the interior and the consequent confinement of development to the coastal fringe.

The European mirage spawned unwillingness to recognize aridity or constraints on carrying capacity, persisting to this day among most figures in politics and business. Ideas such as "progress" and the doctrines of social Darwinism were highly corrosive, generating careless attitudes to the land. Native species and indigenous peoples were treated as "vermin," doomed to extinction. Attempts to promote "closer settlement," through clearance schemes, soldier settlement, and Homestead Acts, generally failed due to small farm size, poor planning, and inadequate infrastructure. Ignorance of the ecological constraints defeated both the social goals and the political aim of breaking or limiting squatter power (Bolton 1992; Dovers 1994, 126). The colonial mentality was fatally indifferent to the Australian environment, natural or social.

During the nineteenth century, the economics of exportable dry-land produce and minerals led to dependence on imports of manufactured goods and continuing identification with Britain. Australia's culture was a derivative transplant in which "Australian" meant inferior: the "cultural cringe," still a very common attitude.

The wholesale expropriation of the indigenous population defined the bulk of the continent as unalienated "Crown" land (Holmes 1996). The colonies were unable to control the spread of pastoral settlement effectively (Gilbert 1981). The stress on "development" led to active government involvement in the promotion of economic growth. Distinctive political institutions and policies emerged and persisted.

Political Economy

Nineteenth-century Australia relied primarily on wheat and wool. Mining began with the gold rushes of the 1850s, but despite its importance, it passed wool in export value only in the 1980s. Declining crop yields were checked by technical breakthroughs, such as development of rust-resistant wheats, application of important trace elements, and novel machinery. But, although "exploitative pioneering" gave way to "national development" posited on "wise use" of resources, the myth of Australia's "unlimited" potential persisted (Frawley 1994). Ongoing flirtation with uneconomic "closer settlement," diversification of European-style agriculture, continued rejection of native species, as well as clearing of land and introduction of exotic species radically modified ecosystems. Despite frequent warnings from those "in the know," these impacts were largely ignored (Bolton 1992).

Aridity lends water political importance. Irrigation emerged in the 1860s; subsidized large-scale schemes followed some 20 years later. Management of the Murray-Darling system, fourth largest in the world in terms of area, but ninth largest by flow, has generated persistent disputes over navigation, access to water, and recently, water quality. The Australian Constitution of 1901 contains explicit provision for institutions designed to help resolve Murray waters problems. However, during the twentieth century, management was based on a series of politically stable but ecologically unsatisfactory compromises (Davidson 1974; Kellow 1992; Walker 1994). Only 15 years after the last of these was concluded, the continuing decline in water quality was causing renewed concern, while conflict over water sharing was intensifying (Hay 2000; Smith 2000; Scanlon 2001a). Calls for further institutional change were emerging (Scanlon 2001b).

As in other capitalist economies, "state-subsidized profit taking" results in substantial infrastructure provision from the public purse. By nineteenth-century standards, it was remarkably extensive: governments invested in communication, transport, public utilities, and even marketing for primary producers. But populations were sparse. Distance so inflated the cost of providing infrastructure, and so reduced returns, that even government provision was scanty. The major cities are coastal precisely because navigation was by far the cheapest form of communication. The arid interior, difficult of access, offered few rewards.

The single-minded focus on "development" meant neglect of environmental and even social problems as irrelevant to nation building. By around 1900 fear of the "Yellow Peril" led to the "populate or perish" slogan and the White Australia policy.

Slow economic growth was punctuated by major depressions, such as those of the 1890s and 1930s. Australia's dependence on world markets meant continuing vulnerability to price fluctuations.

Australia's resemblance to other "dominion capitalist" economies, such as Canada, New Zealand, Argentina, and Brazil, is strong. All displaced indigenous civilizations for European "development." All depend on exports of primary or minimally processed products; secondary industry remains relatively undeveloped, with foreign capital dominant; and their service sectors are relatively large (Head 1986).

Historically, world market prices for primary products, including minerals, have steadily declined; all are far lower than for manufactured goods with "added value." Global market dominance by manufacturing nations means low prices and high tariff barriers to countries dependent primarily on unprocessed exports. Financial dependence and, in particular, reliance on overseas capital leads to technological dependence and a "branch-plant" economy.

Statist Developmentalism

Australian governments have rarely addressed this predicament; attempts to do so have provoked violent opposition from the pastoral, industrial, and financial sectors (Connell 1977; Sexton 1979; Cochrane 1989; Carden 1999). Instead, undifferentiated economic growth was promoted through tax relief, services at low prices, free land, and other inducements. Governments created marketing boards, engaged in overseas promotion of Australian products, and became involved in the direct marketing of services.

The consequent interlocking of industrial and governmental interests was expressed in "colonial socialism"—a pattern of colonial government borrowing from Britain for infrastructure, especially telegraph lines and railways, to stimulate growth. Governments became investors, involved themselves in marketing, and, because of their control of Crown land, were "the landlords of most of the Australian continent" (Butlin, Barnard, and Pincus 1982, 13).

"Colonial socialism" was a misnomer. The motivation of this policy complex was developmentalist, unconnected with traditional socialist or welfare-statist objectives, such as income equality, social welfare, and a "fair go" for working people. It was well established long before the rise of a strong Labour movement and decades before the Labor Party had any significant input into policy; the term *socialism* is inappropriate. Rather, it was a distinct episode in a more durable pattern of development, which can be termed *statist developmentalism*, to capture the Australian state's strong role in development (Walker 1999).

Statist developmentalism will certainly persist well into the twenty-first century, yet it remains a prisoner of its past. Ingrained nineteenth-century attitudes and ideas continue to dictate the "correct" way to develop. Even present-day "state-assisted marketization" is sponsored and subsidized by government (Bell and Head 1994, 21; Davis 1995). Undifferentiated economic growth, careless of ecological impact, remains a tenet of faith, leading to avoidable conflict between "growth" and "environment," both crudely defined.

Policy History: Major Recent Issues

Australia was among the earliest of "Western" nations to address environmental problems. Even in the nineteenth century, scholars, scientists, and even artists bemoaned the impacts of "development" (Bonyhady 1993). From the 1860s on, measures for fauna conservation, reservation of state forests and natural parks, and imposition of a strict quarantine regime steadily emerged. However, the scale was not extensive, nor was public awareness widespread. Attempts to organize extensive soil conservation took place from the 1930s (Bolton 1992).

Nowadays, both "green" and "brown" issues engage public concern (Lothian 1994; Crook and Pakulski 1995). Many brown issues are critical, especially soil and nutrient depletion in agriculture, coastal-zone management, air and water pollution, toxic waste, and sewage disposal. Both types of issue can be immediate, long-term, or both. However, "long-term" issues are frequently neglected or displaced in Australian politics, often breeding problems for the future (Lowe 1994).

Among the serious long-term issues is Aboriginal land rights. The High Court's *Mabo* decision of 1992 overturned the doctrine of *terra nullius*, under which the Crown had seized the whole continent as "uninhabited." Instead, it declared that indigenous land rights existed where land was unalienated and a continuing connection could be shown. In a further decision, *Wik*, in 1997, the Court ruled that native title could coexist with Crown leasehold. Because many claims are over areas of conservation importance, a barely dormant conflict exists between environmental management, in terms of ecological imperatives, and land use by partially modernized indigenes with access to modern technologies. However, claimant groups generally promise responsible management, and the potential for innovative land use is exciting (Ross, Young, and Liddle 1994).

The 1970s

At the federal level, a divided and dispirited Liberal–Country Party coalition government, which had ruled since 1949, was replaced by the Whitlam Labor government in 1972. This short-lived government achieved a good deal, in the face of violent opposition, before its removal in 1975. Exploiting the Commonwealth's constitutional powers, it expanded its participation in social welfare, development, and environmental policy, deriving its authority both from existing powers and from treaty obligations. A series of landmark High Court decisions underpinned these moves. Important environmental measures included the *Seas and Submerged Lands* (1973), the *Australian Heritage Commission* (1975), and the *National Parks and Wildlife Conservation* (1975) *Acts.* The Great Barrier Reef Marine Park Authority (GBRMPA) was set up in 1975, and the Australian Institute of Marine Science at James Cook University in Townsville was established. The early 1970s also saw the first legislation requiring environmental impact assessment and two major inquiries into uranium mining and the exploitation of the mineral sands of Fraser Island, a large ecologically unique sand mass near Rockhampton in Queensland. The Liberal government of 1975–1983 continued the Whitlam Labor government's environmental agenda where it was established and institutionalized—for example, GBRMPA—but developed few new policy directions of its own. To antinuclear forces strongly urging a "leave uranium in the ground" policy, familiar "energy shortage" arguments were counterposed by the mining lobby. A decision favoring limited uranium mining—which ignored stringent warnings in the report of the Ranger Uranium Environmental Enquiry of 1976–7—was offset by a ban on sand mining of Fraser Island, widely thought to be a sop to environmentalists.

The Liberals lost the 1983 federal elections at the height of one of Australia's most celebrated environmental confrontations, the Gordon-below-Franklin hydroelectric dam dispute in Tasmania. Worsening economic conditions, unemployment, and a general perception of lackluster governmental style were also critical issues, but opposition to the dam project was significant.

The Gordon-below-Franklin Dam

Hydroindustrialization in Tasmania was fatally delayed by conservative opposition. Finally adopted in 1915, over 20 years after it was first proposed, it never delivered expected benefits, such as small-scale industrialization. It consequently depended on attracting large industrial consumers, which led to demand-matching problems and tariffs that penalized smaller customers. Its political orthodoxy, however, was so great that, in 1944, the state government legislated to give the Hydro-Electric Commission (HEC) near independence. Between 1953 and 1962 alone, the state's grid capacity was trebled and major new bulk consumers were attracted. But by the 1960s, escalating interest rates aggravated rising capital costs; "footloose" bulk consumers consumed nearly 70 percent of all electricity produced without significant benefit to the state; demand-matching problems persisted; and industrial expansion stalled, one bulk consumer closing down in the 1970s. The immensely powerful HEC's rolling construction program absorbed 54 percent of all Tasmania's loan funds in 1969–70, while other states averaged 23 percent (Tighe 1992).

The HEC's 1972 flooding of Lake Pedder, in Tasmania's hitherto untouched Southwest, sparked a controversy that raged from 1967 to 1976, radicalizing Tasmanian environmentalists. The HEC's 1979 recommendation to the Tasmanian government for an immediate start on the proposed Gordon-below-Franklin dam instantly provoked strong opposition. An enlightened Labor premier sought to reexamine the proposals and break the HEC's stranglehold on energy planning, but the attempt split his party and led to the government's defeat in the Tasmanian Parliament. In the ensuing election Labor was defeated (Lowe 1984). The continuing commitment of both major political parties to the orthodoxy of hydroindustrialization forced an extraparliamentary campaign, led by the Wilderness Society, culminating in a spectacular blockade of the dam site in December 1982. In March 1983, the newly elected Hawke (federal) Labor government immediately banned construction. The High Court subsequently ruled the Commonwealth's use of its external affairs and corporations powers valid. The Tasmanian government applied compensation of $A276,500,000 to the construction of an alternative hydroelectric dam, creating serious overcapacity with consequent inflationary effects for ordinary (noncontractual) users (Blakers 1994). Tasmanian governments continue to pursue industrial development and resource exploitation, despite repeated failure and viable alternatives (Crowley 1989; Lowe 1991; Davis 1995).

The High Court's ruling placed powerful sanctions in the Commonwealth's hands, including the ability to prohibit a wide range of activities by state governments or their instrumentalities (Tighe and Taplin 1985). These supplemented earlier decisions in the 1970s and were, in turn, augmented by subsequent decisions. These had the effect of expanding Commonwealth powers, mainly by reference to treaty obligations, the corporations power of the Constitution, and, recently, by the doctrine of "implied rights" (for example, to free speech in a democracy). However, the Commonwealth has been sparing in its use of its newly validated powers. The doctrine of "new federalism," emphasizing cooperation rather than conflict, has frequently been used to justify inaction. Conservative forces still emphasize states' rights, often as an excuse for nonintervention in environmental damage issues. During the 1990s, a deliberate policy of withdrawal of federal government from numerous areas of regulation and oversight has led to significant underuse of these powers (Economou 1999).

Other Conflicts

A second proposal, for a pulp mill at Wesley Vale in northern Tasmania, was dramatically abandoned in 1989 by the joint venturers, Noranda–North Broken Hill (NNBH), who claimed that the environmental conditions imposed were too onerous. The project had

> promised to make an invaluable contribution to Australia's balance of payments problem by earning export income and displacing costly kraft-pulp imports. . . . the promise of a transfer of state-of-the-art manufacturing technology through the establishment of "value added" production coincided with the Government's policy of encouraging the revitalisation of an internationally efficient manufacturing sector. For the Tasmanian Liberal Government, whose political ascendancy in 1982 was based on a policy platform of aggressively pursuing large-scale development and rapid resource exploitation, the project's promise of 2,400 construction, 400 permanent operational, and 711 indirect employment positions in a region suffering an unemployment rate of 20 percent was irresistible. (Economou 1992a, 43)

But intense opposition stemmed both from middle-class "greenies" and a powerful coalition of local interests. Farmers and residents were fearful of the effects of airborne pollution; the fishing industry feared contamination of Bass Strait by discharges of chemicals, especially organochlorines (and possibly dioxin), from the bleaching process (McEachern 1991). Timber supply limitations threatened existing timber quotas in southern Tasmania, provoking opposition from other sectors of the timber and woodchipping industry (Economou 1992a). The normally pro-Liberal constituency elected Christine Milne, leader of Concerned Residents Opposing Pulp-mill Siting (CROPS), to the Tasmanian Parliament as a Green Independent. The Gray Liberal Government's attempt to consolidate its political support through traditional "brokerage" politics (Sharman 1977) had backfired.

The joint venturers blamed government "indecisiveness" for changing environmental requirements, preventing effective planning. But major planning decisions in Australia are frequently taken in an unstructured, casually authoritarian way; unforeseen changes are inevitable consequences (Walker 1992b; Bolton 1982). At Wesley Vale, business and government policies converged; approval at both federal and state levels took local support for granted; the Environmental Impact Statement (EIS) was merely tacked on to appease potential opponents (Economou 1992a; Chapman 1992). The EIS technique in Tasmania, as elsewhere, has proved deficient, susceptible to bias, wide open to self-interest, and vulnerable to poor or inadequate scientific data (Formby 1989; Walker 1994). The Tasmanian government, in any case, eventually caved in to all the joint venturers' demands; it was the scale and power of opposition from other vested interests, and possibly the prospect of poor profits, that determined the project's abandonment (McEachern 1991). Environmentalist opposition alone was not enough.

Though intensively studied, Tasmania's environmental problems are not unique. Conflicts over conservation of the Great Barrier Reef are among the most enduring and have focused global attention on Australia, as well as generating important initiatives, such as the Marine Park (Bowen 1994). The extensive continental shelf is rich in minerals and fisheries; the northern reaches generate ongoing boundary and jurisdictional conflicts. The Antarctic, 43 percent of which is claimed by Australia, presents unique management problems (Herr, Hall, and Davis 1982; Doyle and Kellow 1995). Proposals during 1980–81 for large aluminium smelters in the lower Hunter Valley in New South Wales generated strong opposition, mainly from the wine industry, which feared the effects of water pollution and fluoride fallout from the smelting process. One major proposal was abandoned and another postponed, due to adverse economic conditions. Environmentalists were blamed, but the available scientific data on fallout and its effects remain inadequate and, consequently, indecisive (Simpson 1992).

Such conflicts often sharpen differences between states and the federal government. The aggressively developmentalist Country (later "National") Party regime (1957–1989) in Queensland was persistent in exploiting natural resources with the sole aim of profit, with little concern for the conservation of the environment, the careful management of resources, or the devastation that might be left as a legacy for future generations (Bowen 1994, 238–239). Especially under Bjelke-Petersen, premier from 1968 to 1989, this regime picked fights with conservationists and the federal government, indulged in "spoiling" behavior, and misused its police powers against political opponents (Brennan 1983; Coaldrake 1989; Whitton 1989). Bitter conflict over environmental issues persisted for some three decades and had important formative impacts on the conservation movement. The federal government frequently found itself mediating and arbitrating; often, it appears to have deliberately avoided confrontation. Queensland's environmental performance, especially in the area of land clearance, remains far behind that of other states.

Cases such as these underline the importance of natural resources to the Australian states. Their exploitation is one of the few means of expanding revenues without a fiscal penalty. State plans for natural resources therefore frequently conflict with conservationist pressure, often markedly stronger at the federal level (Walker 1992b). Australian states depend on the federal government for about half their revenues; in the smaller states, these subventions underwrite pork-barrel politics. In Tasmania, for example, "brokerage" politics extends selective favors—mostly "development" expenditures—to specific areas as electoral bait. Though it collapsed dramatically at Wesley Vale, it was a major explanation for Labor's long period in office from 1934 to 1969 (Sharman 1977; Davis 1986).

Ecologically Sustainable Development

Unless environmental management is conceived purely as conservation, and conservation itself as mere prohibition, governments must undertake planning and regulation, as well as encouragement of relevant research, promotion of environmentally responsible behavior, assistance to innovation, and, most importantly, coordination of action where individual efforts are inadequate. Sustainability, in short, demands more than mere conservation or management.

The Whitlam government (1972–1975) responded to the first UN Conference on the Human Environment in Stockholm in 1972 and the setting up of the United Nations Environment Programme (UNEP) (Dovers 1997). It adopted the then-fashionable environmental impact statement (EIS) methodology, and the marine park organizational structure for the Barrier Reef reflected overseas thinking (Bowen 1994). Impacts on domestic policy increased after the Brundtland Report in

1987 and the 1992 United Nations Conference on Environment and Development (UNCED) in Rio de Janeiro (Dovers 1997). Jolly and McCoy (1994) explicitly attribute the emergence of the Ecologically Sustainable Development (ESD) process in Australia to international developments. However, given Commonwealth reliance on High Court interpretations of its treaty powers, this emphasis may reflect domestic constitutional needs more truly than enthusiastic international cooperation.

In seeking to develop an environmental management capability, the Hawke Labor government (1983–1992) sought to defuse tensions. Hawke's political style relied on negotiating skills developed in his long trade union career. Successive "Accords" on wage restraint were negotiated with the trade unions; cosy relationships with some business interests evolved (McEachern 1986, 1991). Three initiatives of the late 1980s—the Ecologically Sustainable Development (ESD) Working Parties, the Resource Assessment Commission (RAC), and "resource security" legislation—were important extensions of "Accordism" (Economou 1992b).

The ESD working parties, set up in 1990, covered Agriculture, Energy Use, Energy Production, Fisheries, Forest Use, Manufacturing, Mining, Tourism, and Transport. Each had between 12 and 18 members, representing the Australian Council of Trade Unions (ACTU), industry, relevant governmental authorities (state and federal), the Australian and New Zealand Environment and Conservation Council (ANZECC—an intergovernmental body), Commonwealth Government Departments, the CSIRO, and, in one case, two independent scientists. Though environmental groups were represented on most committees, industrial and bureaucratic interests were dominant; Greenpeace walked out of the Forest Use Working Party at an early stage, claiming irretrievable bias. Nonetheless, consensus on a number of issues emerged between industry and the environmental movement. But the resulting reports, submitted in November 1991, were widely criticized as stressing economically, rather than ecologically, sustainable development. Major omissions included a consideration of alternative energy sources and energy conservation. Fragmentation by sectors, with a concomitant neglect of intersectoral issues and coordination, was also seen as a weakness (Jolly and

McCoy 1994). Many of the ESD initiatives were "taken on board" by the Commonwealth and state bureaucracies, and a network of cooperating agencies emerged, supported by appropriate legislation. But although ESD was adopted, or at least deferred to, in many bureaucratic agencies, it was less popular politically. Dovers and Lindenmayer (1997, 73) view as tragic the bureaucratization of the ESD process after the inclusive working groups delivered their reports and the failure to institutionalize and continue the productive and promising ESD discourse.

The RAC replaced ad hoc decision processes, such as judicial inquiries and Royal Commissions—unsatisfactory alike to environmental groups, business, and industry—by extension of "Accordism" to land-use concerns. By achieving even partial consensus through examination of issues in a quasi-judicial tribunal, it sought to bind the various parties to enduring policy outcomes (Economou 1992b). Three major inquiries—Kakadu Conservation Zone, Forest and Timber, and Coastal Zone—each helped to break new ground in defining, refining, and quantifying important resource-use issues. They gained the active participation of scientists, economists, interested pressure groups, business and industry, Commonwealth government departments, and, significantly, the states. The RAC was abolished in 1993. Some thought that the transparency of its deliberative processes was unpalatable to the central government; others suggested that its activities displeased the states; a third explanation pointed to shifts in cabinet priorities (Stewart and McColl 1994; Economou 1996; Dovers and Lindenmayer 1997, 74).

The third leg of the Hawke government's environmental policy process was "resource security" legislation, "guaranteeing" resources to extractive industry, especially in the forests. This depended on an assumption that the RAC and the ESD process could identify those resources that could be safely exploited and those that needed conservation. But the 1991 legislation was defeated in the Senate.

The RAC and ESD were casualties of economic "rationalism." Betraying widespread expectations that it would reverse a drift to unemployment and deindustrialization, the Hawke government continued tariff reductions, floated the currency, deregulated banks, and privatized some public enterprises. This process, driven more by the convictions of civil servants and by global trends than by voter desires or Australian needs, assigns low priority to the environment (Pusey 1991, 1992; Falk 1992). Government progressively abdicated responsibility for regulation and management, with particularly serious implications for sustainability (Dovers 1995, 1997).

The Hawke government's limited environmental policy success had depended on a single senior minister—Richardson—who articulated a vision of cooperation between industry and environmentalism between 1987 and 1990 (McEachern 1991). This staved off the traditionally stronger "economic" departments, which viewed environmental concerns as trivial, but Richardson has been criticized for emphasizing short-term gains at the cost of enduring institutional change (Wilkinson 1996). Hawke's government was certainly inconsistent and selective, approving major trade-offs for small environmental gains in the forests and refusing to intervene over important issues, such as the Daintree road in northern Queensland. It released a "comprehensive" environment statement, which was widely agreed to evade the "hard" issues, suggesting, for example, that standards on emission of greenhouse gases might have to be relaxed if exports made it "necessary."

A rare early success was negotiation, in 1985, of a new Murray Waters Agreement, establishing the Murray-Darling Basin Ministerial Council and Commission. The previous Murray Valley Commission's limited powers had become inadequate. A fortunate conjunction of Labor governments in all states, plus major changes in attitude and some innovative moves by strategically placed policy bureaucrats and ministerial advisers, broke a 70-year-old deadlock. The Ministerial Council explicitly provided for political decision making; the Commission enjoyed extended powers. For the first time, quality criteria were included in water allocations and allowance was made for seasonal variability (Kellow 1992; Doyle and Kellow 1995). Plans for amelioration of salinity, management of natural resources, and nutrient management followed. Local groups were successfully drawn into regional policy design and implementation. But the new arrangements remain consultative and subject to veto by participating states; the resources deployed, especially for the Natural Resource and Nutrient Management strategies, are inadequate to the task (Crabb 1988, 1991). Demands on water remain high, the nutrient and natural resource management programs are making slow progress, and the critical issue of land retirement—removal from production of waterlogged and saline land—has not been squarely faced (Walker 1994; Simons 1996). By the late 1990s, exploitation of additional land, often with water-hungry crops such as cotton, was leading to renewed conflict over allocations, while salinity continued to worsen (Lowe 1999).

The 1990s

Keating, formerly Hawke's treasurer, ousted him as prime minister in 1992. He was more confrontational, less popular, and more closely tied to Treasury and the "development" ministries. ESD was bureaucratized; the RAC was shut down; little attention was devoted to other environmental issues. Policymaking became less comprehensive and consensual (Economou 1996, 1999). It was often made on the run; a promise to "phase out" woodchipping of native forests over five years was conveniently forgotten within a year, and implementation of regional forestry agreements was widely thought to be biased toward industry (Lindenmayer 1994).

Failure to follow through on the Hawke government's initiatives led to repeated embarrassment as issues "blew up" in the face of a government that had failed to do its homework. Forest policy was the most serious example; a National Forest Policy Statement was agreed with the states in 1992, but deadlines for Regional Forest Agreements in conformity with its modest conservation objectives were not met (State of the Environment Advisory Council, 1996). Although a National Greenhouse Strategy incorporated Australia's obligations under the Framework Convention on Climate Change, reduction of emissions to 1990 levels by 2000 was not achieved, and use of fossil fuels, emission of greenhouse gases, and air pollution had all worsened (Taplin 1994; State of the Environment Advisory Council, 1996; Lowe 2001). Measures such as energy conservation and use of abundant solar and wind power were not actively promoted by either level of government.

The Liberals Return

Labor progressively lost control of the states throughout the late 1980s and early 1990s, with only New South Wales remaining by 1995.

However, all states had swung back to Labor by early 2002. Federally, the Liberals returned in a lackluster poll in 1996, and barely squeaked in at the 1999 election with a bare majority of seats and fewer votes nationally than the Opposition. Their fortunes improved with the "khaki" election of 2001, though control of the Senate still eluded them. Their broad policy orientation was toward globalization, trade liberalization, and economic "rationalism." Withdrawal of government from many areas of former responsibility was a major policy. Industry self-regulation was espoused.

Hostility to trade unions led to major confrontations. Environmental issues were not so much rejected as ignored. The dismantling of the Hawke-era institutions continued. Woodchipping quotas were immediately and substantially increased (Grattan 1996). Restrictions on uranium mining were abandoned. Proposals to set up a major conservation fund were tied to the partial sale of the publicly owned telecommunications utility, a move widely seen as a Trojan horse (Gordon 1996). Conclusion of Regional Forest Agreements remained tardy, and devolution of powers to the states placed the carrying out of treaty obligations and recognition of indigenous rights in question (Dargavel 1998). Biodiversity issues, such as control of invading species and quarantine, were seen as less than adequate (Groves and Willis 1999; Sharp 1999; Thresher 1999).

At successive international summit conferences on climate change, Australia sought exemption from the Climate Change provisions on the ground of heavy fossil-fuel dependence (which suggests that, in addition to caving in to foreign mining interests, the Liberal Party Howard government had quite failed to see the incentive function of binding targets for emission reduction in the first place). Research funding for alternative energy sources dried up almost totally (Blakers 2000). One minister dismissed the greenhouse issue as a passing fad ("Greenhouse Effect? No Worries, Says Parer," 1997). April 1997 saw a complaint, by the retiring chairman of the State of the Environment Advisory Committee, that scientists were becoming frustrated because the government ignored their advice (Woodford 1997). Hodgson and Barns (1998) conclude that the Howard government was excessively influenced by major corporate interests, to the detriment of national policy development.

The Howard government's response to *Wik* was to introduce a "ten-point plan" that would permit the extinction of native title on leaseholds by proxy and by stealth; a dissatisfied National Party continued to demand outright extinction. Yet most leasehold land is overexploited and much of it severely degraded. Policing of leasehold conditions is so inadequate as to invite derision; security of tenure would aggravate the situation. Massive subsidies from the public purse to "drought-stricken" farmers are still the rule, and sustainable management practices are not generally required.

In February 2001, Labor gained Western Australia and retained Queensland in landslide victories within one week of each other; the Liberals had again lost control of all but one state. The shift in voter sentiment almost certainly reflected, primarily, dissatisfaction with the Goods and Services Tax (GST), introduced by the Howard government, which took effect on July 1, 2000, and with high petrol prices. The latter were widely perceived as a breach of the government's promise that fuel prices would fall after the introduction of the GST. The events of September 11, 2001, presented the Howard government with a heaven-sent opportunity to campaign on "khaki" issues and to whip up xenophobic sentiment against refugees. "Boat people," many of Afghan origin, became a major red herring in the December 2001 federal election, returning the Liberals to power. Labor "me-tooism" ominously neglected important economic, social, and environmental issues. Small gains for the Greens and Democrats were the only indication that environmental issues continued to resonate with some of the electorate.

No decisive policy changes can now be anticipated for some years to come.

Policy Process and Institutions

Australia possesses a hybrid "Westminster" system, with cabinet government both centrally and in the six states, but with provision for judicial review by a High Court. The 1901 Constitution confers specific powers, primarily in defense, foreign affairs, and related matters, on the central government, residual powers remaining with the states. Two Territories are now self-governing, under delegated power. Commonwealth fiscal power led to progressive increases in its control, especially over management of the economy, and High Court interpretation has further extended its power. The states retain major responsibilities and are most directly concerned with resource exploitation in particular; however, they are increasingly constrained by intergovernmental agreements (most importantly, the Intergovernmental Agreement on the Environment (IGAE)) and central fiscal power. While constitutional constraints are of great importance, their impact on policy tends to be constraining rather than empowering (Bonyhady 1992; Ramsay and Rowe 1995).

The study of Australian government is difficult. Cabinet secrecy is not affected by recent Freedom of Information legislation; consequently, decision processes have to be inferred. Committee systems are underdeveloped in the federal Parliament and almost unknown in the states; there is little public, formalized scrutiny of policy proposals. In a small country, personal and informal factors are of great importance. Thus, explanatory models of personal political networks and influence, such as McEachern's "business mates," vie with depictions of a more structured and principled political process, such as that advanced by Economou.

Political appointments to senior bureaucratic posts have become common since the 1970s, but public service advice was never as neutral as the "Westminster" model presupposes. Some departments, such as Treasury, always had firm views, often determined by theoretical, rather than pragmatic, considerations, while others were either captured by, or devoted to, particular sectors of the economy. Outcomes of internecine disputes between powerful central departments and the more junior "policy" departments frequently reflect relative power, rather than the merits of the case. Small-scale politics often lead to large-scale corruption. "Help" for friends is a political tradition; many personal fortunes have been made at public expense.

Jurisdictional disputes and federal-state division of responsibility are worsened by a fiscal system that divorces the raising of revenue from its disbursement, undermining the independence of the states. As the major service providers, responsible for education, policing, and numerous other labor-intensive functions, state budgetary flexibility is small (Groenewegen 1983, 1994). Because their constitutional responsibilities include control of resources, the states are major agents of development.

Table 7.1
Australia: A Statistical Profile

Land	
Total area (1,000 km^2)	7,682.00
Arable and permanent crop land (% of total) 1997	6.90%
Permanent grassland (% of total) 1997	54.0%
Forest and Woodland (% of total) 1997	19.40%
Other land (% of total) 1997	19.70%
Major protected areas (% of total) 1997	7.7%
Nitrogenous fertilizer use (t/km^2/arable land) 1997	1.70%
National figure excludes Great Barrier Reef	
Population	
Total population (100,000 inhabitants) 1998	187.00
Growth rate % (1980–1998)	27.30%
Population density (inhabitant/km^2)	2.40
Gross domestic product	
Total GDP (billion U.S. dollars) 1997	$463.70
Per capita (1,000 U.S. dollars/capita) 1997	$20.70
GDP growth 1980–1998	81.40%
Current general government	
Revenue (%/GDP) 1996[b]	35.40%
Expenditure (%/GDP) 1996[b]	34.80%
Government employment (%/total) 1997[b]	14.80%
Energy	
Total supply (MTOE) 1997	102.00
% change 1980–1997	44.40%
Total consumption (MTOE) 1998[c]	69.14
% change 1988–1998[c]	28.10%
Road-vehicle stock	
Total (10,000 vehicles) 1997	1,099.00
% change 1980–1997	51.20%
Per capita (vehicles/100 inhabitants) 1997	59.00
Air profile and greenhouse gases	
CO_2 emissions (tons/capita) 1998[a]	16.60%
CO_2 emissions (kg/USD GDP) 1998[a]	.84
Nitrogen oxide emissions (kg/capita) 1998[a]	118.00
Nitrogen oxide emissions (kg/USD GDP) 1998	6.10
Sulfur oxide emissions (kg/capita) 1998[a]	101.00
Sulfur oxide emissions (kg/USD GDP) 1998	5.20

Table 7.1
(continued)

Pollution abatement and control	
Total (household excluded) expenditure (%/GDP) 1997	.80%

Sources: OECD, *OECD Environmental Data, Compendium 1999* (Paris: OECD, 1999), except as noted.
[a] OECD, The OECD Observer, *OECD in Figures* (Paris: OECD, June 2000).
[b] OECD, *National Accounts of OECD Countries* (Paris: OECD, 2000).
[c] IEA/OECD, *Energy Balances of OECD Countries, 1997–1998* (Paris: IEA/OECD, 2000).

The importance of resource taxes as sources of additional revenue explains their exploitative attitudes. Most, too, behave more like plebiscitary dictatorships than democracies. Decisions once made are rarely reversed, and objectors are ridiculed, often with media support.

Federalism, however, often brings environmental issues out into the open; few recent major cases have lacked an element of federal-state conflict or public rehearsal of disputed issues. Additionally, the High Court's pivotal role in constitutional interpretation has tended to expand Commonwealth powers.

Both at Commonwealth and at state level, poor coordination results in policy fragmentation. Departments conflict; often, issues are not fully explored before being decided; available data is ignored. Some exercises, such as State of the Environment Reports, are repeatedly commissioned and pigeonholed. Economic irrationalism downplays administrative coordination in favor of markets, seriously impeding regulation. Most states now have departments of Environment in some form; most, however, are junior and lack influence. Their initiatives are often compromised or watered down. The Commonwealth's Environment Department has frequently been reorganized or merged.

During the 1990s, devolution of responsibility has marked "small government" policies. Responsibility for major environmental issues has been "handed back" to the states and to local government. However, adequate funding has rarely accompanied them, despite some easing of state fiscal constraints, due to increased revenues from gambling and a promise of revenue from the Goods and Services Tax introduced in 2000 (Crowley 1999). Nor do state and local government typically have the clout—even when they have the will—to cope with major economic actors, such as multinational corporations (Doyle 1999).

Models of Interaction

The capability of government in a pluralist system to achieve "economic transformation," as well as to coordinate policy effectively, depends on its relative power. The performance of its "defining" functions entails some capacity to decide on overall policy and to plan and implement effectively (Rueschemeyer and Evans 1985).

Nations can be classified as "strong" or "weak" by their relative capability. The "Anglo-American" states, such as Britain, the United States, Canada, Australia, and New Zealand, with political institutions enjoying limited powers, have typically been poor at planning. Their policymaking tends to be reactive, without overall policy direction, often responding to crisis by seeking a quick, temporary "fix." Stronger states, such as European autocracies, have fewer controls and restrictions on government and can adopt more anticipatory measures (Atkinson and Coleman 1989).

Australian policymaking often seems to be "pressure pluralist," with weak government and fragmented, disorganized business with multiple policy agendas. Planning is skimpy and ineffective, bureaucratic conflict common, and policy reactive, without any evident plan. Coordination routinely suffers. Government relies on intergroup competition to expose policy issues and, hence, to define its own policy goals; it is disorganized, with departments and agencies pursuing their own, at times conflicting, objectives. The most common decision strategy is incrementalism.

The Australian Paradox

Yet pressure pluralism is not universal. In Queensland, persistent developmentalism has embodied elements of clientele pluralism—to the point of rubber stamping some proposals—but has also displayed state direction, most notably in policies for central Queensland mineral development. Out-of-favor groups, including nonconforming scientists and conservation interests, were deliberately excluded (Mullins 1986). In Victoria, a short-lived, formalized forest policy network, representing all major interests, could have been been characterized as weakly corporatist (Kellow 1986; Wright 1988). The Franklin dam case saw a head-on collision between a well-developed *concertation* network, composed of the HEC and its clientele of large-scale power consumers, and the emerging Green Independents.

The central government's strength was displayed both in the regime of tariff protection erected in the 1950s and 1960s and the recent switch to "free trade" under the influence of economic "rationalists" in the Commonwealth Treasury (Bell 1989). In the states, powerful developmentalist premiers have been common: Playford, in South Australia, Court, in Western Australia, and Bjelke-Petersen, in Queensland. Often, a dominant bureaucrat was present: Wainwright, in South Australia, and Hielscher in Queensland, for example.

Yet state developmentalist planning has been sectorally constrained. Most comprehensive was South Australia, paradoxically under a conservative government from 1933 to 1965, where industrialization was supported with housing and price control policies designed to keep the cost of living low and, hence, attract investment. But South Australia is critically unable to control its economic environment. Victoria performed poorly in irrigation prior to 1909, but New South Wales, which took account of the Victorian experience, encountered near-identical problems (Walker 1994).

Even central government in Australia, though historically more powerful than in other Anglo-American nations, still has a record of failure in interventionist planning. Commonwealth Government attempts, in 1946–1949, the early 1970s, and the 1980s, were fairly rapidly abandoned and were in any case ineffective.

The Australian state's weakly anticipatory capability is limited by a (partly ideological) preference for an arm's-length relationship with industry (Bell 1989). This doubly inhibits development of a capacity for detailed, planned economic intervention. Firstly, government establishes only the broad outlines of "macro" policy, depending on private investment for implementation. Capital's significant veto power includes freedom to resist or transform policy. Secondly, state capability exists only where state intervention was necessary in the past, confining the significant capabilities and skills, often embodied in administrative institutions such as statutory corporations, to a few sectors.

Hence the paradox of a "free market" economy with high levels of government intervention. Except where private investment failed, as in transport and communications, no significant government capability developed. In most fields—for example detailed control of industry policy—it has few skills and poor resources.

This incapacity directly constrains environmental policy and especially ecosystem management. Despite their large-scale and long-term implications, they also require sensitive, "fine-grained" implementation. Thus, to support salination and land management policies in the Murray-Darling Basin, it was necessary to establish numerous local bodies for consultation and implementation (Burton 1990; Simons 1996). The Tasmanian government's failure to consult stiffened local opposition to the Wesley Vale pulp mill proposal. And lack of administrative capability at the "micro" level may reduce some classes of policy to impracticability. Regulatory policies are especially likely to be affected; failure to regulate sewage and industrial effluents in Sydney seriously affected public health and amenity (Beder 1989). A toxic waste scare in Brisbane eerily recapitulated Love Canal, though less dramatically (Walker 1994).

Yet, with progressive withdrawal of government activity in many sectors, further decline in capacity, if not power, can be expected.

Communities and Networks

Just as government lacks power due to sectoral cross-currents, so the major political party groupings are the mere tip of a complex iceberg, and can obscure linkages that cut across traditional party lines. Labor has difficulty reconciling demands from roughneck unions for access to threatened natural resources—such as timber and minerals—with concerns over wilderness and biodiversity. The Liberals grapple with long-standing conflicts between manufacturers—tending to demand high tariffs and resist cuts—and the importing and exporting sectors. Rural party allegiances are not monolithic. Pastoralists reliant on leaseholds do not share all their interests with freeholding wheat, sheep, or dairy farmers; irrigation farmers have special interests in water and subsidies; and so on. Some groups, including miners, road hauliers (large trucking firms), media proprietors, and the financial community, are exceptionally powerful. Most not only possess institutions and publications for defining and expressing their orthodoxies, but also have preferential access to the mass media.

Among Australia's policy communities is one of the world's oldest and most powerful trade union movements. Big business, though often poorly organized or disunited, has always had a major influence on government, "representative" bodies frequently being less influential than individual major firms or industry groupings. Both the Labor and the Country (now blandly labeled "National") parties have direct, historically strong links with their supporting interest groups, the latter through groups such as the National Farmers' Federation and the Cattlemen's Union. The weaker links between the Liberals and industry pressure groups are explained both by the historical discontinuity of business organization and by the highly diverse nature of business demands. Recently, parties such as the Greens and the Democrats, the latter representing a broadly social liberal perspective, have gained some Parliamentary representation. Only in Tasmania, and then only briefly, has it been decisive.

Nongovernmental organizations (NGOs), such as the Australian Conservation Foundation, The Wilderness Society, Greenpeace, some religious bodies, and state and regional conservation councils, also play an important part in identifying and defining areas of policy concern. The neoclassicist economist community in the Australian bureaucracy accounts for the dominance in both major parties' policies of deregulation, "free market," and anti–welfare state ideas: so-called economic rationalism. While concentrated in Treasury, their influence spreads throughout the most powerful departments. Crucial support is derived from the international financial community, most academic economists, and their supporters and sycophants in the media (Pusey 1991, 1992; Goldfinch 1999).

Policy communities frequently have orthodox constructions of their central problems, acceptance helping to determine who is "in." Attention and sympathetic consideration to community members' proposals confer much greater chances of adoption, even if "outs" exist who offer better analyses and better policies. Frequently, the media reinforce such orthodoxies.

In Australia, the pluralist assumption that vested interests are "legitimate," in the sense of having just claims on government's time and attention directly proportionate to their financial interest, is well embedded. Government reliance on them for information and feedback on policy options confers privileged standing. Reduced attention to other individuals or organizations inevitably results in policy bias.

Australia's environmental groups are, at best, marginal to the dominant policy communities. Media attention is fitful and consultation by government incomplete. Their construction of the environmental problem contradicts growthism and developmentalism; their full message is rarely heard or comprehended. Their exclusion was underlined very dramatically when Greenpeace withdrew from the ESD Working Parties in 1992; no attempt was made to reintegrate it nor to address its criticisms. Environmentalist and independent scientific opinion both remained significantly underrepresented.

Epistemic Communities

Epistemic (knowledge-based) communities play an important role in policy. The international scientific community enjoys particular authority. Frequently, there is a defined body of scientific opinion on a particular issue, emanating from a specialist group, but backed by the prestige of the community as a whole. Well-developed national scientific communities have important impacts on policy (Haas 1989, 1990, 1992).

In Australia, science is revered, but not always heeded. Its findings are eagerly exploited to support desired courses of action, but as often rejected or ignored. The Wesley Vale proponents were highly critical of scientific evidence for unacceptable contamination of Bass Strait and its fisheries. Technological factors rarely, if ever, determine technical decisions in the electricity industry in Australia and New Zealand; rather, arguments from technological necessity justify decisions reached on other grounds (Kellow 1986). Western Australia's Government Entomologist had discovered, as long ago as 1945, that all known major cotton pests were found in the Ord River area. Yet government persistently ignored these findings, adopting cotton as the preferred crop on the initial Ord River Dam project of 1963. Cultivation collapsed in 1974, due to uncontrollable pest problems (Walker 1992b).

During the 1980s, when fish caught in the vicinity of Sydney's notorious sewer outfalls were proving to have very high levels of pesticide and heavy metal contamination, the government of New South Wales, through its Minister for Agriculture, the State Pollution Control Commission, and the Sydney Water Board, attempted to prevent the publication of discussion and analysis that would make raw scientific information intelligible to laypersons, on the grounds that they might unduly alarm the public (Beder 1990).

The Victorian government's forest policy from 1982 to 1991 incorporated consultation with both local communities and scientists, promising a management regime sensitive both to ecological factors and local community needs. But expedient changes, primarily in order to increase state revenue from timber cutting, drew sharp protest from scientists and conservationists. By 1993, clerical error, in combination with pitifully inadequate reserves, led to loggers destroying an important habitat for the endangered Leadbeater's possum (O'Neill 1993; Lindenmayer 1994; Dovers and Lindenmayer 1997).

In 1979–1981, the Terania Creek rainforest Inquiry in New South Wales dramatized the impediments posed to the rational use of scientific knowledge in policy by the adversarial legal procedures. In his conduct of the Inquiry, a retired judge of the Industrial Court not only oversimplified the issues by insisting on dividing all witnesses into "pro-" and "anti-" logging groups, but also assumed that the State Forestry Commission's witnesses were impartial. His insensitive handling of eminent scientists in the witness box goaded them into indignant public protest. The final report ignored the scientific issues and the environmental values involved (Taplin 1992).

Scientific conservatism, specialization, and tunnel vision often contribute to inability to see the big picture. Institutionalized scientific advice to government risks domination by "old-boy" networks, formal and informal, and time lags reflecting past patterns of scientific research investment. It self-selects for atypical, politically active scientists (Wilson and Barnes 1995).

Incomplete or uncertain scientific information—for example, over the effects of climate modification—is often seized on and exploited, frequently beyond the scope of legitimate scientific disagreement, casting doubt on matters on which all informed parties agree.

Policy Networks, Incrementalism, and Closure

Policy is ultimately determined by policy networks: that subset of a policy community in which all members have a direct role in formulating and implementing policy. Networks are not necessarily coextensive with policy communities and may focus about particular processes, "such as budgeting, auditing or planning" (Wright 1988, 606). Exclusion of some policy communities or their members—"closure"—is essential to the decision process. For example, the ESD Working Groups of 1981, imperfectly representative of the policy communities concerned, were replaced by a bureaucratic ESD Steering Committee charged with implementing the recommendations.

Environmental policy networks in Australian government are fluid. Membership changes by issue, though some groups, such as the pastoralists, banking sector, and mining industry, are exceptionally strong and ubiquitous. The major environmental groups tend to be treated by government as "single-issue" groups and excluded from the "important" business of economic growth.

Closure can lead to policy persistence, even when external conditions are changing. Australian governments have been remarkably reluctant to reexamine major policy regimes. Exclusion of unorthodox views is abetted by Australia's conservative, monopolistic media, themselves pursuing sectional interests. The apparent fluidity of incrementalism, especially under pressure pluralism, might seem to imply great flexibility, but its lack of comprehensiveness reinforces its fatal, ineradicable conservative bias (Dror 1969; Goodin 1982). Thus, the Tasmanian government did not review hydroindustrialization in response to its Franklin dam defeat; its economic strategy remained cargo-cultist, encouraging high-energy, low-labor content industry, while depending on federal financial subventions to take up the slack in the economy (Davis 1986, 1995; Crowley 1989). The Queensland government pursued the Tully Millstream hydroelectric proposals, which would inundate parts of a World Heritage rainforest, and even resurrected the discredited Bradfield scheme for diversion of northern rivers to the interior. The adoption of "market" policies and the "user-pays" principle have barely dented major subsidies to rural producers and to roads.

The most important policy imperative under conditions of change is to find alternatives: new ways of constructing problems, innovative ways of addressing them, and novel solutions and ways to implement them. All require expanded "policy space" (Clay and Schaffer 1984). Governments and their bureaucracies, fearful of upsetting stable patterns of support by antagonizing powerful groups, may actively seek to avoid alternatives. They practice agenda control, hiding it behind the technicalities of "mainstream" decision analysis or justifying it as technological necessity. Closure and agenda control are mutually reinforcing. Coupled with pressure pluralism, vested interests, and incrementalist "garbage can" policymaking, comprehensiveness, in particular, suffers.

Furthermore, decision makers do not necessarily respond to problems by solving them. More often, there is either "displacement" or symbolic activity designed to reassure the public (Dryzek 1987). In Queensland, the Bjelke-Petersen government responded to persistent popular hostility to coastal sand mining with National Parks legislation which failed to preclude mining. Decisions may be based on partial information available through experts or administrators, selectively interpreted and presented and often radically incomplete (Simpson 1992). Implementation is often hampered by distortion or reinterpretation, reflecting the professional biases of those charged with the task, or the interests of favored "clients."

Policy Performance

Evaluation of the effectiveness of government policy, especially in the environmental sphere, involves some assessment of adequacy: the degree to which proposed policies can resolve the problems they address and the degree to which they have done so when implemented. There is no established methodology for this purpose and little interest among political scientists (Walker 1995). Evaluation frequently depends on the opinions of scientists, administrators, and other interested parties.

Australian political scientists are divided over the impact of environmental issues on public policy. Some claim that they are here to stay and have significantly changed Australian politics. Others point to the lack of major institutional change and the general inadequacy of government policy, arguing that environmentalist "successes" to date have been on single issues, requiring enormous energy to defend.

Evidence for the first position is poor. Public environmental concern does not pervade policy processes. Claims that major industrial and pressure groups have responded positively to the Green movement tend to ignore both the inadequacy of policy outcomes and the tenacity with which industry groups have asserted the primacy of economic growth. A full understanding of environmental issues still appears remote (Walker 1992a, 1995). Progress toward sustainability is very slow (Dovers 1997; Dovers and Lindenmayer 1997). Creation of environmental departments or agencies has not engendered major policy innovation (Russ and Tanner 1978). Some intergovernmental agreements have been reached, but implementation has often lagged and some are of dubious adequacy. The durability of the few institutional reforms made is questionable. Those institutions that have survived the Keating and Howard governments tend to be advisory. Politicians and senior bureaucrats do not appear to understand environmental issues, let alone the imperatives of ecology. Regulation has been neglected, and in some cases has actually declined in effectiveness (Falk 1992).

Supporters of the second position point to a continuing lack of concern with long-term issues and the assumption by politicians and economists that the "fundamental" economic and policy measures are reversible. Pointing to irreversibility, threshold effects, and urgency, they reject incrementalism, emphasizing the need for ecological concerns to be built into all policy processes and for the application of the "cautionary principle." If the brief prominence enjoyed by environmental policy during Hawke's prime ministership reflected both the commitment to "Accordism" and an unusual conjunction of ministerial interests, renewed attention to traditional concerns easily obliterated it (Economou 1996, 1999).

Australian public policy continues to be short-term, ad hoc, incrementalist, and heavily influenced by vested interests (Hodgson and Barns 1998). The "winner-take-all" system of cabinet government means that electorally successful strategies may be immune to rational counterargument until long after their usefulness is dissipated (Painter 1996).

When major long-term issues have low electoral weight, they are easily ignored.

The central themes of Australian public policy have been strongly developmentalist. They have been driven by a crude economic pragmatism that has been highly vulnerable to the claims of vested interests, while at the same time being insensitive to public interest, social, and environmental concerns. The short period of policy interest in environmental issues resulted neither in an enduring institutional regime nor in enhanced attention to environmental issues. Indeed, the effective takeover of public policy by economic "rationalists" from the 1970s onward resulted in decreased attention to environmental issues as economic growth became the sole criterion of policy performance.

Australian Environmental Policy: Perspectives on Adequacy

Historically, human societies have tended to devote increased productivity to population expansion; the enhanced living standards resulting from the industrial revolution are consequences of vastly improved productivity, which combined with imperialist exploitation to outstrip population growth in the richer countries (Boyden 1987; McMichael 1995). But this phenomenon is necessarily brief. Globally, extended overexploitation is leading to serious ecosystem overload, threatening a classic Malthusian dead end. In Australia's case, derivative culture, social organization, and political ideas since European settlement have led to the attitude that resources must be developed at all costs: more population, exports and imports, expansion of cultivation, and the bigger, the better.

Contrary to such assumptions, Australia is already overpopulated. There was never a "frontier," in the North American sense, due to the aridity of the interior. The best coastal land was taken up early in the period of European settlement. Environmental problems have been endemic ever since. Water is one of the most serious; soil and nutrient depletion is a major neglected "sleeper," of which present knowledge is seriously inadequate; and present energy-use patterns aggravate pollution and mean that the system is living on borrowed time. The "European" economy is underpinned by minerals, all of which face eventual depletion, threatening "boom-and-bust" consequences in socioeconomic terms.

Australian governments and political parties have been slow to adapt to these challenges. At the end of the 1980s, a process of "policy learning" appeared to be generating wise, "landmark" decisions with increasing frequency, though still at a bare snail's pace. Since that time, however, changes, both of political mood and of government have reversed the process. A recent study, though emphasizing bureaucratic acceptance of ESD, concluded with masterly understatement that it was "poorly understood by government" (Crowley 2000). Economou suggests that environmental issues have passed through a complete issue-attention cycle, a view that Crowley criticizes for underestimation of issue persistence (Economou 1999; Crowley 1999). There is in fact a serious dichotomy between the salience of environmental issues for voters—consistently high since the 1970s—and for the media.

Attention to issues in the Australian media, indeed, is often short-lived and rarely critical. The press in Australia is dominated by two major publishing groups, and only two cities have competing daily newspapers. Television and radio are similarly monopolistic. This near-total lack of independent mass media, plus unrelenting political pressure on such "neutral" outlets as the Australian Broadcasting Commission (ABC) and the Special Broadcasting Service (SBS), has stifled public debate. A continuing lack of media recognition and public discussion of important issues has led, time and again, to hasty solutions, imperfectly thought out and applied. This has been especially true of environmental issues.

The close intertwining of government and corporate interests is of particular importance here. Though there are distinct differences between Labor and Liberal in power, the bipartisan political consensus on development and growth marginalizes environmental issues as "quality-of-life" concerns.

This phenomenon is structural. Dependent on investment for policy success, all Australian governments stress their "reasonableness" to business. Despite their considerable administrative and entrepreneurial capability, notably in organizing and running statutory corporations and QUANGOs (quasi-NGOs), Australian governments have never succeeded in overall planning nor in coordination of industrial activity. Substituting indiscriminate growth as a goal has served the interests of both business and government—a convergence embodied in the ideological prop of "development," deliberately ill-defined. For its success, however, it requires the assumption that natural resources, in particular, are effectively inexhaustible. The near-simultaneous confrontation of resource limits and serious depletion and pollution problems places an exceptionally high stress on this very sweeping assumption.

Australian politics is riddled with sectional demands from vested interests, commonly cloaked in "practical" language, the stalking horse for economic reductionism and self-interest. In this canon, "development" is the unquestioned underpinning of universally desired material prosperity. The conviction of its "goodness" is common: sometimes to the point of near-religious fervor. It pervades attitudes to population, immigration, foreign capital and its role, and perceptions of the role of political leadership. Impracticable myths, such as hydroindustrialization in Tasmania or the Bradfield scheme for diverting East Coast rivers, persist and resurrect themselves with mind-deadening regularity (Horne 1976; Holmes 1963).

Environmental policy in Australia fails five simple tests: institutional change, comprehensiveness, attention to irreversibility, allowance for threshold effects, and concern for long-term issues. The dominance of economic irrationalism since the early 1980s has led to systematic, myopic neglect of noneconomic issues in the narrowest sense. Its inadequacy in the face of problems with invisibility and "sleeper" characteristics reinforces persistent neglect, not only of the long-term self-interest and invisible issues bound up with the environment, but also of ethics, aesthetics, and welfare. Structurally, Australian politics is egg-bound: peripheral and symbolic issues gain wide-eyed attention, but globalization, loss of national control of economy and ecology, and ecological decline are ignored.

Quite apart from their considerable aesthetic (nonutilitarian) value, Australia's ecosystems are a reservoir of biological resources of enormous potential. They present opportunities for novel approaches to "development" that have not been explored. But governments and the major political parties still construct the environmental problem in terms of "jobs versus conservation"; the social cost alone of the resulting disputes is excessive.

Some links between conservation and production are beginning to emerge; in an atmosphere of obsessive, election-driven short-termism, informal and nongovernmental initiatives are outstripping government. A striking collaboration in 1988 between the Australian Conservation Foundation and the National Farmers' Federation extended the Landcare movement from its beginnings in Western Australia and Victoria. Landcare gained federal government financial support in 1990, but remains essentially voluntary, encouraging local programs to enhance sustainability through adoption of "best practice," revegetation, and erosion prevention (State of the Environment Advisory Council, 1996; Dovers and Lindenmayer 1997). The Scientific and Engineering Network on a Sustainable Environment (SENSE) has been joined by a national body to promote sustainable development (Lowe 2001). In another arena, implementation of Agenda 21 by local councils was very impressive during the 1990s. However, limited powers, rate capping, and policy conflict have blunted much of this progress (Adams and Hine 1999). It is therefore crucial that such positive developments inform government policy and are integrated into it; otherwise they are likely to remain isolated showpieces.

This is important because effective environmental management will depend on resolution of the fundamental conflict, at all levels of public decision making, between the rationalities of neoclassical economics and bureaucratic administration, on the one hand, and that of ecology, on the other. That requires revaluation of natural resources, adoption of a "steady-state" approach to materials, emphasis on energy conservation, and the redefinition of the notion of growth to emphasize ecological and social utility. Practical policy implications include increased conservation and recycling of materials and energy; emphasis on energy "capture" rather than creation or generation; and a conservative, stewardship-oriented approach to environmental exploitation. There is ample potential for renewable energy, especially solar and wind, and for novel technologies (Blakers and Diesendorf 1996; Intergovernmental Panel on Climate Change, 2001c). The probability that these would create unforeseen economic opportunities, rather than stagnation, has been widely canvassed (Boyden, Dovers, and Shirlow 1989; Daly 1992; Birch 1993; Intergovernmental Panel on Climate Change, 2001c). Treatment of environmental disputes as single issues, and a failure to systematically address the issues of land use, technology, and economic organization that they raise, are fatal. "Progress toward sustainability requires . . . a willingness to build ecological thinking into all social and economic planning" (State of the Environment Advisory Council, 1996, 10–28). The IPCC, in its Working Group Reports on climate change of March 2001, strongly emphasized the importance of government in making policy choices, given the differing mixes of energy-use patterns and opportunities (Intergovernmental Panel on Climate Change, 2001c).

The major challenge for Australia is not dissimilar to that for other developed states. Staving off a future of overpopulation, starvation, resource shortage, and general misery requires the timely development of an ecologically sensitive policy regime in an aware society. Such a regime would necessarily be capable of understanding and utilizing the best modern scientific knowledge. Only by such means could a genuine "stationary state," a novel community with high, but sustainable, standards of living within a stable, sensitively managed natural environment, be attained. This attractive and exciting prospect for enlightened and intelligent public policymaking demands further, serious exploration.

References

Adams, G., and Hine, M. (1999). Local environmental policy making in Australia. In K. J. Walker, and K. Crowley, eds., *Australian environmental policy 2* (pp. 186–203). Kensington, NSW: New South Wales University Press.

Atkinson, M. M., and Coleman, W. D. (1989). Strong states and weak states: Sectoral policy networks in advanced capitalist economies. *British Journal of Political Science*, *19*(1), 47–67.

Beder, S. (1989). *Toxic fish and sewer surfing: How deceit and collusion are destroying our great beaches*. Sydney: Allen & Unwin.

Beder, S. (1990). Science and the control of information: An Australian case study. *The Ecologist*, *20*(4), 136–140.

Bell, S. R. (1989). State strength and capitalist weakness: Manufacturing capital and the tariff board's attack on McEwenism, 1967–74. *Politics*, *24*(2), 23–38.

Bell, S. R., and Head, B., eds. (1994). *State, economy and public policy in Australia*. Melbourne: Oxford University Press.

Birch, C. (1993). *Confronting the future—Australia and the world: The next hundred years*. New ed. Ringwood, Victoria: Penguin.

Blainey, G. (1983). *The tyranny of distance: How distance shaped Australia's history*. Rev. ed. Melbourne: Sun Books.

Blakers, A. (1994). Hydro-electricity in Tasmania revisited. *Australian Journal of Environmental Management*, *1*(2), 110–120.

Blakers, A. (2000, December). Solar and wind electricity in Australia. *Australian Journal of Environmental Management*, 7(4), 223–236.

Blakers, A., and Diesendorf, M. (1996). A scenario for the expansion of solar and wind generated energy in Australia. *Australian Journal of Environmental Management*, *3*(1), 11–25.

Bolton, G. C. (1982). From Cinderella to Charles Court: The making of a state of excitement. In E. J. Harman, and B. W. Head, eds., *State, capital, and resources in the north and west of Australia* (pp. 27–42). Perth: University of Western Australia Press.

Bolton, G. C. (1992). *Spoils and spoilers: A history of Australians shaping their environment*. 2nd ed. Sydney: Allen & Unwin.

Bonyhady, T. (1993). A different Streeton. *Art Monthly Australia*, *61*, 8–12.

Bonyhady, T., ed. (1992). *Environmental protection and legal change*. Sydney: Federation Press.

Bowen, J. (1994). The Great Barrier Reef: Towards conservation and management. In S. R. Dovers, ed., *Australian environmental history: Essays and cases* (pp. 234–256). Melbourne: Oxford University Press.

Boyden, S. (1987). *Western civilisation in biological perspective: Patterns in biohistory*. Oxford: Clarendon Press.

Boyden, S., Dovers, S., and Shirlow, M. (1989). *Our biosphere under threat: Ecological realities and Australia's opportunities*. Melbourne: Oxford University Press.

Brennan, F. (1983). *Too much order with too little law*. St. Lucia: University of Queensland Press.

Brook, D., ed. (2000). *A. D. Hope: Selected poetry and prose*. Rushcutters Bay, NSW: Halstead Press.

Burton, J. R. (1990). Community involvement in the management of the Murray-Darling Basin. *Canberra Bulletin of Public Administration*, 62, 76–80.

Butlin, N. G., Barnard, A., and Pincus, J. J. (1982). *Government and capitalism: Public and private choice in twentieth century Australia*. Sydney: Allen & Unwin.

Carden, M. (1999). Unsustainable development in Queensland. In K. J. Walker, and K. Crowley, eds., *Australian environmental policy 2* (pp. 81–102). Kensington, NSW: New South Wales University Press.

Chapman, R. J. K. (1992). *Setting agendas and defining problems: The Wesley Vale Pulp Mill Proposal*. Deakin Series in Public Policy and Administration No. 3. Geelong, Victoria: Centre for Applied Social Research, Deakin University.

Clay, E. J., and Schaffer, B. B., eds. (1984). *Room for manoeuvre*. London: Heinemann.

Coaldrake, P. (1989). *Working the system: Government in Queensland*. St. Lucia: University of Queensland Press.

Cochrane, T. (1989). *Blockade: The Queensland loans affair 1920 to 1924*. St. Lucia: University of Queensland Press.

Connell, R. W. (1977). *Ruling class, ruling culture: Studies of conflict, power and legitimacy in Australian life*. Cambridge: Cambridge University Press.

Crabb, P. (1988). *The Murray-Darling Basin Agreement: An examination in the light of international experience* (CRES Working Paper 1988/6). Canberra: Centre for Resource and Environmental Studies, Australian National University.

Crabb, P. (1991). Resolving conflicts in the Murray-Darling Basin. In J. W. Handmer, A. H. Dorcey, and D. I. Smith, eds., *Negotiating water* (pp. 147–159). Canberra: Centre for Resource and Environmental Studies: Australian National University.

Crook, S., and Pakulski, J. (1995). Shades of green: Public opinion on environmental issues in Australia. *Australian Journal of Political Science*, *30*(1), 39–55.

Crosby, A. W. (1993). *Ecological imperialism: The biological expansion of Europe, 900–1900*. Cambridge: Cambridge University Press.

Crowley, K. (1989). Accommodating industry in Tasmania: Eco-political factors behind the Electrona Silicon Smelter Dispute. In P. Hay, R. Eckersley, and G. Holloway, eds., *Environmental politics in Australia and New Zealand* (pp. 45–58). Hobart: Centre for Environmental Studies, University of Tasmania.

Crowley, K. (1999). Explaining environmental policy: Challenges, constraints and capacity. In K. J. Walker, and K. Crowley, eds., *Australian environmental policy* 2 (pp. 45–64). Kensington, NSW: New South Wales University Press.

Crowley, K. (2000, September). *Review of implementation of ecologically sustainable development by Commonwealth departments and agencies* (Productivity Commission, Inquiry Report, Report No. 6. Canberra: AusInfo, 1999). In *Australian Journal of Political Science*, *59*(3), 119–120.

Daly, H. E. (1992). *Steady-state economics*. London: Earthscan.

Dargavel, J. (1998). Politics, policy and process in the forests. *Australian Journal of Environmental Management*, *5*(1), 25–30.

Davidson, B. R. (1974). Irrigation economics. In H. J. Frith, and G. Sawer, eds., *The Murray Waters: Man, nature and a river system* (pp. 193–211). Sydney: Angus and Robertson.

Davidson, I. (1992). Aboriginal Living. In J. Smith, ed., *The unique continent: An introductory reader in Australian environmental studies* (pp. 91–100). St. Lucia: University of Queensland Press.

Davis, B. W. (1986). Tasmania: The political economy of a peripheral state. In B. Head, ed., *The politics of development in Australia* (pp. 209–225). Sydney: Allen & Unwin.

Davis, B. W. (1995). Adaptation and deregulation in government business enterprise: The Hydro-Electric Commission of Tasmania. *Australian Journal of Public Administration*, *54*(2), 252–261.

Dovers, S. R. (1995). Information, sustainability and policy. *Australian Journal of Environmental Management*, *2*(3), 142–156.

Dovers, S. R. (1997). Sustainability: Demands on policy. *Journal of Public Policy*, *16*(3), 303–318.

Dovers, S. R., ed. (1994). *Australian environmental history: Essays and cases.* Melbourne: Oxford University Press.

Dovers, S. R., and Lindenmayer, D. B. (1997). Managing the environment: Rhetoric, policy and reality. *Australian Journal of Public Administration*, *56*(2), 65–80.

Doyle, T. (1999). Roundtable decision making in arid lands under conservative governments: The emergence of "Wise Use." In K. J. Walker, and K. Crowley, eds., *Australian environmental policy 2* (pp. 122–141). Kensington, NSW: New South Wales University Press.

Doyle, T., and Kellow, A. (1995). *Environmental politics and policymaking in Australia.* South Melbourne: Macmillan.

Dror, Y. (1969). Muddling through—"Science" or inertia? In A. Etzioni, ed., *Readings on modern organizations* (pp. 166–171). Englewood Cliffs, NJ: Prentice-Hall.

Dryzek, J. (1987). *Rational ecology.* Oxford: Blackwell.

Economou, N. (1992a). Problems in environmental policy creation: Tasmania's Wesley Vale Pulp Mill Dispute. In K. J. Walker, ed., *Australian environmental policy* (pp. 59–80). Kensington, NSW: New South Wales University Press.

Economou, N. (1992b). Reconciling the irreconcilable? The Resource Assessment Commission, resources policy and the environment. *Australian Journal of Public Administration*, *51*, 461.

Economou, N. (1996). Australian environmental policy making in transition: The rise and fall of the Resource Assessment Commission. *Australian Journal of Public Administration*, *55*, 12–22.

Economou, N. (1999). Backwards into the future: National policy making, devolution and the rise and fall of the environment. In K. J. Walker, and K. Crowley, eds., *Australian environmental policy 2* (pp. 65–80). Kensington, NSW: New South Wales University Press.

Falk, J. (1992). Economic rationalism and environmental regulation. In P. Vintila, J. Phillimore, and P. Newman, eds., *Markets morals and manifestos: Fightback! and the politics of economic rationalism in the 1990s* (pp. 195–204). Murdoch, W. A.: Institute for Science and Technology Policy, Murdoch University.

Formby, J. (1989). Where has EIA gone wrong?: Lessons from the Tasmanian Woodchips Controversy. In P. Hay, R. Eckersley, and G. Holloway, eds., *Envi-*

ronmental politics in Australia and New Zealand (pp. 3–17). Hobart: Centre for Environmental Studies, University of Tasmania.

Frawley, K. (1994). Evolving visions: Environmental management and nature conservation in Australia. In S. Dovers, ed., *Australian environmental history* (pp. 55–78). Melbourne: Oxford University Press.

Gilbert, A. (1981). The state and nature in Australia. *Australian Cultural History*, 9–28.

Goldfinch, S. (1999). Remaking Australia's economic policy: Economic policy decision-makers during the Hawke and Keating Labor Governments. *Australian Journal of Public Administration*, *58*(2), 3–20.

Goodin, R. E. (1982). *Political theory and public policy*. Chicago: University of Chicago Press.

Gordon, M. (1996, July 13–14). Howard alienates Greens. *The Australian*, p. 20.

Grattan, M. (1996, July 12). Green fury over woodchips. *The Age*, p. 1.

Greenhouse effect? No worries, says Parer. (1997, March 14). *Sydney Morning Herald*, p. 8.

Groenewegen, P. (1983). The political economy of Federalism 1901–81. In B. W. Head, ed., *State and economy in Australia* (pp. 169–175). Melbourne: Oxford University Press.

Groenewegen, P. (1994). The political economy of Federalism since 1970. In S. R. Bell, and B. Head, eds., *State, economy and public policy in Australia* (pp. 169–193). Melbourne: Oxford University Press.

Groves, R. H., and Willis, A. J. (1999, September). Environmental weeds and loss of native plant biodiversity: Some Australian examples. *Australian Journal of Environmental Management*, *6*(3), 164–171.

Haas, P. M. (1989). Do régimes matter? Epistemic communities and Mediterranean pollution control. *International Organization*, *43*(3), 377–403.

Haas, P. M. (1990). *Saving the Mediterranean*. New York: Columbia University Press.

Haas, P. M. (1992). Banning chlorofluorocarbons: Epistemic community efforts to protect stratospheric ozone. *International Organization*, *46*(1), 187–224.

Hay, P., Eckersley, R., and Holloway, G., eds. (1989). *Environmental politics in Australia and New Zealand*. Hobart: Centre for Environmental Studies, University of Tasmania.

Head, B. W., ed. (1986). *The politics of development in Australia*. Sydney: Allen & Unwin.

Herr, R. A., Hall, R., and Davis, B. W., eds. (1982). *Issues in Australia's marine and Antarctic policies*. Hobart: Public Policy Monograph, Department of Political Science, University of Tasmania.

Hodgson, N., and Barns, I. (1998, June). Australia's response to the Framework Convention on Climate Change. *Science and Public Policy*, *25*(3), 142–154.

Holmes, J. M. (1963). *Australia's open North*. Sydney: Angus and Robertson.

Holmes, J. (1996). The policy relevance of the state's proprietary power: Lease tenures in Queensland. *Australian Journal of Environmental Management*, *3*(4), 240–256.

Horne, D. (1976). *Money made us*. Ringwood, Victoria: Penguin Books.

Hoy, A. (2000, August 15). Rivers of no return. *The Bulletin*, pp. 40–42.

Intergovernmental Panel on Climate Change. (2001c). *Working Group III Report, Climate Change 2001: Mitigation*. Cambridge: Cambridge University Press.

Jolly, B., and McCoy, E. (1994). Ecologically sustainable development and the fragmentation of policy: A critical review. *Policy Organisation and Society*, 7(Summer), 34–39.

Kellow, A. (1986). Electricity planning in Tasmania and New Zealand: Political processes and the technological imperative. *Australian Journal of Public Administration*, *40*(1), 2–17.

Kellow, A. (1992). *Saline solutions: Policy dynamics in the Murray-Darling Basin*. Deakin Series in Public Policy and Administration No. 2. Geelong, Victoria: Centre for Applied Social Research, Deakin University.

Lindenmayer, D. B. (1994). Timber harvesting impacts on wildlife: Implications for ecologically sustainable forest use. *Australian Journal of Environmental Management*, *1*(1), 56–68.

Lines, W. J. (1991). *Taming the Great South Land: A history of the conquest of nature in Australia*. North Sydney: Allen & Unwin.

Lothian, J. A. (1994). Attitudes of Australians towards the environment, 1975 to 1994. *Australian Journal of Environmental Management*, *1*(2), 78–99.

Lowe, D. (1984). *The price of power: The politics behind the Tasmanian Dams Case*. South Melbourne: Macmillan.

Lowe, I. (1991, August). Towards a green Tasmania: Developing the "Greenprint." *Habitat Australia*, pp. 12–14.

Lowe, I. (1994). The greenhouse effect and the politics of long-term issues. In S. R. Bell, and B. Head, eds., *State, economy and public policy in Australia* (pp. 315–333). Melbourne: Oxford University Press.

Lowe, I. (1999, July 17). Trans Tasman call to arms. *New Scientist*, *163(*2195), 55.

Lowe, I. (2001, March 10). Living within our means. *New Scientist*, *169*(2281), 55.

Marshall, J., ed. (1966). *The Great Extermination: A guide to Anglo-Australian cupidity, wickedness and waste*. Melbourne: Heinemann.

McEachern, D. (1986). Corporatism and business responses to the Hawke Government. *Politics, 21*(1), 19–27.

McEachern, D. (1991). *Business mates: The power and politics of the Hawke Era*. Sydney: Prentice Hall.

McMichael, A. J. (1995). *Planetary overload: Global environmental change and the health of the human species*. Cambridge: Cambridge University Press.

Mullins, P. (1986). Queensland: Populist politics and development. In B. W. Head, ed., *The politics of development in Australia* (pp. 138–162). Sydney: Allen & Unwin.

O'Neill, G. (1993, May 12). Rare possum's habitat destroyed by mistake. *The Age*, p. 5.

Painter, M. (1996). Economic policy, market liberalism and the "end of Australian politics." *Australian Journal of Political Science, 31*(3), 287–299.

Ponting, C. (1992). *A green history of the world*. Harmondsworth, Middlesex: Penguin.

Pusey, M. (1991). *Economic rationalism in Canberra: A nation-building state changes its mind*. Melbourne: Cambridge University Press.

Pusey, M. (1992). What's wrong with economic rationalism? In D. Horne, ed., *The trouble with economic rationalism* (pp. 63–69). Newham, Victoria: Scribe Publications.

Pyne, S. J. (1992). *Burning bush: A fire history of Australia*. Sydney: Allen & Unwin.

Ramsay, R., and Rowe, G. C. (1995). *Environmental law and policy in Australia*. Sydney: Butterworths.

Rose, R. (1976). On the priorities of government: A developmental analysis of public policies. *European Journal of Political Research, 4*, 247–289.

Ross, H., Young, E., and Liddle, L. (1994). Mabo: An inspiration for Australian land management. *Australian Journal of Environmental Management, 1*(1), 24–41.

Rueschemeyer, D., and Evans, P. B. (1985). The state and economic transformation: Toward an analysis of the conditions underlying effective intervention. In P. B. Evans, D. Rueschemeyer, and T. Skocpol, eds., *Bringing the state back in* (pp. 44–77). Cambridge: Cambridge University Press.

Russ, P., and Tanner, L. (1978). *The politics of pollution*. Camberwell, Victoria: Visa.

Scanlon, J. (2001a). The National Action Plan on Salinity and Water Quality and the Murray-Darling Basin Initiative. *Environmental Politics and Law Journal, 18*(2), 105–107.

Scanlon, J. (2001b). The need to reform the Murray-Darling Basin Commission. *Environmental Politics and Law Journal, 18*(2), 230–234.

Sexton, M. (1979). *Illusions of power: The fate of a reform government*. Sydney: George Allen & Unwin.

Sharman, G. C. (1977). Tasmania: The politics of brokerage. *Current Affairs Bulletin*, *53*(9), 15–23.

Sharp, R. (1999, September). Federal policy and legislation to control invading species. *Australian Journal of Environmental Management*, 6(3), 172–176.

Simons, M. (1996, March 2–3). River of no return? *The Australian*, p. 26.

Simpson, R. W. (1992). Air pollution control strategies: A case study in the Hunter Region of NSW. In K. J. Walker, ed., *Australian environmental policy* (pp. 103–123). Kensington, NSW: New South Wales University Press.

Smith, D. (1990). *Continent in crisis: A natural history of Australia*. Ringwood, Victoria: Penguin.

Smith, J., ed. (1992). *The unique continent: An introductory reader in Australian environmental studies*. St. Lucia: University of Queensland Press.

Smith, R. (2000, June). Crisis in the Lower Murray. *Trees and natural resources*, *42*(2), 22–24.

State of the Environment Advisory Council. (1996). *Australia: State of the environment 1996*. Collingwood, Victoria: CSIRO.

Stewart, D., and McColl, G. (1994). The Resource Assessment Commission: An inside assessment. *Australian Journal of Environmental Management*, *1*(1), 12–23.

Taplin, R. (1992). Adversary procedures and expertise: The Terania Creek Inquiry. In K. J. Walker, ed., *Australian Environmental Policy* (pp. 156–182). Kensington, NSW: New South Wales University Press.

Taplin, R. (1994). Greenhouse: An overview of Australian policy and practice. *Australian Journal of Environmental Management*, *1*(3), 142–155.

Thresher, R. E. (1999, September). Diversity, impacts and options for managing invasive marine species in Australian waters. *Australian Journal of Environmental Management*, 6(3), 137–148.

Tighe, P. (1992). Hydroindustrialisation and conservation policy in Tasmania. In K. J. Walker, ed., *Australian environmental policy* (pp. 161–195). Kensington, NSW: New South Wales University Press.

Tighe, P., and Taplin, R. (1985). Lessons from recent environmental decisions: The Franklin Dam Case and rainforest in New South Wales. In D. Jaensch, and N. Bierbaum, eds., *The Hawke Government: Past, present, future*. Adelaide: Australasian Political Studies Association.

Walker, K. J. (1989). The state in environmental management: The ecological dimension. *Political Studies*, *37*, 26–39.

Walker, K. J. (1992a). Business, government and the environment: Unresolved dilemmas. In S. R. Bell, and J. Wanna, eds., *Business-government relations in Australia* (pp. 243–254). Sydney: Harcourt Brace Jovanovich.

Walker, K. J., ed. (1992b). *Australian environmental policy*. Kensington, NSW: New South Wales University Press.

Walker, K. J. (1994). *The political economy of environmental policy: An Australian introduction*. Kensington, NSW: New South Wales University Press.

Walker, K. J. (1995, September). Environmental policy: Adequacy of government response. Paper presented at the 37th Annual Conference of the Australasian Political Studies Association, University of Melbourne.

Walker, K. J. (1999). Statist developmentalism in Australia. In K. J. Walker, and K. Crowley, eds., *Australian environmental policy* 2 (pp. 22–44). Kensington, NSW: New South Wales University Press.

Whitton, E. (1989). *The hillbilly dictator: Australia's police state*. Crows Nest, NSW: ABC Enterprises.

Wilkinson, M. (1996). *The Fixer: The untold story of Graham Richardson*. Port Melbourne: Heinemann.

Wilson, S., and Barnes, I. (1995). Scientists' participation in environmental policy. *Search*, *26*(9), 270–273.

Woodford, J. (1997, March 31). MPs ignoring environment warning bells, say scientists. *Sydney Morning Herald*, p. 1.

Wright, M. (1988). Policy community, policy network and comparative industrial policies. *Political Studies*, *36*, 593–612.

8

Japan's Environmental Regime: The Political Dynamics of Change

Jeffrey Broadbent

Pollution and Politics in Japan

Japan has often been strongly praised for its rapid economic growth—called, by some, a "miracle." It has been just as strongly criticized for the massive pollution and severe health problems caused by that growth—a pollution "debacle." Yet Japan cleaned up some of this pollution with an unprecedented rapidity. These contrasts make Japan the eminent case for exploring the political dynamics of industrial and consumer pollution. With its dense population and scarcity of resources, Japan represents the future. If, under these circumstances, it has managed to solve even a few of its pressing environmental problems, these solutions should offer instructive lessons for the rest of the world.

Japan's environmental regimes have passed through four major phases. Scholars roughly agree on the timing and content of these phases, but not on the dynamics—how and why the changes took place (Broadbent 1998, 36–37; Funabashi 1992; Hasegawa 1998; Iijima 1993, 29–66; Pharr and Badaracio 1986).

In phase 1, a "polluter's paradise" for extraction, industrial manufacturing, power generation, and urbanization, the government imposed virtually no restrictions on pollution from the start of industrialization to the mid-1960s. In the 1950s and early 1960s, few people paid attention to environmental problems. But as industrial manufacturing and consumer consumption rose dramatically, the quality of air, water, land, and the living environment declined quickly. Japan's "economic miracle" produced a "pollution debacle." During that era, Japan produced some of the world's most terrible pollution-caused illnesses, symbolized by brain-destroying and disfiguring mercury pollution in Minamata and Niigata. Terrible air pollution in cities brought asthma, bronchitis, and other respiratory problems. Chemical pollution of food poisoned unsuspecting consumers.

In phase 2, the mid-1960s to the mid-1970s, facing massive protests and international pressure, the government achieved a remarkable turnaround in industrial pollution. New pollution control laws, regulations, and "guidance" from ministries went into effect, producing what some called a "polluter's hell," but also an environmental minimiracle. Massive investments in pollution technology and equipment reduced waste output. Air and water pollution improved significantly.

In phase 3, the mid-1970s to the late 1980s, some scholars see business's acceptance of industrial pollution controls, while others see regulatory stasis or retreat. Protests quieted and attention turned back to economic issues. However, other types of environmental disruption expanded and accumulated: industrial toxic waste and consumer waste, as well as from public works and other large-scale construction projects (such as golf courses). During this period, the Japanese government resisted international agreements concerning newly emerging global environmental problems, such as depletion of the stratospheric ozone layer or trade in endangered species.

In phase 4, though, from the 1990s through the early 2000s, Japan took new stances. The ill effects of toxic waste from industry and nuclear power had become increasingly apparent. Moreover, global environmental problems and pressures intensified greatly. International governmental and nongovernmental actors increasingly pressured Japan to act. This situation lead Japan to embark on a series of new global environmental initiatives.

Seen from its rather desperate initial economic situation, Japan looks like an unlikely place to find even a shred of successful environmental protection. The land is very low in the mineral resources necessary for industrial production. Japan has to import over 90 percent of its fuels. Arable level land is scarce, and the population is very dense. Accordingly, population and industry crowd closely together. These conditions did not deter economic growth. By around 1980, with only 0.3 percent of the world's surface area and 3 percent of the world's population, Japan accounted for about 10 percent of the world's economic product (Johnson 1982, 6) and about 10 percent of total world exports. This growth gave Japan the highest density (per square kilometer) of industry, energy use, and population of any industrial nation (figure 8.1) (Broadbent 1998, 12–18). Such resource and trade dependency would seem to preclude effective pollution control. On the other hand, the objective situation—Japan's high intensity of urban pollution and vast exposed population—might, at some point, "kick in" to propel it toward remedies.

This book and chapter focus on how institutions—in particular, business, government, and global politics—affect environmental policy and practice. The formal political institutions of Japan consist of a bicameral elected legislature (the Diet) with a prime minister (selected by the ruling party in the Diet) and cabinet, a set of administrative ministries and agencies, and a judicial system culminating in a Supreme Court. Local government consists of similar institutions on a smaller scale; the governor is elected, but the vice-governor (and usual successor) is appointed by

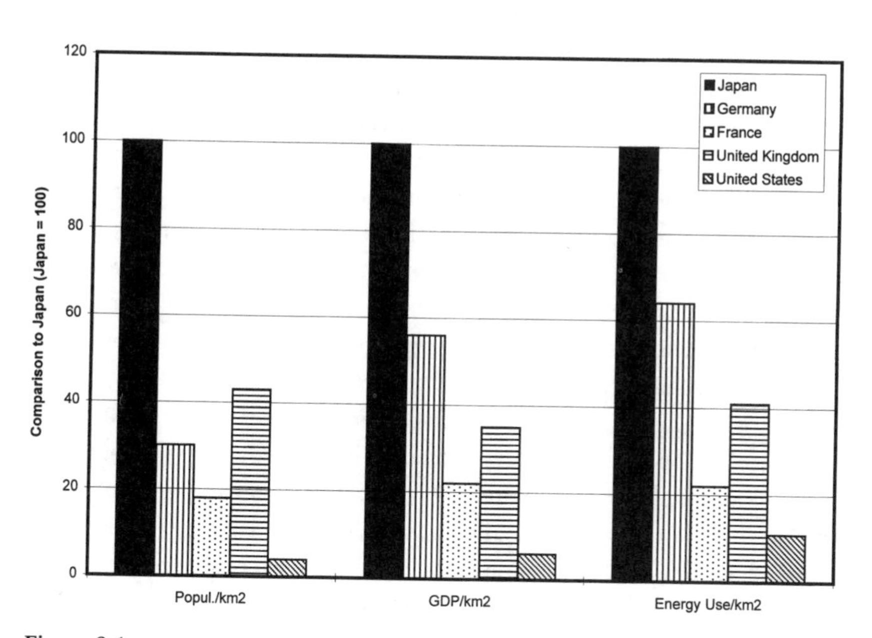

Figure 8.1
Comparative population and industrial density, 1970. *Source:* OECD 1993.

the national government. Most of the local governments' budget is doled out by the national government, limiting local fiscal autonomy but not discouraging all initiative (Muramatsu 1997).

These formal institutions did not fully define the effective loci of political power. Over most of the post–World War II years, the macropattern of power took the shape of a "T." The horizontal bar at the top of the T: a "ruling triad" composed of career ministry bureaucrats, Liberal Democratic Party (LDP) politicians, and business leaders with roughly equal veto power. The vertical bar under the T: Japan's "vertical society," wherein the ruling triad dominated the rest of society, except during times of upheaval. Japanese scholars often call this T pattern the "1955 system" because the LDP was established in that year. Despite occasional crises and policy compromises, the T held until the early 1990s. Until the 1980s, it was strongly supported by the U.S. government (including large financial contributions to the LDP from the U.S. Central Intelligence Agency) (Smith 1997, 17; Weiner, Engleberg, and Sterngold 1994). In 1993, one corner of the triad, the LDP, lost control of the Diet, with important consequences for social change.

The popular press has often conveyed an image of the Japanese state as unified, of a single will (Fallows 1989). At first glance, the T image might seem the same. However, within that elite arena, the ruling triad was not a harmonious unit. Its members split into different policy "tribes" (*zoku*) that defended different turf and interests (Inokuchi and Iwai 1987). Among these were, for instance, the "construction tribe" (*kensetsuzoku*), the "industrial growth tribe," and the "environment tribe." Each tribe was a minitriad composed of party politicians, government bureaucrats, and interest groups (supporters from business, unions, or other sectors of society).

Nor were they equal in power. Scholars have portrayed the Japanese state and its policies as "plan rational"—integrated under the Ministry of International Trade and Industry (MITI) around the goal of rapid national economic growth (Johnson 1982). While certainly more coordinated than the U.S. economy, MITI hardly exercised unchallenged authority. MITI had its "growth tribe" of supporters in the LDP and among manufacturing businesses. But other "tribes," such as the "construction tribe," had other interests, sometimes undercutting MITI's goals. Their struggles occurred within the corridors of Kasumigaseki, Tokyo's political district. They rarely emerged into public view. Yet their results determined the actual (as opposed to stated) outcomes of policy.

Certainly, MITI was a major player in setting national policies. Indeed, MITI tried with military fervor to organize the entire society around this goal. MITI tried to set up favorable conditions for Japanese businesses, so they could grow rapidly at home and then beat foreign competition in world markets (Johnson 1982; Okimoto 1989; Sakakibara 1993). MITI officials worked closely with business representatives in the Industrial Structure Advisory Council and Industrial Sector Associations (*gyokai*) to develop sector-specific plans. MITI and other ministries set up collaborative research institutes to develop basic technology that businesses then put to commercial uses.

The LDP, however, operated as a nationwide political "machine." Its political power depended on delivering patronage to local voters: local construction projects, jobs, expensive help and services from local party representatives, outright bribes at election time. This patronage depended on the party's power and money. The LDP, in control of the Diet, could decide the fate of policies desired by government ministries. It could, for instance, approve massive budgets of the ministries of construction and transportation. The LDP then demanded huge government outlays for public works construction. Through collusion (*dango*), the construction companies (*zenecon*) submitted rigged, highly inflated bids. They sent a portion of their huge profits back to the helpful LDP politicians (usually leaders of Diet factions) (Broadbent 1998; Masumi 1995; Woodall 1996). By doling this money out to junior LDP politicians, LDP leaders built their own compliant factions within the party. The politicians then pumped this money (and public works) into local communities, as various kind of patronage, to secure votes.

Japanese business organized along corporatistic lines under a peak association, the Federation of Economic Organizations (FEO, Keidanren). Under the FEO, numerous industrial-sector associations, such as the steel producers' association, represented the collective interests of that sector. This structure gave business the capacity for coordinated action in response to challenges. Labor unions in the business sector largely supported business policies.

Against these pro-growth "tribes" stood a shifting "environmental "tribe" or coalition. This coalition was usually weak, but not always. Propitious political opportunities occasionally gave members of the environmental coalition the chance to affect national or local environmental policy. As the story shows, those were the times of major change. Coalition members included a few sympathetic LDP politicians; pro-environment government agencies (sometimes); (rising and falling) victims' groups, protest movements, and citizens' organizations; (perpetually out-of-power) opposition political parties (notably, the Japan Socialist Party or JSP); (sporadic) court decisions; and (occasionally) innovative local governments. The mass media, often sympathetic to citizen environmental concerns, played a crucial role in spreading public awareness of these issues.

Unions representing government workers, though, such as teachers, post office workers, government clericals, and railroad employees (before privatization), at times took more pro-environmental stances. Until the 1990s, the JSP represented these unions and, as a party-out-of-power in the Diet, often criticized government growth and environmental policies. On a smaller scale, the Japan Communist Party (JCP) played a similar role, though rarely cooperating with its long rival, the JSP. In the 1960s, when worsening pollution propelled citizens into forming small community environmental protest movements, they often allied with the JSP and/or JCP to attain local electoral success.

Despite affecting government pollution policy, however, these grassroots efforts petered out without leaving much institutional legacy (such as the national environmental public interest groups in the United States). This failure reflected the barriers local movements faced, in terms of local conservative resistance, a parochial popular culture, and the lack of supportive laws and institutions (such as incorporation, nonprofit status, and nurturant foundations) (Broadbent 1998).

The weakness of the environmental coalition reflected its context. Elite government and business assumed a vertical, paternalistic stance toward society (Nakane 1970), producing an "administered society" or "nationalist-paternalist capitalist state" (Bellah 1985; Fukui 1992). Ordinary members of society had to contend with the influence of a bureaucratic state at every turn, throughout their lives. Government and business bureaucracies strongly shaped individuals' options in life. They shepherded the public, in ordered fashion, toward maximizing national productivity. The LDP, in contrast, solicited and depended on voters. But the LDP promoted voting not as an act of individual critical choice, but as one of uncritical loyalty exchanged for personal and community patronage.

This institutionalized domination discouraged the formation of an autonomous civil society, the crucial source of democratic debate. Certainly, some unions, opposition political parties, media, and educators injected leavening into this flattened civil society. But, on the whole, the situation allowed ordinary citizens little experience with self-generated voluntary organizations, critical policy discussions, or the exercise of political will. As a result, the public had little capacity to reflect on its encompassing structures or to resist them.

The T-shaped power structure, given minimal citizen input, long ignored environmental and other social problems. At crucial junctures, though, when citizen discontent erupted, its intensity threatened to disrupt both LDP political dominance and economic growth. Then the triad made strategic policy compromises to deflate the opposition (Calder 1988). The 1971 "Pollution Diet," described below, is a case in point. Ironically, when so motivated, the T structure enabled elites to rapidly and effectively implement selected environmental improvements. Under such circumstances, policy moved rapidly to change industrial practice, and changes in practice could be effective. But the triad assiduously avoided procedural changes, such as writing pollution law so as to legally empower ordinary citizens, since that would weaken the basic T structure.

In the 1990s, though, a number of factors, including the Kanto earthquake, severely shook this T formation. The resulting instabilities dispersed some power, allowed greater political pluralism, and ushered in policy and institutional change. New environmental policies recognized the need for radical social change to meet the challenge of sustainability. What this portends in practice remains to be seen.

Japan's particular mix of environmental success and failure, emphasis and neglect, emerged from a shifting combination of factors. At any given point in the process, a particular combination of domestic and foreign factors was at work. These factors included pollution intensity;

material interests and strategies; formal and informal institutions, patterns, and practices; and the subjective world of culture (beliefs, ideals, moralities) formed shifting causal patterns (Broadbent 1998). Foreign, especially U.S., pressures played major roles in forming Japanese policies. In changing combinations, these various factors shifted the goals and strength of Japan's various policy tribes. These shifts produced new policies, which, in turn, caused further shifts. To understand the dynamics of change in Japan's environmental policies, we need to examine these political processes.

Phase-Change Dynamics of Japan's Environmental Regime

Phase 1: "Polluter's Paradise" (1800s to mid-1960s)

During phase 1, government and business largely collaborated for the sake of economic growth, ignoring the effects of pollution on ordinary people. Accordingly, this phase saw minimal environmental regulation (the first in Osaka, 1887) (Organization for Economic Cooperation and Development, 1977, 7; Kato, Kumamoto, and Matthews 1981, 77). In the late 1800s, runoff and smoke from the Ashio copper mine severely poisoned farmers and their land (Crump 1996; Notehelfer 1975). The famous environmentalist politician, Tanaka Shozo, pleaded their case to the national Diet, but to little effect (Strong 1977). In 1897, after years of appeals with no redress (Shoji and Sugai 1992), over 2,000 local farmers organized protest marches in Tokyo (Strong 1977). The government finally ordered the mine to reduce its pollution, but it did not repair the land or compensate the farmers (Stone 1975, 395). In 1900, the government violently repressed continuing protests but provided partial compensation (McKean 1981; Stone 1975, 398; Shoji and Sugai 1992, 28). In the 1970s, Ashio farmers again demanded compensation and received a partial settlement (McKean 1981).

During the pre–World War II years, many other pollution cases, such as the Besshi and Hitachi copper mines, followed this pattern, sometimes with positive results, but often with violent conflict and repression (Broadbent 1998, 138; Gresser Fujikura, and Morishima 1981, 10–11; Tsuru 1989, 18). The issues at stake in these conflicts were local economic welfare and health, not protection of the environment for its own sake. These issues and interaction patterns continued to characterize Japanese environmental politics (Dimitrov 1997, 5; Stone 1975, 391). At this time, the government handled each case on its own and did not develop an overall pollution policy (Gresser, Fujikura, and Morishima 1981, 15). During World War II, government and industry paid even less attention to pollution control, brutally suppressing potential protests and dumping toxic waste wherever it was most convenient (Ui 1992a). These practices bequeathed a number of pollution time bombs to postwar Japan (Iijima 1993, 19).

In the first decades after the war, during frenetic economic rebuilding, companies stepped up production and introduced new, untested products and production processes with little or no attention to pollution. At first, this situation vastly increased traditional forms of pollution, such as smoke. During the 1950s, coal-fired power plants provided electricity to a vast array of manufacturing plants, from synthetic textile mills to steel refineries. These power plants belched forth vast quantities of black smoke (from coal); the steel refineries produced red smoke (from iron oxide particles) (Hoshino 1992). A textbook, seeing the smoke as a symbol of industrial power, proudly proclaimed Osaka the "capital of smoke" (*kemuri no miyako*) (Kawana 1987, 132; Tsuru 1989, 19). The authorities and mass media evinced little awareness that the smoke could also bring harm. The Tokyo Factory Pollution Prevention Ordinance of 1949 set no standards and no penalties, but only required permits (Gresser, Fujikura, and Morishima 1981, 16). From the 1950s on, rural depopulation and urban overcrowding, undeterred by government rural development plans, intensified human exposure to pollution (Broadbent 1998).

Rapid industrialization, pushed by MITI and eagerly pursued by businesses, ignored the potential for pollution. In 1955, the Morinaga Milk Company distributed milk accidentally laced with arsenic, which caused many cases of subacute arsenic poisoning. Victims of methylmercury pollution in the city of Minamata, Kumamoto Prefecture, were first officially recognized in May 1956. By 1958, Tokyo's Edogawa River was seriously polluted by effluents from the Honshu Paper Company's factory. A few years later, rice oil from the Kanemi Soko Company, in northern Kyushu, seriously poisoned consumers.

In 1953, the Ministry of Health and Welfare (MHW)—the main environmental protagonist until the creation of the Environmental Agency—conducted the nation's first national survey of pollution and drafted a bill to protect the living environment. However, other ministries, the LDP, and business leaders ignored this proactive, pro-environmental ministry (Johnson 1982, 284; McKean 1981, 215). Rather, social protest brought about the passage of the earliest pollution laws. By the late 1950s, industrial pollution had already reduced the fishing catch in Tokyo Bay. Local fishing people broke into and rioted inside a polluting factory. In 1958, prompted by these protests and local governments, the national government passed two laws to protect water and air quality in Tokyo Bay. But like the Smoke and Soot Regulation Law of 1962, these laws had no enforcement provisions and little effect (Gresser, Fujikura, and Morishima 1981, 17; Iijima 1993, 20; Matsubara 1971, 156). As a result, air quality worsened rapidly. The average big-city atmospheric sulfur dioxide concentration went from 0.015 parts per million (ppm) in 1960 to 0.060 ppm in 1965. Nitrogen oxide, likewise, rose sharply (Barrett and Therivel 1991, 34). In cities, like Yokkaichi, this air pollution quickly caused an epidemic of asthma and other respiratory diseases (Tsuru 1989, 19). It made Tokyo air so murky that Tokyoites could only see Mt. Fuji a few days per year.

In the early 1960s, public protest in Yokkaichi led the government to pass a law controlling air particulate levels (Huddle, Reich, and Stiskin 1975). This law had little impact, in part because the pro-growth Ministry of International Trade and Industry (MITI) administered it (McKean 1977, 216).

From 1955 to 1970, industrial waste output increased eightfold. In the 1960s, much waste went into shore landfills, adding to estuary pollution. Highly toxic waste, encased in concrete, was dumped into the deep ocean (Hiraishi 1989, 326). New synthetic fabric factories discharged two-thirds of their raw materials as waste into rivers, lakes, and bays (Hoshino 1992). Farmers made intensive use of fertilizers and other agricultural chemicals that also flowed into the waters. Cases of "red tide," massive blooms of red algae caused by these water pollutants that killed off fish, rose from 60 in 1968 to 300 in 1977. Water sources not meeting human health standards went from under 50 in 1960 to 583 in

1970 (Barrett and Therivel 1991, 36; Kelley, Stunkel, and Wescott 1976, 85). Vast industrial water use led to ground subsidence, with about 5 percent of Tokyo subsiding at 10 centimeters or more per year (Barrett and Therivel 1991, 38). Industries and ports encased virtually the entire urban coastline (and much of the rural coast) in concrete, denying beaches and the coast to the local residents. Noise and vibration from construction and ground and air traffic impinged ever more on people's everyday lives (Funabashi et al. 1985).

At first, many of these forms of pollution did not stir up much public protest. Effects on human health worsened, though. Arsenic in milk (1955), untested drugs, and PCB contamination of cooking oil (1968) poisoned large numbers of people (George 1996; Gresser, Fujikura, and Morishima 1981; Huddle, Reich, and Stiskin 1975; Ishimure 1990; Kelley, Stunkel, and Wescott 1976; McKean 1981; Mishima 1992; Uchino 1983, 169; Ui 1992b).

It has been said that people did not realize the damaging effects of pollution at this time. However, the evidence suggests that the dominant politicians and bureaucrats were aware of this danger, but largely disregarded it so as not to slow down Japan's rapid growth (Iijima 1993, 22; Johnson 1982, 284). MITI and certain other ministries, along with the LDP and organized big business, did their best to discourage, discredit, and demoralize pollution victims and dampen protest (McKean 1981; Upham 1987).

Four big pollution cases symbolized the situation: the Minamata mercury poisoning (first publicly acknowledged in 1954), Niigata mercury poisoning (1965), Yokkaichi asthma (1973), and Itai Itai cadmium poisoning (1968). The "Big Four" cases went through similar social dynamics. When residents complained of acute illnesses, they hit a political wall: government and corporate denial, inaction by once-trusted politicians, and even social rejection by their own communities, especially by those who worked in the polluting factories (Iijima 1992; Ui 1968; Upham 1987). The "growth coalition" buttressing this political wall consisted of economic ministries like MITI, the ruling LDP, and big business. Organized labor working in the private sector largely followed their lead.

As a result, in phase 1, Japan achieved the unenviable reputation of being the world's most polluted country—a veritable "kingdom of pollution" (Iijima 1993, 20–21; Matsubara 1971, 158). Drawing a lesson from this example, ecologist Paul Ehrlich likened Japan to a "miner's canary"—a warning signal to other industrial nations—heed the dangers of industrial pollution or suffer this fate! (McKean 1977, 204). During phase 1, sporadic citizen protest, not international pressures or a proactive state, brought about what minimal environmental protection occurred. Elites did not repress protest violently, but neither did they try to ameliorate the severe pollution.

Phase 2: "Polluter's Hell" (mid-1960s to mid-1970s)

In phase 2, some scholars argue that severe government regulations created very difficult times for industrial polluters, justifying the label "polluter's hell" (Pharr and Badaracio 1986). However, other scholars would disagree with the punitive regulatory connotations of that label. They argue the definite pollution reductions resulted from a consensus among growth-coalition members of the political and economic necessity of restraining domestic pollution (Broadbent 1998).

Imperiled citizens, caught between intensifying pollution and unresponsive government, created new, effective styles of protest. In the mid-1960s, pushed by a rising tide of information and pressure, the mass media took up their cause. The crippled hands and disabled children of fishing people in Minamata, poisoned by mercury-laden wastewater from the Chisso petroleum refinery, became widely accepted symbols of the horrors of pollution and the callousness of elites (Gresser, Fujikura, and Morishima 1981; Huddle, Reich, and Stiskin 1975; McKean 1977; Szasz 1994, 84; Ui 1972). In other words, the public reframed pollution from a symbol of progress to a symbol of a severe social problem.

In 1964, for the first time, the residents of Mishima and Numazu tried to prevent pollution, rather than just seeking compensation after the fact. Responding to large protests, the town councils rejected government plans for a highly polluting petrochemical complex, thereby stopping it (Hashimoto 1988, 68; Lewis 1980). Emboldened by that example and stimulated by the horrific icons of pollution broadcast in the news, the drumbeat of popular antipollution complaint and protest rose steadily throughout the country (see figure 8.2). Victims' groups persevered.

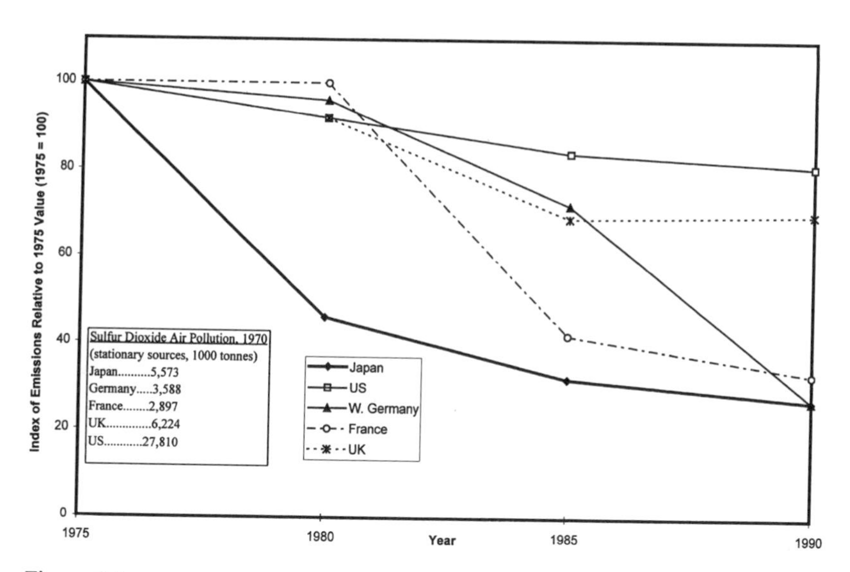

Figure 8.2
Reduction in sulfur dioxide emissions, 1975–1990. *Source:* OECD 1993.

Eventually, they received support from grassroots groups and academic scientists, who verified their claims and helped them file lawsuits.

The localness of Japanese culture, however, encouraged movements to take a "not in my backyard (NIMBY) stance toward the pollution issue. They complained about pollution in their own community, but paid little heed to problems in other communities, let alone to the fate of other species, "Nature," or "the planet" (Broadbent 1998, 287; McKean 1981, 131–136). Officials in the MHW and a few LDP politicians expressed concern over the damaging health effects of pollution. But they were no match for the powerful pro-growth coalition of economic ministries, big business, and the LDP.

By the mid-1960s, though, rising protest, along with criticism from the media, the public, and international agencies, pushed the government toward a response. The MHW and the LDP urged the creation of an official governmental Advisory Council on Pollution (Hashimoto 1970; Hashimoto 1988, 72; McKean 1977, 216). Hoping to intensify public protest, the MHW installed pollution monitoring stations at busy city corners that flashed up-to-the-minute reports of air pollution levels right there (Kawana 1988). Building on this lead, Japan eventually introduced

Table 8.1
Japan: A Statistical Profile

Land	
Total area (1,000 km^2)	376.00
Arable and permanent crop land (% of total) 1997	11.40%
Permanent grassland (% of total) 1997	1.70%
Forest and Woodland (% of total) 1997	66.80%
Other land (% of total) 1997	20.10%
Major protected areas (% of total) 1997	6.8%
Nitrogenous fertilizer use (t/km^2/arable land) 1997	11.50%
Population	
Total (100,000 inhabitants) 1998	1,264.00
Growth rate % 1980–1998	8.20%
Population density (inhabitant/km^2) 1998	334.60
Gross domestic product	
Total GDP (billion U.S. dollars) 1998	$2,544.00
Per capita (1,000 U.S. dollars/capita) 1998	$20.10
GDP growth 1980–1998	64.60%
Current general government	
Revenue (% of GDP) 1997[b]	31.70%
Expenditure (% of GDP) 1997[b]	28.80%
Government employment (%/total) 1997[b]	5.90%
Energy	
Total supply (MTOE) 1997	515.00
% change 1980–1997	48.60%
Consumption (MTOE) 1998[c]	336.54
% change 1988–1998[c]	23.20%
Road-vehicle stock	
Total (10,000 vehicles) 1997	6,921.00
% change 1980–1997	86.70%
Per capita (vehicles/100 inhabitants)	55.00
Air profile and greenhouse gases	
CO_2 emissions (tons/capita) 1998[a]	9.30
CO_2 emissions (kg/USD GDP) 1998[a]	.45
Nitrogen oxide emissions (kg/capita) 1998[a]	11.00
Nitrogen oxide emissions (kg/USD GDP) 1998	.60
Sulfur oxide emissions (kg/capita) 1998[a]	7.00
Sulfur oxide emissions (kg/USD GDP) 1998	.40
Pollution abatement and control	
Total (household excluded) expenditure (%/GDP) 1997	1.60%

Sources: OECD, *OECD Environmental Data, Compendium 1999* (Paris: OECD, 1999), except as noted.
[a] OECD, The OECD Observer, *OECD in Figures* (Paris: OECD, June 2000).
[b] OECD, *National Accounts of OECD Countries* (Paris: OECD, 2000).
[c] IEA/OECD, *Energy Balances of OECD Countries, 1997–1998* (Paris: IEA/OECD, 2000).

1,700 ambient-air-quality monitoring stations in severe areas, providing excellent statistics on urban air quality.

In 1965, MITI itself began using its administrative guidance to persuade industries to install tall smokestacks to diffuse smoke more widely, to burn low-sulfur fuels that would produce less sulfur dioxide pollution, and to build green belts between new industrial zones and residential areas (usually in close contiguity). Helped by the "horizontal" cooperative relationships existing among members of the ruling triad, MITI brought about an effective consensus among business leaders. Viewing the wave of protest, business leaders quickly grasped the emerging political necessity of reducing air pollution. MITI's research institutes worked cooperatively with business to develop new pollution control technologies. Within the cooperative context guaranteed by MITI and supported by government subsidies and technical aid, power plants implemented the suggested measures. Sulfur dioxide air pollution quickly began its precipitous drop in Japan. In this instance, MITI's regulatory style contrasts greatly with the bureaucratic "arm's-length" command-and-control regulatory style typical of the United States.

Furthering this change, in 1964, Finance Minister Tanaka Kakuei created the Pollution Control Service Corporation (PCSC, Kogai Boshi Jigyodan) to subsidize pollution equipment investment by industry. In 1965, the Diet approved this corporation, set up a Pollution Advisory Council (PAC), and established Diet committees on pollution in both upper and lower houses—probably the first in the world (McKean 1977, 216). The PCSC provided technical and financial support to industry and local government for pollution abatement.

From 1965, under MHW guidance, prefectures began taking surveys of and entering into informal Pollution Control Agreements (PCAs) with local polluting factories (Hashimoto 1970). Under a Socialist Party mayor, industrial Yokohama City concluded Japan's first PCA, which had considerable success in reducing local pollution. The quickness of these administrative responses contrasts with government reluctance to pass formal public regulatory laws.

In 1966, under MHW insistence, the PAC urged "radical" measures: human health should have priority over economic growth, industries should be held "strictly liable" for pollution damage even when they had not been legally negligent (no-fault liability), industries should pay for pollution prevention measures, and an Environmental Agency should be established (Hashimoto 1970; McKean 1977, 217). At first, top growth-coalition members—MITI, the Ministry of Construction, and the Federation of Economic Organizations—flatly rejected the PAC's proposals (Nishimura 1989, 19). They advocated the exact opposite: priority to economic growth without industrial liability for pollution (McKean 1981, 218). The seesaw battle within the government between pro-growth and pro-environment factions has continued ever since. In the mid-1960s, though, the LDP's worsening political situation—a shrinking Diet majority, a rising tide of pollution protest, and local opposition party victories (figure 8.2)—forced it to compromise (Broadbent 1998; McKean 1977). In 1966, hoping to defuse protest, the LDP prime minister asked the MHW to draft a Basic Law for Pollution Control (hereafter, the Basic Law or BL). As seen here, external challenge to the ruling triad may make the voice of a sidelined ministry suddenly audible.

Once again, the MHW's draft recommended the "radical" measures advocated a few months prior by the PAC (McKean 1981, 19; Hashimoto 1988, 112). And again, bowing to growth-coalition resistance, the prime minister removed some of those measures from the draft and substituted the infamous "harmony clause"—"Protecting the people's health from pollution shall be carried out in harmony with healthy economic progress." This clause gave priority to growth over public health and environmental protection (Hashimoto 1988, 112; Gresser, Fujikura, and Morishima 1981, 24). Yet the MHW's recommendations provided the Basic Law's basic policy concepts (Nishimura 1989, 18). Perhaps due to its ambiguity, the Diet passed the Basic Law in 1967 as Japan's first national framework for pollution control. The BL did identify the basic types of pollution, state government intent to control and reduce them, and, most important, increase local government response capacity. But the BL did not specify environmental standards, assign responsibility for pollution payments, or give relief to pollution victims. Moreover, the pro-growth Prime Minister's Office, not the MHW, controlled the BL's enforcement (Hashimoto 1970; McKean 1977). Therefore, the BL itself had little immediate effect on pollution. However, as a result of the law, opposition local governments passed local pollution ordinances and PCAs that required stricter standards. In 1968, based on the BL, the Diet passed the Air Pollution Control Law, which set emission standards for sulfur dioxide. But opposition from the petrochemical, power, and other industries forced the responsible ministry to largely neglect the enforcement of these standards (McKean 1977, 221).

While intended to coopt and defuse citizen worries, to the contrary, the BL legitimized citizen concerns and sparked even more protests (Broadbent 1998; Matsubara 1971, 157; McKean 1977, 220; Pempel 1982, 231). Still not seeing real pollution reductions, movements adopted more unruly means: demonstrations, sit-ins, rallies, shutdowns, and lawsuits (Broadbent 1998, chap. 3). In other words, the situation pushed more and more citizens from a petition to a protest mentality.

Taking many forms, environmental protest mounted and peaked in the early 1970s (figure 8.2). After delays, the Big Four lawsuits by pollution victim's groups neared the Japan Supreme Court. Between 1971 and 1973, the Supreme Court, breaking precedent, decided in favor of the victims in all four cases and forced polluting companies to pay large damage settlements (Gresser, Fujikura, and Morishima 1981; McKean 1981, 45–57; Ui 1989b, 565–566; Upham 1987). At the same time, protest-movement demonstrations and rallies peaked in 1971 and again in 1973 (according to mentions in the *Asahi Newspaper*). With protest-movement support, opposition party mayors grew from 20 in 1947 to 138 (out of 643) in 1973 (Broadbent 1998; McKean 1977, 227; McKean 1981, 222; Steiner, Krauss, and Flanagan 1980, 326) (figure 8.2). These mayors popularized important innovations: citizen participation, pollution control agreements with local industry, and the concept of environmental rights (Broadbent 1988, 1998; McKean 1977, 222–224; McKean 1981, 221–224; Reed 1986; Shibata 1989b; Ui 1989a, 114).

Foreign pressures also mounted. The United States passed its National Environmental Protection Act in 1969; in 1970, it started negotiations with Japan over standards for the Pacific Ocean environment (Kato 1989, 3). President Richard Nixon criticized Japan for getting an unfair trade advantage by neglecting pollution. In March 1970, the International Social Science Council held an international conference in Tokyo concerning the global aspects of environmental decline (Tsuru 1989, 33).

The conference's "Tokyo Resolution" proclaimed universal "environmental rights," further legitimizing this new concept in Japan (Tsuru 1989, 33).

This convergence of domestic and foreign pressure threatened the LDP and business leaders, who feared it would damage Japan's economic growth (interview, Michio Hashimoto, April 3, 1990) (McKean 1977, 226–227; Pempel 1982, 231). Moreover, Japanese companies bought their oil through the "Seven Sisters" foreign-owned oil companies, so they were not heavily dependent on profits from the sale of oil in Japan. Measures for pollution reduction and conservation did not impinge quite as so heavily on business pocketbooks in Japan as they did in the United States.

The Central Pollution Countermeasures Headquarters (CPCHQ), created by Prime Minister Sato in July 1970, concluded that BL's "harmony clause" had become "an inflamed appendix and had to be surgically removed" (*Asahi Newspaper*, August 10, 1970; cited in Matsubara 1971, 156–166, and McKean 1977, 228). In a panic, without the usual broad consultation, the CPCHQ hurriedly designed amendments to the Basic Law (Energii to Kogai, No. 140, December 3, 1970, 1129). The amendments removed the "harmony" clause, gave priority to health, designated the financial responsibilities for pollution cleanup, clarified national and prefectural division of regulatory powers over pollution, and greatly strengthened measures against air and water pollution (Kato 1989, 3; Matsubara 1971, 163).

In November 1970, the Diet passed the amendments as 14 new antipollution laws (Nishimura 1989, 27; Pempel 1982, 231, 244–247). This "Pollution Diet" marked the major turning point in Japanese environmental law. The new laws made polluters financially responsible to their victims under civil law, determined cost allocation for pollution control, and provided for mediated dispute resolution. They amended existing laws concerning air and water pollution, traffic pollution and noise, hazardous material transport, waste disposal and sewage, toxic waste, and Natural Parks. They also set up new institutions, including the Environmental Agency.

On paper, these new laws constituted the strictest antipollution regime in the world (Huddle, Reich, and Stiskin 1975; Pempel 1982, 231). At this point, Japan, the United States, and Sweden were the world leaders in comprehensive environmental legislation and institutions, influencing Germany, France, and other countries (Schreurs 2000). As is typical of Japanese laws, though, the new laws did not specify standards or sanctions.

In 1971, government agencies set up "regulatory instruments" to implement the new laws. These included recommended emission standards for sulfur oxides, carbon monoxide, water quality, and noise. Government worked closely with the business community's industrial-sector associations (ISA, *gyokai*) to develop specific standards tailored to the conditions of each industrial sector. Each ISA represented a specific industrial sector, such as steel manufacturing or consumer electronics. As the conduits for securing business participation and cooperation, the ISAS defined Japan's style of implementation (Ren, forthcoming). Through negotiation, they apportioned the economic risk faced by member firms from costly pollution control investments. They also trained technicians and helped firms exchange related technical expertise and experience.

In some scholars' estimates, these were the "world's most innovative regulatory instruments." For instance, they set standards according to area levels of pollution, rather than just restricting individual factory smokestack emissions. They also set up a way to fine polluters to pay compensation to victims—the so-called polluter-pays principle or PPP (Weidner 1989c, 489). The 1971 Law on the Development of Pollution Control Organization for Special Factories required that certain factories have their own pollution control managers and supervisors. In 2000, they numbered 40,000 and 23,000, respectively. By then, 470,000 had passed the national certificate program for pollution control managers.

However, the conventional term *regulatory instrument* may be misleading in the Japanese case. The new laws did not define compulsory emission standards, only voluntary or suggested levels. They did not require agencies or local governments to enforce any particular standards. The laws required no impact assessment and provided no funding for enforcement (Imamura 1989, 44; Matsubara 1971, 167; Upham 1987). They gave citizens no right to sue. As in other legal areas, citizens could not sue companies for violation of regulatory standards, use class action suits to sue for personal damages, or sue the government for nonenforcement of standards (Feinerman and Fujikura 1998, 260–262). Instead, the instruments left regulation and enforcement up to, and divided among, many national and prefectural government agencies.

The Japanese negotiated approach becomes clear when contrasted with the U.S. method of "command and control"—imposing standards by legal fiat and sanctions. As shown by the Superfund case, the U.S. approach invited lengthy litigation, often proving inefficient. Such problems eventually pushed the United States toward market-based "emissions trading." But neither command-and-control nor market approaches suited Japan's culture of interelite negotiation. The U.S. 1969 National Environmental Protection Act (NEPA) gave courts leverage, allowed citizens and industry to sue the EPA concerning enforcement problems, and encouraged citizen participation in policymaking. In contrast, the Japanese approach minimized the voices of local people and hampered their legal leverage. This situation pushed Japanese victim groups away from the courts and toward mediation—a process controlled by the government (Upham 1987).

The Diet established the Environmental Agency (EA) in May 1971, mandating it to coordinate the diverse environmental laws and conduct research toward new laws and standards. The EA set up the Central Environmental Advisory Council, with members mostly from business and academia, with specialized subcommittees for advice on a range of environmental issues. However, the laws gave the EA almost no regulatory power and little funding (Gresser, Fujikura, and Morishima 1981; Kato 1989, 3; Reich 1983; Shibata 1989a, 44). Environmental programs and policies remained spread out among many government ministries and agencies, weakening their impact (Gresser, Fujikura, and Morishima 1981, 234–242). Despite these limits, the EA's first director general, Oishi Buichi (a well-known environmental advocate), took dramatic action. He attended the 1972 United Nations Conference on the Environment, in Stockholm. In response to harsh criticism there, he signed a joint pledge that Japan would greatly reduce its air pollution. Japan achieved this goal much faster than other counties (Broadbent 1998, 15; Organization of Economic Cooperation and Development, 1993, 17) (figure 8.2).

The new pollution standards used an area-wide allowable total, rather than just individual factory limits as before. This allowed the government to demand that all factories in an area reduce their pollution output, a much more efficient approach to regulation. The new system also apportioned the costs of the newly required abatement equipment to help firms meet the new standards (Industrial Pollution Control Association of Japan, 1983, 14–15; Kato, Kumamoto, and Matthews 1981, 80–81; Ministry of Foreign Affairs, 1977, 13; Nishimura 1989, 29; Organization for Economic Cooperation and Development, n.d., 29; Weidner 1989a, 1989c).

The Pollution-Related Health Damage Compensation Law of 1973 made compensation automatically available to victims suffering from any of four air pollution–related and three water pollution–related symptoms (Industrial Pollution Control Association of Japan, 1983, 14–15). Compensation included free medical care, rehabilitation, and physical-disability compensation. The compensation fund was based on a European idea—the polluter-pays principle. In Europe, the PPP was supposed to be drawn entirely from polluting factories, not subsidized by the government (public) treasury. Officially, the Japanese government created the compensation fund this way, through two types of fees: 80 percent from factories emitting smoke and soot, assessed according to their pollution emissions, and 20 percent from car owners as a tonnage tax (Industrial Pollution Control Association of Japan, 1983, 15; Weidner 1989a, 146–52).

Some scholars say the law worked this way. They call it the "punish-polluters principle" and contend that the onerous costs it imposed convinced air polluters (in places like Yokkaichi) to reduce their pollution (Imura, forthcoming). Other scholars, though, reject this analysis of the PPP. They say the government heavily subsidized the polluters' payments to the victims' funds, so as not to excessively burden industry (Yoshida, forthcoming; Ueta, forthcoming). In any case, by 1991, the compensation fund held about 24.4 billion yen ($190 million) (Environmental Agency of Japan (Kankyocho), 1992, 3).

The new pollution laws empowered prefectures to set their own desirable standards for air, water, and noise pollution appropriate to local conditions (Gresser, Fujikura, and Morishima 1981, 25). For the first time, these standards could be stricter than those set by the national government. The intention of this measure was to give prefectures the flexibility to combat severe local pollution. The EA urged prefectures to set standards stricter than the national ones (Broadbent 1998, chap. 5; Pempel 1982, 232), to use the PPP to force local polluting businesses to pay victims directly, and to conduct impact assessments. Westerners should not think of Japanese law, regulation, ordinances, and agreements according to their own universalistic standards. All forms of law remain heavily administered through informal agreement and bureaucratic fiat, giving little leverage to ordinary people (Upham 1987).

Faced with this opportunity, local governments went in two directions. Progressive (opposition-party-affiliated) local governments made stronger use of the PPP law, concluded aggressive Pollution Control Agreements (PCAs), passed many pollution regulations, and set up pollution complaint bureaus (Broadbent 1988; Broadbent 1998, chap. 8; Wallace 1995, 243–244; Weidner 1989b, 473; Weidner 1989c, 490). The more conservative local governments, though, especially those dominated by the LDP, probably tended to negotiate less demanding PCAs with local businesses (Broadbent 1998, 284; Ui 1989c, 563–565). In the final analysis, two-thirds of the PCAs had no effect on local pollution. But some, especially those targeting large central pollution sources, were effective (Matsuno, forthcoming). Around Tokyo (then run by Socialist governor Rxōkichi Minobe), PCAs set the strictest standards on power plants, often demanding zero emissions (a switch to natural gas). All told, 149 new pollution-related local regulations and national laws were passed during the 1970s (Ui 1992c, 74). Some of the stricter prefectural regulations embarrassed the central government into revising the national laws. Such feedback from the "states" is typical in the United States but was new in Japan. Of course, PCAs deal only with local, single-firm problems, so they cannot provide a comprehensive national solution.

Attempts to further "democratize" pollution regulation, however, met stiff resistance. On the national level, the Environmental Agency, following the U.S. NEPA provisions, tried to make mandatory an environmental impact assessment (EIA) based on local citizen participation. It wanted to make this EIA with provisions for public disclosure, public hearings, and citizen participation, and make it mandatory for all large-scale public and private projects. The EA proposed this idea a number of times but was rejected. MITI, LDP, and business leaders (the ruling triad) objected to the participatory provisions and repeatedly quashed the EIA proposal (Barrett and Therivel 1991, 98; Broadbent 1998, 294; Hashimoto 1987, 77; Tsuru 1989, 39). Some called the EIA defeat the Environmental Agency's "most notable failure" (Miller and Moore 1990). The triad members disliked citizen participation because it ran the risk of empowering ordinary citizens to reject an economic development plan. This was anathema to a bureaucracy and political-economic elite desirous of maintaining arbitrary authority toward ordinary citizens.

Despite their limits, though, the new national and prefectural regulations reportedly shocked business leaders. In some accounts, they converted Japan's "polluter's heaven" into a "polluter's hell" (Nishimura 1989, 261–265). Nevertheless, the business community preferred predictable administrative regulations to unpredictable court-ordered ones, as long as they did not empower local citizens to demand impact assessments (Shibata 1989a, 37).

The new regulatory regime channeled real money into environmental protection. From 1970 to the mid-1970s, government expenditures on environment policy and pollution control went from 1.2 to 2 percent of the total government budget. Local government expenditures on environmental protection increased threefold, producing a uniquely thorough local monitoring of pollution (Ministry of Foreign Affairs, 1977, 14; Weidner 1989c, 492). Over the same period, business environmental protection investment increased 40 percent annually (Pempel 1982, 232). By the mid-1970s, corporations spent an average of 6.5 to 8.5 percent of their total capital investments for pollution control (Organization for Economic Cooperation and Development, 1977, 89–93), the highest ratio in the OECD (Organization of Economic Cooperation and Development, 1977, 69–73).

The OECD said that Japanese businesses applied new pollution control technologies irrespective of their cost (unlike typical U.S. businesses). At that point, Japan's environmental expenditures, as a proportion of its GNP, were three times as large as the next closest OECD country (Environmental Agency of Japan (Kankyocho), 1987a, 170–177; Organization for Economic Cooperation and Development, 1977, 71; Organization for Economic Cooperation and Development, 1993). "Smoke-scrubbing" units to desulfurize smokestack gases increased from 102 in 1970 to 1,134 in 1976 (Environmental Agency of Japan (Kankyocho), 1987b, 97), giving Japan 80 to 90 percent of all such units in the world by the 1980s (interview, Tokyo Electric Company, 1990). The financial burden pollution control placed on businesses varied by industry. Overall, the government did not want to suddenly burden business with heavy costs. Government "policy-based financing" certainly mitigated business's burden (Environmental Agency of Japan (Kankyocho), 1992, 5; Meyerson 1980; Ministry of Foreign Affairs, 1977, 178; Organization of Economic Cooperation and Development, 1977, 86; Reich 1983, 204). Some scholars contend that government loans and subsidies far outweighed businesses' own investments in pollution control (Ueta, forthcoming). Yet the Environmental Agency contends that pollution control investments placed heavy financial burdens on Japanese firms, especially the smaller ones (Environmental Agency of Japan (Kankyocho), 1992, 173–175).

The government subsidized business pollution control through two financial institutions. The Japan Development Bank (JDB) provided low-interest loans for pollution control to big corporations, and the Pollution Control Service Corporation (PCSC, later JEC) did the same to small and medium-sized enterprises (SMEs) as well as local government. In 1975, the peak year, such loans accounted for up to 50 percent (JDB) and 80 percent (PCSC) of pollution control investments. In that year, government subsidies to business covered about 10 percent of the costs of such investment (Imura, forthcoming). In addition, the government provided tax breaks and rapid depreciation. PCSC loans drew from the Fiscal Investment and Loan Program (FILP), run by the Ministry of Finance. (FILP funds came from the massive public postal savings accounts.) These loans often stimulated more loans from private banks as well. PCSC also taught SMEs about good pollution technology and practices, drawing on the experience of larger firms. During the mid-1970s, PCSC loans covered about 50 percent of pollution control investment by SMEs. In later decades, this amount declined and stabilized at around 15 to 25 percent (Ren, forthcoming). Going into the 1990s, JDB loans accounted for about 30 percent of big-business pollution control investment.

The examples of JDB and PCSC/JEC show the need for government to set up special, centralized institutions to help industries tackle pollution control through both technical advice and loans. These key institutions played central roles in Japan's pollution control successes. They serve as important models for other industrialized countries, as well as for developing countries (Ren, forthcoming). In the United States, government financial support for business pollution control investment has mainly taken the form of (sporadic and disorganized) tax breaks.

These programs still left some burden of investment on business and required their active participation in the effort. That government and business could work together to target investment this way reflects the integrative capacities of Japan's ruling triad. Under external pressure, the triad could apply an effective "technical fix" to a pressing social problem. (A technical fix differs, however, from a social reorganization—as Japan's incapacity to reorganize banks in the 1990s illustrates.)

Japan's response to the 1973 and 1979 "oil shocks," as the Japanese called the OPEC dramatic oil-price rises, also illustrates this capacity. The oil shocks prompted an amazingly effective energy belt tightening in Japan. After the 1973 oil shock, the government passed the Law to Promote Rational Use of Energy. Worked out with business cooperation, this law promulgated better heat waste management conservation practices. It required a certified energy-use manager in each plant or big office building. These measures helped business become more energy efficient, causing a drop in energy per GNP by 30 percent from 1973 to 1995. Japan reduced energy demand by 35 percent from 1973 to 1997. Conservation in fuel use (reduction per unit of GNP) accounted for 40 percent of the reduction in sulfur dioxide air pollution from 1975 to 1980 (Imura, forthcoming). The second oil shock prompted a similar wave of energy-saving investment. The 1979 revision of the Rational Use of Energy Law had detailed guidelines for energy saving and offered loans from the JDB, plus tax exemptions and accelerated depreciation.

Japan's facility in implementing these dramatic changes in business practice compared favorably to U.S. ineptness. The key to Japan's facility lies in the dense reciprocal ties between ministries and business (Broadbent 2000). Through these ties, a talented technical bureaucracy (MITI) could work with business representatives to formulate a "rational" common plan, reducing defection by ensuring common rules (and burdens). Under external pressure or crisis, this collective system overcame the "anarchy" of capital. The degree to which these changes in pollution output and energy use became institutionalized—a taken-for-granted reality—in business practice remains a crucial question.

Despite industry worries, the new pollution regulations did not have an overall negative impact on the Japanese economy. Certain industries, such as textiles and steel, and some smaller firms claimed a 90 percent loss of profits in the peak years of pollution equipment investment, but these leveled off after a few years (Ren, forthcoming). On the positive economic side, the pollution regulations spurred the emergence of an industry making air and water pollution control equipment, which became very profitable, and an export sector. These good economic results inclined business toward more active cooperation in technical and organizational reform to control pollution. The enormous social costs of pollution, measured in medical, compensation, and other costs, would have been greatly reduced by even earlier pollution control. In any case, as a result, Japan became the world's most efficient user of energy and resources, with recycled water use up significantly and certain pollutants—sulfur dioxide air pollution and chemical oxygen demand (COD) discharge by the paper-and-pulp industry—down significantly (the latter from changes in the production process).

These measures and investments bore environmental fruit. Visible environmental conditions improved dramatically. By the 1980s, Mount Fuji once again became a common sight from Tokyo (Pempel 1982, 235). Sulfur dioxide air pollution declined much more rapidly than in other OECD countries (Broadbent 1998, 15). Total sulfur dioxide discharge went from 5.5 million tons in 1975 to 1.6 million tons in 1990 (figure 8.2). The highest recorded value of sulfur dioxide air pollution was 0.059 ppm in 1967, but only around 0.01 ppm in 1986. During the same period, water quality related to human health improved dramatically. Japan also reduced pollution by moving noisy factories out of Tokyo. A 1977 OECD report concluded that Japanese "trends in environmental quality . . . are on the whole . . . more favorable than in other countries" (Organization for Economic Cooperation and Development, 1977, 67). Even environmentalists praised Japan's 1970s pollution regime as the most thorough and effective in the world (Huddle, Reich, and Stiskin 1975).

After 1973, the wave of pollution protest and progressive local governments rapidly ebbed. The worst air and water pollution had rapidly improved; local governments had set up offices and channels to handle pollution disputes. The 1973 OPEC oil shock and accompanying recession drew public attention away from environmental issues. But more important, local elites—using "soft social control"—had dampened most community movements. Local conservative political bosses (*bosuteki sonzai*) usually controlled the traditional village and neighborhood associations. When a protest movement started to well up, they "gnawed away" (*nashi kuzushi*) at protest leaders and members, dissuading their participation with bribes and threats (Broadbent 1998, 217). The parochialism of village culture aided this effort. The average citizen had little interest in environmental issues beyond the confines of their own villages or neighborhoods. The protest wave was a collection of NIMBY droplets. Citizen parochialism, in part, resulted from Japan's long history of soft social control, which inhibited the formation of a proactive citizenry.

Unlike the U.S. and European experiences, Japan's environmental protest wave did not leave behind strong, national, institutionalized environmental NGOs (Broadbent 1998, 289; Schreurs 1997a). The few organizations, based in Tokyo, and addressing national or international environmental issues, suffered poverty and marginality. NIMBYism, inexperience with volunteerism, and distrust of charities reduced popular donations to national environmental organizations (Broadbent 1998, 286–292; Dunlap, Gallup, and Gallup 1992; Miller and Moore 1990). At the same time, the ruling triad kept many barriers in the way of civil society. Nongovernmental organizations (NGOs) could not incorporate freely, meaning they had no legal existence as an organization. The forms of incorporation available to NGOs required subordination to a government ministry, to the point of accepting a retired official onto their board of directors, often as managing director, and paying that official's salary. This move effectively surrendered autonomy. NGOs could not obtain tax-exempt status. Foundations did not offer grants to citizen groups (Broadbent 1998; Schreurs 1997b, 326–327). Moreover, the legal system was designed against empowering citizens. To dissuade lawsuits, environmental (and other) laws provided no explicit standards (Broadbent 1998; Upham 1987); Japanese law did not permit class action suits (Feinerman and Fujikura 1998, 262).

These local and national forms of soft social control blocked the formation of a lively civil society and public sphere. They denied ordinary citizens experience with autonomous and effective voluntary organization (not found in traditional neighborhood organizations). They severely weakened the popular environmental activist community in Japan (Broadbent 1998, chap. 6). In its place, government and corporations filled the public sphere with their own environmental QUANGOs (quasi-NGOs) (Broadbent 1998, 292; Ui 1989a, 115; Ui 1992c, 219). These "tame" NGOs created an ersatz civil society and lukewarm public debate.

Due, in part, to the weakness of civil society, the 1970 pollution regime had distinct limits. It neglected long-range, slow-acting, invisible pollutants. It paid little heed to preventive measures, citizen participation, the protection of other species, and the provision of environmental amenities (pleasing, relaxing qualities of the environs) (Gresser, Fujikura, and Morishima 1981, p. 25 and n. 119 on p. 423). It failed to establish effective recycling programs. It did not impose impact-assessment requirements on new, gigantic projects, some of which devastated aboriginal Ainu lands (Iijima 1998). Starting in the 1960s, government plans repeatedly failed in redistributing industry to the hinterlands, where they would have diluted pollution, provided jobs, and drawn population from the crowded cities (Broadbent 1998, chap. 3).

Furthermore, the amenities or quality of life of Japan's cities continued to deteriorate with crowding, noise, vibration, lack of parks, and loss of local ambience and culture through standardized national building practices by giant contractors (*zenecon*). In such a crowded society, compared to the United States, for instance, amenities dependent on open space were inherently more difficult to provide. But this was not the only barrier. Without a strong, active civil society, government and corporate bureaucracies often rode roughshod over local voices that defend local amenities—a historic building, a wooded ridge, a unique species (Scott 1998). The impulse for protection was there—following the U.S. lead, Japan set up its first national parks in the early 1900s, before England or France. By 1997, Japan had 28 national parks and 6.8 percent of its land loosely protected (Schreurs 2000). But strong protection was more difficult to obtain: conservation laws designate only 5,600 hectares as wilderness areas and only 30,000 hectares for wildlife protection (Barrett and Therivel 1991, 44). Reflecting this situation, the laudatory 1977 OECD report added that "Japan has won many pollution-abatement battles, but has not won the war for environmental quality" (Organization for Economic Cooperation and Development, 1977, 83). Lack of environmental amenities, the report said, was the main source of continuing popular discontent.

Japan's new environmental regulatory regime of phase 2 used informal persuasion over business as well as some measure of regulatory control, it subsidized business's adoption of pollution control equipment, and it compensated victims. But, as much as possible, it left the input for policy decisions to the ruling triad elite and the implementation of policy in bureaucratic hands. Accordingly, the resultant policies responded to sources of severe, electorally threatening levels of public criticism, such as air, water, and obvious disease-causing pollution. But it did not take proactive responsibility for more subtle, long-term, or refined types of environmental-quality matters. The ruling triad focused strategically on the most visible kinds of pollution—of the air and water—and on the compensation of victims to reduce and blunt the wave of popular (and foreign) protests and lawsuits. For these matters, the triad gave "voice" to the Ministry of Health and Welfare and, later, the Environmental Agency. The triad did little for parks, noise, and crowding, however. The elite traditionally gave little thought to such public amenities. Moreover, the triad neglected invisible and long-term pollutants, such as toxic waste, because such problems did not then impact enough on public awareness to arouse threatening protest. In sum, the Japanese response to the pollution crisis took a "technocratic" path, addressing certain substantive problems, but retaining maximum state autonomy. This "closed" response differed sharply from the "opening" response in the United States that empowered citizens, leading to more debate, more litigation, and a different pattern of outcomes.

Phase 3: "Maintenance or Retreat?" (mid-1970s to late 1980s)

One would expect, given the strength of Japan's pro-growth coalition, that once popular and foreign environmental pressure declined, so, too, would pollution control (Sabatier 1975). Some scholars argue this decline occurred (Funabashi 1992, 7; Kawana 1995a, 7). Others argue that, during phase 3, industry accepted the new pollution control regime and cooperated in its maintenance (Pharr and Badaracio 1986).

During phase 3, both protest and foreign environmental demands steeply declined. Recession focused people's worries on their jobs and focused government's policy efforts on reducing the national debt. With less protest, courts reduced their pro-environmental stance (Upham 1987). Conservative Western regimes (Reagan, Thatcher, Kohl) put little pressure on Japan for environmental improvement.

Inexorably, though, Japan's domestic and global environmental impact continued to expand. Japan's expanding exports, along with its growing population, consumerism, and urbanization, increased its intake of the world's resources and its output of waste. Its assiduous factories "added value" to these resources by turning them first into basic stocks (steel, oil, naphtha, and so on), and then into radios, televisions, cars, and other consumer goods. What was this small country to do with the enormous piles and tanks of waste materials, often toxic, left behind after the finished goods went off? The first strategy had been dumping. The land area alongside many Japanese bays has been expanded by massive landfill, consisting of consumer and industrial waste. The fate of toxic waste is less clear—much was dumped, sometimes illegally, in local rivers and seas, landfills, islands, and the offshore ocean. Phase 2's domestic pollution regulations, though, made visible pollution increasingly difficult.

To escape domestic regulations, in the 1970s polluting Japanese companies began to "export" their most polluting plants (Ui 1989b; Watanuki 1979, 112). Japanese oil, aluminum, and other refineries set up shop in Indonesia, Venezuela, and Brazil. Japanese chemical companies exported DDT and other toxic pesticides to Third World countries in great quantities (Iijima 1998; Watanuki 1979, 122).

To "green" its own archipelago, Japan moved its wood sourcing to foreign countries. The Japanese government supported intensive planting of cedar forests in mountainous areas (Japan is now 70 percent forested). At the same time, Japanese trading companies systematically stripped Southeast Asia's rain forests of their exotic hardwoods. In 1986, Japan consumed two-thirds of world imports of timber and half of world imports of tropical hardwoods (Dauvergne 1997; Iinkai 1995; Pearce and Warford 1993, 291–293).

Likewise, Japan's increasingly prosperous population could afford a better diet. With domestic fisheries partly damaged and contaminated, Japan culled new supplies from the world's oceans. In the 1980s, Japanese companies deployed 60 huge mechanized fishing ships using gigantic drift nets. Along with vast quantities of usable fish, these nets killed an enormous number of unwanted fish, turtles, dolphins, and other species. These foreign resource extractions often destroyed not only local ecologies, but also the habitats and livelihoods of indigenous peoples, such as the Australian Aborigines and Malaysian Panang (Iijima 1998).

Despite its growing global impact, during most of the 1980s the Japanese government displayed great reluctance to join in and abide by international environmental agreements (Feinerman and Fujikura 1998, 253; Weidner 1989c, 519–521). Despite signing the Convention on International Trade in Endangered Species of Wild Fauna and Flora (CITES) Treaty in 1973, Japan only ratified it in 1980 after getting the largest number of exemptions, allowing it to import 40 percent of the world's traded ivory (Feinerman and Fujikura 1998, 270–271; Weiss 1998, 109; Mofson 1997, 174). Despite a 1985 moratorium on whaling by the International Whaling Commission, Japan continued to kill about 300 Minke whales per year (Weber 1994, 55). Japan signed the London Convention on Marine Pollution by Dumping Wastes and other Matter in 1973, but refused to practice it. Japan dumped much more waste into the ocean than any other nation (4.55 million tons in 1991), including low-level radioactive waste (Feinerman and Fujikura 1998, 281).

So, too, with ozone-depleting chlorofluorocarbons (CFCs). Japan used about 10 percent of the world's CFCs. Due to objections from MITI and Japanese companies, however, the government refused to attend or sign the 1985 Vienna Convention on reducing CFC use and production (28 countries did, including the major producers). This type of behavior caused some observers to label Japan the world's "ecooutlaw" (Begley and Takayama 1989, 70; Miyaoka 1998, 167).

However, a few concerned government ministries and politicians became actively engaged with global environmental issues. After the United States published the *Global 2000 Report on the World Environment*, in 1981 (Barney 1981), the Japanese Environmental Agency set up a committee to study such problems as deforestation, desertification, and global warming. In 1982, the Japan EA director, Bunbei Hara, persuaded the UN to establish the World Commission on Environment and Development (Schreurs 2000). Moreover, Japan's energy vulnerability, revealed by the oil shocks of 1973 and 1979, sparked MITI into support for conservation and alternative sources of energy, from solar to atomic. In 1974, MITI started its "Sunshine Project," promoting alternative sources of energy. In 1978, its "Moonlight Project" promoted energy efficiency and nuclear power (Schreurs 2000). Backed by government subsidies and loans, industry consortiums developed better ways to reuse waste heat and monitor fuel usage. They even found the new technology a commercial success (Choy 1989, 5–6; Flavin and Young 1993, 193; Hashimoto 1987, 76–77; Matsuoka 1989, 447; Miller and Moore 1990; Ministry of International Trade and Industry, 1988). As a result, between 1973 and 1988, Japan's GNP grew 81 percent, but its energy use only grew 16 percent, putting its efficiency ahead of other OECD nations (Schreurs 2000).

Alternative energies, including nuclear, rose from 3.3 percent in 1973 to 26.4 percent of the total in 1994 (Schreurs 2000). As its main "alternative" energy, MITI advocated nuclear power (Matsuoka 1989, 446). These plants started to come online in the 1970s and quickly generated severe problems of radioactive waste disposal. In 1976, the Japanese government began dumping tens of thousands of barrels of low-grade radioactive waste into the Mariana Trench in the Pacific Ocean. This behavior shows how, without domestic and foreign resistance, the ruling triad takes shortcuts—by dumping waste into the "commons."

During phase 3, many other neglected pollutions accumulated. Companies and municipalities dumped vast quantities of industrial and consumer solid waste into shoreline landfill areas—including one in Tokyo

Bay, ironically called the "Island of Dreams" (Huddle, Reich, and Stiskin 1975)—and into the ocean (Feinerman and Fujikura 1998, 281; Ishino 1989, 326; Weidner 1989c, 521). Japan's farmers used more pesticides per hectare than almost any other nation. Pesticide runoff caused a rising number of red tides (destructive algal blooms) along the coasts (Vig and Kraft 1990). Developers turned forests and beaches into ski and golf resorts and yacht clubs with little restriction (Miller 1989, 2).

In Japan's "construction state" (*kensetsu kokka*), politicians fed high-priced public works contracts to general contractors without competitive bidding (the *dango* system) (Woodall 1996). Conservative politicians received kickbacks from these projects to finance their patronage to voters. This system constituted a core policy motivator between the construction sector and the LDP, within the ruling triad. Its preferred policies often ran counter to the rational preferences of the triad ministries, such as MITI and the Ministry of Finance. But, working with the Ministry of Construction and the Ministry of Transportation, these "irrational" resource uses often triumphed. The construction state built many environmentally damaging projects of dubious public utility, such as the Nagara River Dam and the coral reef–destroying airport at Ishigake Island. It harbors and buildings reconstructed local communities to fit a single, reinforced-concrete central vision, thereby erasing regional culture and environs (Yoshida, forthcoming).

Taking advantage of the political lull, in 1975 auto industries forced the EA to postpone its strict 0.25 g/km nitrogen oxide auto emission standard. In its place, until 1981, the EA adopted a higher, interim 0.6 g/km standard—still the world's strictest (Tsuru 1989, 34–36; Weidner 1989c, 486). Nonetheless, upstart Honda and another auto company saw a sales potential in efficient cars. By the mid-1970s, they brought out car engines that met the 0.25 standard, setting a world benchmark (Hashimoto 1987, 73–74; Shibata 1989a, 108). This gave them an edge that greatly increased their world sales, especially in the United States. This example, and that of air pollution equipment, cited earlier, illustrates a general point: government regulation can be better than the market in calling forth needed technological innovation from business.

This industrial society, with growing consumer affluence, produced an ever-growing stream of waste. In 1976, the national government passed a new waste-handling law requiring local government to supervise recycling, incineration, and the safe disposal of toxic waste (Ishino 1989). Already, the shortage of landfill and dump sites was apparent. By 2000, there were 1892 landfill sites, with declining total capacity.

Recycling was seen as a way to reduce the waste stream. By 1987, some 83 percent of all cities had private or governmental recycling programs (Herskowitz and Salerni 1987, 26–33). But they were not very effective. By 2000, municipal waste had only a 3.9 percent recovery (recycle) rate—mostly cans and bottles, not plastics (Imura, forthcoming). Both businesses and municipalities refused to bear the costs of recycling. Industrial waste did better, with a 40 percent recycling rate. In 2000, industrial waste totaled 430 millions tons per year and growing, eight times consumer waste in weight. Of that, an estimated 1 million tons, including toxics, were dumped illegally (Imura, forthcoming; Weidner 1989c, 494).

Frustrated with dumping and recycling, the government turned to incineration. By the late 1980s, Japan incinerated 68 percent of its consumer waste in about 1,800 incinerators, while the United States put 90 percent into landfills (Herskowitz and Salerni 1987, 15). This seemed like a cleaner solution, for a while, until the public realized the amount of toxic chemicals, especially dioxin, contained in incinerator smoke (Weidner 1989c, 529).

The latent effects of toxic waste accumulated throughout Japan's society and ecology (Ariyoshi 1979; Feinerman and Fujikura 1998, 281; Ishino 1989, 326; Watanuki 1979; Weidner 1989c, 521). By the 1980s, the effects began to manifest themselves. Carcinogenic organochloride compounds seeped into some municipal wells, closing them down (Hashimoto 1987, 80; Ishino 1989). Newspapers reported rising rates of unusual human fetal deformities.

The government did not deal with these problems, however. Instead, it backtracked on existing regulations. In 1986, under industry pressure, the Central Council for Environmental Pollution Control ruled that industrial air pollution was no longer a main cause of respiratory disease in Japan. On that basis, in 1987, the government canceled all 41 Class 1 areas—areas in which the PPP forced polluters to pay into a victims' compensation fund—and decided to not certify any new people as victims of air pollution (Environmental Agency of Japan (Kankyocho), 1987b, 193; Miller and Moore 1990; Nakemata and du Florey 1987; Shibata 1989a; Weidner 1989a). Again, during this period, the EA's perennial request for an EIA met defeat.

Only the nuclear power issue stirred up much new popular protest during phase 3. Public support for the nuclear power program declined precipitously over the 1980s. In a 1990 government survey, 90 percent felt "uneasy" about it ("Unease about Nuclear Power 90%," 1990). Through the 1990s, this unease intensified into outright opposition in many places. This popular opposition led to local protests, slowing Japan's nuclear trend. No new reactor sites were approved after 1986, though already-contracted ones kept coming online and MITI held to its expansionist plans (Brown et al. 1992, 61; Hasegawa 1996; Takubo 1997).

At the end of this period, the Japanese government still exhibited great reluctance to recognize and act on many global environmental problems (Kawashima 1997, 114–119). Only U.S. and European bans on ivory, coupled with intense criticism from international NGOs, sufficed to get Japan, in 1989, to implement the CITES Treaty and ban the import of ivory (Miyaoka 1998, 176) ("Japan Joins Ban on Ivory Trade," 1989). Likewise, in 1989, only great foreign pressure (*gaiatsu*) forced Japan to cut its drift-net deep-sea fishing ships to 20 ("Stripmining the Seas," 1989).

Finally, in 1993, when the UN was about to adopt a moratorium on drift-net fishing, Japan ceased it altogether (Miyaoka 1998, 177) Under threat of U.S. trade sanctions, Japan finally signed the 1987 Montreal Protocol on Substances That Deplete the Ozone Layer (CFCs), and then ratified both the Vienna and Montreal acts in 1988 (Kawana 1995b, 53; Schreurs 1997a, 148). The JDB introduced loan schemes to help reduce freon gas (1988), an ozone-depleting chemical, and nitrogen oxides (1989). Despite MITI resistance, some ministries began planning the implementation of the Montreal Treaty on ozone depletion. But despite some initial steps, the Japan government still largely denied the thornier problem of global warming.

In its 1988 Environmental White Paper, the Environmental Agency produced Japan's first official recognition of global warming and set up a research group on the problem (Environmental Agency of Japan

(Kankyocho), 1988, 43–121). Japan hosted the 1989 United Nations Environmental Program (UNEP) Conference on Global Environmental Protection. But Japanese government representatives refused to go along with the Netherlands, Germany, and the United Kingdom in setting CO_2-reduction targets. The government also excluded NGO participation from the conference.

In response, Japanese NGOs simultaneously held an alternative conference, calling the official one a "PR exercise." The NGOs labeled Japan the "number one destroyer of the environment in the Third World" (Schreurs 1997a, 196). This brought international attention to the role of Japan's Official Development Assistance (ODA) in destroying tropical forests ("Charging Japan with Crimes against the Earth," 1989; Cross 1989; Iinkai 1995; Schreurs 1997a).

By then, Japan accounted for about 40 percent of the world trade in tropical timber. It also imported about 40 percent of the world's traded fish, increasingly from coastally destructive fish farms in Southeast Asia. Japan also exports a great deal of waste, such as used car batteries, to that area. Japan largely ignored its global environmental "shadow."

In sum, during phase 3, Japan maintained many of the environmental institutions (regulatory regimes) achieved during phase 2. But it backtracked on some of these, refused to recognize new, growing domestic environmental threats, and was not very cooperative on international environmental issues. Citizen activism subsided while many people enjoyed unprecedented wealth, prosperity, and satisfaction, sensing Japan had, at last, surpassed the United States in many ways.

Phase 4: "New Global and Local Demands" (1990s to early 2000s)

Momentous changes ushered in phase 4. The Cold War ended, reconfiguring Japanese opposition parties (Takeuchi 1998). Japan's "bubble" economy collapsed in the early 1990s, ushering in a decade of stagnation and gloom. Shrinking capital and imports led to a U.S.-demanded increase in deficit spending on often-dubious public works—more roads, tunnels, and harbors. These projects further damaged the environment but failed to restart the economy. The pro-growth ruling triad proved unable to make the necessary structural changes in the political and economic institutions that propped it up. It could not abandon Japan's mutually protective economy ("convoy capitalism") by letting many major banks and businesses fail.

At the same time, growing global and domestic pollution raised the pressure for effective response. Public concern about toxic waste, including radioactive waste, continued to strengthen. By 1990, the public ranked the environment third in importance among all issues (Environmental Agency of Japan (Kankyocho), 1992, 129). Yet the government record did not inspire confidence. Local governments did not reliably enforce laws on toxic waste disposal. Illegal dump sites proliferated (Hiraishi 1989). Even existing laws had been inadequately implemented—some victims of the Minamata and Niigata mercury poisoning had not received government aid more than 20 years after their court victory. According to a government study, a majority of firms attempted to save costs on industrial waste disposal expenses, with 25 percent using improper means to dispose of waste (Environmental Agency of Japan (Kankyocho), 1992, 132). Among Japan's infrequent corporate prosecutions, the most frequent was for violation of industrial waste laws. Resulting public concerns interacted with a changing political landscape.

By 1990, Japanese leaders were feeling embarrassed and hampered by international criticism, particularly for their refusal to comply with reductions in ozone-eating CFCs. The OECD criticized Japan for its poor international compliance and failure to include NGOs. It implied that if Japan wanted to be a player in international politics, it would have to meet OECD norms. Concerning climate change, then, the Japanese government decided to keep closer pace with the international community (Kawashima 1997, 116).

In rapid succession, the Japanese government declared the need to include environmental protection among Official Development Aid (ODA) goals. MITI started a Green Aid Plan to transfer environmental technology to ODA recipient countries (1991) (Schreurs, forthcoming) and issued industrial CO_2-reduction guidelines (1993). In 1992, the Environmental Agency created the Japan Environmental Corporation (taking over the loan functions of the old Pollution Control Service Corporation). In addition to the PCSC's loans to small and medium-sized businesses, JEC began to give loans to environmental nongovernmental organizations (ENGOs).

The prime minister appointed the head of the EA to a ministerial Council on Global Environmental Protection, which formulated the 1990 Action Plan to Arrest Global Warning. This plan reflected ministerial conflicts over the issue: MITI argued that Japan had already achieved greater greenhouse-gas reductions through energy-use efficiency than other industrial countries. Therefore, stabilizing greenhouse gases at year-2000 levels would suffice (Schreurs 1996; Schreurs 1997a, 151). The EA, though, wanted to return Japan to 1990 levels.

At the 1992 United Nations Conference on Environment and Development (UNCED) in Rio, Japan joined 154 countries in signing the new international agreements on global environmental problems: Agenda 21, the Framework Convention on Climate Change (FCCC), and Conventions on Biological Diversity and Combating Desertification. Agenda 21 urged action by citizens and local governments to combat global warming. The FCCC suggested stabilizing greenhouse gases at 1990 levels (Japan Center for a Sustainable Environment and Society, 1996, 15). FCCC signatories agreed to work toward a mutually agreeable, specific, binding, global greenhouse-gas-reduction plan and to produce their own National Action Plans. At UNCED, Japan announced it would massively increase its ODA. Director Miyashita, of Japan's EA, announced—without Japanese government decision—that Japan "might possibly" host the Third Conference of the Parties (COP) (the UNCED meeting being COP1; *parties* refers to signatories of the FCCC). His announcement gave rise to international expectations that helped produce that outcome.

On returning home, the EA produced plans to fulfill FCCC and Agenda 21 proposals. The EA urged domestic reduction of greenhouse gases, research, technology, education, and international cooperation to attain a "sustainable development"—type society. MITI remained skeptical of these measures and opposed them. Japan's 1990s recession, by tightening government and business budgets, worked against the effective implementation of carbon dioxide reduction (Miller and Moore 1990). An EA study in the early 1990s found low corporate interest in compliance (Environmental Agency of Japan (Kankyocho), 1992, 130). As a result, from 1990, Japan's carbon emissions climbed more rapidly than those of other ACID countries, surpassing Denmark, the Netherlands, and Germany by 1997 (Flavin and Dunn 1998, 115–129). Similarly, though Japan's automobile fuel was entirely lead free (Organization for Economic Cooperation and Development, 1994, 101), its average auto fuel economy diminished (Flavin and Tunali 1996, 39). In sum, during the early 1990s, Japan did little new to combat greenhouse-gas emissions.

On the other hand, political changes shook up the dominance of the ruling triad, which helped civil society and environmental policy. In 1989, opposition parties won unprecedented control of the Upper House of the Diet, keeping it into the 2000s. Then, in 1993, for the first time in postwar history, opposition parties won control of the more powerful Lower House. This electoral loss shocked not only the LDP, but the whole ruling triad, because it threatened their institutionalized dominance. Once in power, the opposition party coalition started important new environmental and civil initiatives. The domestic NGO community mushroomed and gained legitimacy; international NGOs entered Japan in greater force. These sudden changes made clear how central LDP control of the Diet had been to the "1955 regime" and its conservative policies.

In November 1993, the opposition party ruling coalition, led by Prime Minister Morihiro Hosokawa, passed a revised Basic Environmental Law (the first revision of the 1967 Basic Law for Pollution Control since the 1970 Pollution Diet). As usual, the EA and MITI fought over the content. The EA wanted to include strong substantive measures toward a sustainable society: mandatory environmental impact assessment, environmental taxes and surcharges to pay for environmental protection, and freedom of information. MITI and the Japan Federation of Economic Organizations forced the removal of such measures (Schreurs 2000). The intention of the Basic Law was to provide general direction and a framework for future environmental administration ("How Can We Bring the Environment Basic Law to Life?", 1993). In tone, it differed from its predecessor by containing many expressions of concern about sustainability and the global environment (Nihon Kankyo Kaigi (Japan Environmental Council), 1994). However, critics said it did not give sufficient weight to key issues like environmental assessment, citizen participation, and freedom of information about corporate pollution.

The 1994 Basic Environmental Plan was more substantive, systematically requiring ministries to devise regulatory means to attain the goals of the Basic Law. In addition, the 1994 plan created the Institute for Global Environmental Strategies (IGES), under the EA, to find practical ways to implement solutions.

This plan called for a "network" style of environmental governance, based on the joint efforts of government, citizens, NGOs, and business (Ren, forthcoming). This approach sounds like unrealistic idealism from the EA. However, it built on earlier forms of Japanese business-government cooperation. Industrial-sector associations had always negotiated the technical provisions of new regulations, including pollution control, for their own industry. The rising levels of education and awareness among Japanese citizens made such cooperation more likely to be effective than in less developed countries. In the past, though, as the preceding analysis has shown, getting the business-government-LDP "ruling triad" to solve environmental problems required massive public protest and electoral threat, as well as foreign pressure. Business had not proactively engaged in voluntary cooperation. At best, it was "quasi-voluntary cooperation," under threat of government and public sanction (Ren, forthcoming).

Some observers claim that in the 1990s, business cooperation became more genuinely voluntary. They point to business efforts at Voluntary Action Plans, environmental auditing, life-cycle analysis, clean production (CP) technology, and other measures (Imura, forthcoming). As advocated by the United Nations Environmental Program, CP technologies redesign production processes to eliminate toxic raw materials, reduce pollution and waste, and make products more ecofriendly with long life and recycling. Japanese companies have made important advances in cost-effective CP, but since CP contributes to company competitiveness, companies keep their innovations secret.

Environmental NGOs, however, worry that, if not backed up by popular pressure, "network environmental regulation" could end up as little more than "cooptation." This might be especially true for environmental protection that cannot, like CP, be made to pay for business. The ideal of cooperative network regulation has spread from Japan to the United States and other countries. But seasoned environmental regulators in those countries remain skeptical of business intentions and wary of possible degeneration into cronyism and regulatory "capture."

In 1995, the Japan Development Bank started making low-interest loans to help factories reduce their output of greenhouse gases and to support recycling. The government convened many advisory councils on its environmental performance, and local governments developed environmental plans. Critics charged, though, that these plans produced no substantive improvements ("How Can We Bring the Environment Basic Law to Life?", 1993).

In 1994, the LDP regained control of the Lower House, but at an odd price: a coalition government with its old enemy, the Japan Socialist Party. The LDP-dominated cabinet, led by Socialist Party Prime Minister Tomiichi Murayama (1994–1996), produced some specific environmental programs, but these only covered conservation measures by government agencies. In its 1994 Environmental White Paper, the EA called for the use of market forces (emission trading) rather than regulation ("Use Market Methods on the Environment," 1994), a measure at odds with Japan's administered-market economy.

In 1995, a huge earthquake devastated the Kobe area. The government failed to adequately help quake victims, while volunteer groups did an effective job. This stark contrast greatly strengthened popular support for nongovernmental, nonprofit volunteer organizations (Bestor 1998; Yamaoka 1998; Yamauchi 1998).

Continued economic failure brought the LDP closer to the brink of severe political decline (Katz 2001). The precarious situation made the party more open to policy compromise. Reliance on NGOs during the earthquake, for instance, legitimated the 1998 "Special Nonprofit Activities Law," or NPO Law (Pekkanen 2000). While not measurably improving the NGOs' opportunity for tax-exempt status (Deguchi 1998), by allowing NGO incorporation and removing some bureaucratic oversight, the NPO law weakened some of the barriers facing NGOs in Japan.

Small special-topic domestic environmental groups had continued to work at the national level since the 1970s. Groups such as Japan Tropical Forest Action Network, JATAN, and People's Forum 2001, supported by a few hundred subscribers and led by dedicated, but impoverished, activists, kept up a ferment of concern and critique. In the 1990s, these groups contributed to significant environmental victories: stopping Mitsubishi's plans to build a salt plant in a Mexican bay used by gray whales as a nursery and getting the Japanese government to withdraw its support for World Bank funding for the Narmada Dam in India. New research-oriented domestic NGOs appeared in the 1990s, such as the Japan Center for a Sustainable Environment and Society (JACSES). Branches of international environmental groups, such as Greenpeace Japan and World Wildlife Fund Japan, also increased their presence. These NGOs slowly attained greater legitimacy in Japan and built stronger communication links with government and business-based environmental organizations.

In the mid-1990s, public concern about toxic chemicals swelled and reached crisis levels. Dioxin, PCBs, and other toxins were identified as "environmental hormones" (*kankyo horumon*, endocrine disrupters) ("Environmental Hormones: National Survey Starts," 1998; Yoshida and Iguchi 1998).[1] Scientists attributed increasing rates of fetal deformity, cancer, skin disease, and fish deformities to these sources (Hasegawa 1998; Kawana 1998; Nagayama 1998, 76; Risaikuru, n.d.; Ueda 1998, 76). Fish caught near Japan exhibited extremely high dioxin concentrations (Nagayama 1998, 65). Between 1967 and 1987, skin allergy diseases (atopii) attributable to toxic pollution increased sevenfold (Nagayama 1998, 23).

Harm to human health turned public attention toward the likely sources. Many illegal dump sites, such as Teshima's 510,000 tons of illegal industrial toxic waste, came to light ("Cleanup Order to Waste Hauler," 1996). The smoke from 1,854 local trash incinerators (compared to 148 in the United States) spread dioxin in surrounding communities ("Control Dioxin Emissions," 1997; Nagayama 1998, 124–130; Ueda 1998, 43–50). Nuclear accidents at several nuclear plants intensified public fears (Sawai 1998). Japan became a "risk society" (Beck 1992).

These conditions, worsened by government inaction, set off a new wave of environmental protest ("Citizens' Referendum," 1997). Protests against nuclear power had been simmering since the 1980s. Environmental movements and groups widened their campaigns to include chemical toxins (Kajiyama 1995). Between 1990 and 1997, 717 distinct groups conducted 944 protest incidents against nuclear and toxic pollution (Taguchi 1998, 242). In 1997, the village of Maki, using Japan's first binding local referendum, rejected a nuclear plant (Takubo 1997, forthcoming). In the same year, Mitake City residents used a referendum to reject a proposed dump site and processing facility for industrial toxic waste ("Mitake Industrial Waste Site 80% No," 1997). The dynamics of this growing wave of protest resembled that of the 1960s: pollution buildup, official neglect and denial, accidents and health damage, intensifying citizen protest, and community political resistance.

The government responded to this new wave of protest with new laws. More standards for waste disposal, including marine disposal, are being added. The government enacted the Container Recycling Law in 1995, aimed, in part, at creating waste material stockpiles for future resource mining. In June 2000, the government passed the Basic Law for Formation of a Resource Recycling Society to respond to this general need. Also, businesses organizations, such as the Federation of Economic Organizations, came forth with Voluntary Action Plans for environmental protection. However, businesses and municipalities still resist paying for recycling. This casts doubt on the depth of any business "voluntariness" toward environmental protection.

On a global scale, evidence for global climate change continued to accumulate. In 1995, signers of the 1992 Rio FCCC agreement met again in Berlin (Second Conference of Parties to the Rio Declaration—COP2). At that time, Japan agreed to host the next meeting (COP3) in Kyoto in 1997. During the ensuing preparations for COP3, as host, the Japanese government felt that it should present a specific standard for greenhouse gases reduction (Kawashima 1998). Yet, as always, the EA and MITI disagreed on this standard. The Environmental Agency wanted Japan to go 5 percent below 1990 greenhouse-gas levels by 2010, but MITI demanded a 3 percent increase in greenhouse-gases beyond 1990 levels by then, plus many new nuclear power plants.

Despite government ambivalence, the Japanese public increasingly took the issue seriously. By the mid-1990s, about 80 percent of the public reported high concern over global environmental issues (Schreurs

1996)—a sea change in parochial Japanese culture. Emboldened, Japan-based NGOs convened a Climate Forum in 1996 and made policy proposals. There, the Japan Federation of Lawyers argued for a 20 percent greenhouse-gas reduction by 2010.

Facing a conflux of international criticism and domestic protest similar to 1970, the top Japanese business association—the Federation of Economic Organizations (FEO), or Keidanren—wanted to prevent another harsh regulatory regime. In June 1997, the FEO announced a voluntary industrial greenhouse-gas-reduction plan (Kawashima 1998). The voluntary agreements made by Japanese industrial associations covered 60 percent of manufacturing firms. These agreements aimed for a 10 percent reduction in greenhouse gases by 2000 and a 10–20 percent reduction by 2010 (Flavin and Dunn 1998, 123). Also, many Japanese companies have achieved ISO 14001 certification, testament to good environmental practices.

Alone among the 29 countries of the OECD, Japan lacked an environmental impact assessment (EIA) law. In 1997, the Upper House, ruled by an opposition party coalition, finally passed a weakened version of the EA's long-sought EIA law ("Environmental Assessment Law Passed," 1997). The law required taking the opinions of local citizens "into account," but left all decision-making power with the government. The law is not likely to be effective, since it only concerns large-scale projects and the assessment is done too late in process.

The Kyoto Conference (COP3) convened in December 1997. Concerning greenhouse-gas-reduction targets, the EA found allies in the Europeans, while MITI aligned itself with the conservative U.S. position. In the end, MITI conceded to the EA, partly to head off the even more stringent standards advocated by Germany. The Japanese government proposed that the industrial nations reduce greenhouse gases by 5 percent below 1990 levels by 2008–2112. This proposal fell below the European Union proposal of a reduction of at least 7.5 percent below 1990 levels by 2005 and 15 percent below by 2010, but it was stronger than the U.S. proposal to return to 1990 levels by 2008 to 2012 ("Meeting Reaches Accord to Reduce Greenhouse Gases," 1997). MITI argued that 20 more nuclear power plants would be needed for Japan to attain this goal (Hasegawa 1998).

As in the past, at the Kyoto Conference, the Japanese government tried to ignore the noisy NGOs. However, after the Kobe earthquake, the government had lost its popular aura of infallibility. The government appears to be including NGOs more in the policymaking process on an informal basis (Schreurs 1997b, 329).

The 1998 Law for Promotion of Measures to Prevent Global Warming—the world's first law specifically for this purpose—required disciplined energy conservation throughout the whole of Japanese society, so as to reduce greenhouse-gas emissions. Many Japanese communities have adopted the 1992 Rio Agenda 21 scheme, stipulating many forms of local conservation and pollution reduction.

In 2000, government reorganization raised the status of the Environmental Agency to that of a ministry, a boost in its status. In his 1998 address to the 142nd Diet, the Director General of the Environmental Agency stressed seven areas of environmental policy focus. As its leading idea, the address stressed that global warming will raise air temperatures and sea levels, "shaking the foundations of human society." The origin of this problem, the address continued, lies in "mass production, mass consumption and mass waste." The solution must lie, not in piecemeal policies to protect the environment, but in rethinking all systems and the bold construction of an "environmental protection–style society" (*kankyo hozengata shakai*).[2] This kind of bold systemic thinking goes beyond prior formulations of policy response. If the past politics of environmental policy are any indication, such systemic reforms will not be realized until the Japanese public strongly demands them. Once that threshold is passed, however, if Japan's environmental reforms of the 1970s still provide a relevant example, Japan may respond effectively to control the pollution targeted by public protest.

Conclusion: The Interplay of Politics, Policies, and Practices

What conflux of institutions, interests, cultures, actors, and ecomaterial conditions brought about Japan's particular pattern of environmental politics and policies? Ecological, demographic, cultural, and social conditions interacted with foreign, elite, and grassroots actors. How does the preceding narrative help us assess their relative importance?

During the postwar half century, Japan's voracious extraction and import of the world's raw materials supplied its economic miracle. This torrent of extraction cast a heavy ecological global "shadow." Moreover, only a portion of this torrent became marketable goods. The rest was discarded as manufacturing and consumer waste (as even the goods eventually were). In effect, given its concentrated industry and dependency on import of resources, Japan became one of the world's largest waste handlers. Much of this waste, improperly disposed, fed back into the air, food, and water of Japan's dense population. Japan suffered a high social intensity of pollution—vast numbers of people polluted. This social density intensified popular discontent with pollution illness and environmental degradation (Broadbent 1998).

These social conditions of pollution were not inevitable, however, even given such a dense population, industry, and resource-import regime. The successful aspects of Japanese pollution policy resulted from a conflux of interests and institutions. The efficiency and targets of pollution control and environmental protection were strongly affected by the immediate clash of interests. The priorities of economic growth and pollution control were largely defined by the pro-growth coalition within Japan's ruling triad. In addition, public protest, electoral threat, and the structure of business interests also facilitated those policies.

The strengths of the Japanese economy defined the immediate priorities of businesses. Given Japan's lack of resources and relatively weak control over global resources, especially through the 1970s, mining and oil companies constituted less of a barricade against reform than in the United States. Rather, the strengths of the Japanese economy and, hence, the major interests of its businesses lay in manufacturing, for both domestic and international sales. Accordingly, getting electrical power plants to increase efficiency or all businesses to conserve energy use did not trample on the interests of oil companies in selling more oil, as it has in the United States.

Beyond the immediate clash of interests, though, we have to consider how formal and informal relational patterns and institutions channeled their effects. Japan's pro-growth elite coalition created a Japanese-style "treadmill of production"—a social situation that pushed for ever more productivity, regardless of its social need. In Western theory, the polluting "treadmill" arises from a class agreement between capital and labor to increase productivity for mutual profit (Schnaiberg 1980; Schnaiberg and Gould 1994). In Japan's "Asian-style capitalism," though, along with classes, the state played a much larger role in organizing the economy (Sakakibara 1993), and hence in guiding the "treadmill." The economic ministries, especially MITI, helped coordinate the ruling party (LDP), business leadership, and organized private-sector labor to focus with military intensity on rapid national economic expansion. Elsewhere, I have described their organization as "communitarian elite corporatism" (Broadbent 1998).

Through loans, subsidies, and sponsored research, the government extended considerable support to industry. Japanese economic officials wanted to control pollution without damaging economic productivity. U.S. government officials see U.S. businesses as independent actors pursuing their own, not national, interests. In contrast, Japanese officials treated Japanese businesses as national treasures, producing the lifeblood of the nation. To the pro-growth coalition, national economic growth became a patriotic mission. Accordingly, the Japanese ministries did not so much impose rules on businesses as partner with them under this nationalistic mantle. In the same way, government ministries cooperated with businesses to find the most effective ways to improve the targeted pollution problems without seriously hampering business productivity.

The state was only able to guide business effectively toward environmental protection, though, when a wave of mobilized protest movements and elected opposition officials crashed against the gates of power. The state did not have much inherent power. MITI, the EA, and other ministries needed a countervailing social pressure behind them to become effective persuaders of business. In other words, the state attained maximal persuasive power under specific structural conditions. I liken this structure to the shape of a bow tie: a knot between two wings. In my "butterfly" theory of the Japanese state, the state can exert real regulatory pressure only when it is the "knot," "broker," or "bridge keeper" between two opposed social and political "wings" (Broadbent 2000).

In the 1960s, the lack of an effective countervailing bloc, coupled with highly centralized state authority, set the stage for Japan's environmental tragedies. The triad treated the ordinary citizenry as its ruling child, to be shepherded along toward the higher end of growth. Under those conditions, the government ignored environmental problems and hampered protest mobilization until the late 1960s. The triad changed these policies only when protest reached disruptive and electorally threatening levels, and then only on the targeted issues. This is one example of the "crisis-and-compromise" pattern of LDP rule (Calder 1988). In general, much of the power of social movements comes from their disruptive, factory-stopping, street-blocking potential (Piven and Cloward 1971; Schwartz 1976). The periodic waves of protest seen in Japan resulted partly from the frustration due to the "punctuated equilibrium" of episodic LDP compromise.

Within the administrative government, some ministries were not part of the pro-growth coalition. The EA often clashed with MITI on environmental issues and usually lost. However, the MHW and the EA achieved a greater policy "voice" when powerful external criticism rattled the gates. Only when domestic or international pressures threatened electoral or economic loss to the pro-growth coalition (LDP, economic bureaucracies, and big business) did the government make major changes in environmental policy. This conclusion is well supported by the historical cases noted above: the environmental laws passed by the 1970 Pollution Diet had their roots in MHW proposals. Control of the powerful House of Representatives (Shugiin) by the opposition parties allowed EA proposals to shape the 1993 Basic Law. Getting Japan to sign the Montreal Protocol required massive U.S. pressure. The structural conditions necessary for these policy changes further supports my "butterfly" theory of the Japanese state.

Under such circumstances, when the Japanese government decided to "compromise" and act, it did not need to rely on U.S.-style "command-and-control," arm's-length imposition of regulations. In the United States, this style of implementation sometimes worked, but often led to legal challenge and stalemate. Rather, in Japan, once attaining a workable consensus, the ministries, business leadership, and the LDP negotiated more consensual ways to reduce the worst pollution.

At the same time, the pro-growth triad imposed "soft social control" on the ordinary society to weaken protest movements and opposition political parties. The larger goal of this soft social control was to weaken societal impulses toward an autonomous civil society, so as to fend off sources of future challenge to and compromise by the ruling triad.

Japan's environmental countermeasures stressed elite-controlled technical solutions to pollution threats to human health. Rather than prevention—assessing and rejecting potentially damaging projects—environmental policies preferred postpollution technical and administrative solutions—tall chimneys and smoke scrubbers for power plants, cleaner engines for cars, compensation to victims. Energy conservation and reductions in carbon dioxide emissions, as well as some alternative energy technologies, resulted from Japan's chronic search to reduce foreign energy dependency. The government repeatedly wrote new laws vaguely, without teeth, so as to avoid empowering ordinary citizens to challenge projects and decide issues (Broadbent 1998; Upham 1987).

Less visible, more slow-acting types of pollution, such as toxic waste or global issues, did not call forth public outcry for decades, so the government ignored them. Similarly, due to a lack of public demand, policy largely ignored environmental amenities—protection from noise, vibration, crowding, the lack of greenery, and preservation of other species. Once pollution controls were implemented, it remains unclear how well the changes "stuck." Were they highly dependent on continued political pressure, as the theory of "regulatory decay" would argue (Sabatier 1975)? Or did they become institutionalized in their own right? This is a cutting-edge question about environmental regulation around the world (Mol and Sonnenfeld 2000) and requires more research.

In sum, Japan's objectively hazardous environmental conditions set up the potential for recognizing and reacting to them as environmental "problems." The density of the affected population intensified that potential. The conversion of these conditions into recognized problems, however, and, beyond that, into responsive policies and changed practices depended on social mediated processes. That is, a particular array of institutional patterns, power distributions, and cultural frames channeled Japan's environmental recognition and response into certain pathways. Until the 1990s, local and national forms of soft social control confined public recognition of pollution as a "problem" largely to the local, not national or global level. Relatively few individuals made the leap from local criticism to local resistance, and even fewer adopted universal environmental norms and worked for national or international environmental NGOs (Hannigan 1995; Janicke 1992; Reich 1991; Schnaiberg 1980; Schnaiberg and Gould 1994; Jacobson and Weiss 1998).

Through the 1980s, no strong national civil society emerged around these, or any, issues. Coupled with "communitarian elite corporatism" at the top, this social formation resulted in a narrow range of government environmental policies. The resulting policies dealt mainly with the obvious sources of pollution, those known of and complained about by such movements: air and water pollution that clearly damaged human health. But the policies ignored the more subtle and long-range health dangers posed by insidious pollutants, such as toxic and nuclear waste. Consequently, environmental reforms focused on improving immediate public health, not on long-term threats. Policies gave little weight to public amenities—quality-of-life issues, such as noise, vibration, and park space. They showed much less concern for the preservation of natural areas and other species.

If Japanese democracy had instead enjoyed an alternation of political parties competing for the votes of a more active civil society, the policy outcomes would have been much more sensitive to environmental problems, including long-term toxics and public amenities. When party alternation finally occurred, during opposition party control of the Diet (1993–94), the resulting policies—such as the 1993 Basic Environmental Law and 1994 Plan—support that conclusion. The ruling triad's continuing strategy of resolving crises by policy compromises while rigorously excluding citizen involvement seemed to be reaching a limit. The environmental problems of advanced industrial societies, like Japan, have gotten very intense, complex, and pervasive. Under these circumstances, technocratic solutions have decreasing effectiveness. Solutions increasingly require strong citizen involvement. In the United States, leverage by citizen groups through institutionalized channels, such as lobbying, NGO formation, public mobilization, and referenda, have been crucial to the evolution of environmental policy (Wellock 1998). The solution of the intensifying environmental crisis calls for a vigorous civil society, freedom of information, and positive participation by all "stakeholders.".

Glossary of Acronyms

BL	Basic Law
COD	Chemical oxygen demand
COP	Conference of the Parties (signatories to the FCCC)
EA	Japan Environmental Agency
FCCC	Framework Convention on Climate Change
FEO	Federation of Economic Organizations (Keidanren)
FILP	Fiscal Investment and Loan Program
JEC	Japan Environmental Corporation (successor to PCSC)
LDP	Liberal Democratic Party
MAFF	Ministry of Agriculture, Forests, and Fisheries
MHW	Ministry of Health and Welfare
MITI	Ministry of International Trade and Industry
MOF	Ministry of Finance
MOFA	Ministry of Foreign Affairs
PCA	Pollution Control Agreement
PCSC	Pollution Control Service Corporation
UNCED	United Nations Conference on Environment and Development

Notes

I would like to express my heartfelt appreciation to the sources of support that made this chapter possible: the University of Minnesota, the Center for Advanced Study in the Behavioral Sciences (1998–99), the Asia/Pacific Research Center of Stanford University (1998–99), and a grant from the Pacific Basic Research Center (2000–01) for related research. Comments by Uday Desai, Thomas Hargrave, Miranda Schreurs, Hidefumi Imura, and others greatly improved the chapter. Any remaining flaws are my responsibility.

1. A 1998 survey by the Japan Environmental Agency found 11 suspected endocrine disrupters in varying levels at 123 of 130 sites. Endocrine disrupters can impair sexual development and immune functions and can cause malignant tumors. They include chemicals used in detergents, resins, and plastics. Nonylphenol, found in 76 percent of the sites, is used in detergents and polystyrene plastic and inhibits testicle growth in fish. Bisphenol A was found at 68 percent of the sites. Diethylhexyl phthalate (DEHP), which causes cancer in lab animals and is used to soften plastics and in toys, was found at 55 percent of the sites ("Endocrine Disrupters Found at 90% of Sites," 1998).

2. This was a speech given to the 142nd Diet by the director general of the Japan Environmental Agency. Available at EIC Netto, the Environmental Information and Communication Network (http://www.eic.or.jp) set up by the National Institute for Environmental Studies, a government agency (http://www.nies.go.jp/index-j.html).

References

Ariyoshi, S. (1979). *Fukugo Osen* (Complex pollution). Tokyo: Shinchosha.

Barney, G. (1981). *The global 2000 report to the president.* Obertshansen, Germany: Production Greno.

Barrett, B., and Therivel, R. (1991). *Environmental policy and impact assessment in Japan.* London: Routledge.

Beck, U. (1992). *The risk society.* London: Sage.

Begley, S., and Takayama, H. (1989, May 1). The world's eco-outlaw. *Newsweek*, p. 68.

Bellah, R. (1985). *Tokugawa religion: The cultural roots of modern Japan.* 2nd ed. New York: Free Press.

Bestor, V. L. (1998). Reimagining "civil society" in Japan. Available at http://www.us-japan.org/dc/civil/cspaper.bestor

Broadbent, J. (1988). State as process: The effect of party and class on citizen participation in Japanese local government. *Social Problems*, *35*(2), 131–142.

Broadbent, J. (1998). *Environmental politics in Japan: Networks of power and protest.* Cambridge: Cambridge University Press.

Broadbent, J. (2000). *The Japanese network state in US comparison: Does embeddedness yield resources and influence?* Occasional paper. Stanford: Asia/Pacific Research Center, Stanford University.

Brown, L., ed. (1992). *State of the world 1992.* New York: Norton.

Calder, K. E. (1988). *Crisis and compensation: Public policy and political stability in Japan.* Princeton: Princeton University Press.

Charging Japan with crimes against the Earth. (1989, October 9). *Business Week*, pp. 108–112.

Choy, J. (1989, May 12). MITI to revise Energy Demand Outlook. *JEI Report*, *19B*, 5–6.

Citizens' referendum (Jumin Tohyo Ni Muke Kessoku). (1997, July 13). *Asahi Newspaper*, morning edition, Miyazaki page.

Cleanup order to waster hauler (Gyosha E Sanpai Tekkyo Meirei). (1996, December 26). *Asahi Newspaper*, evening edition, p. 14.

Control dioxin emissions (Daiokishin Hassei O Yokusei). (1997, May 22). *Asahi Newspaper*, morning edition, p. 1.

Cross, M. (1989, September). Tokyo nods its head toward the environment. *New Scientist*, *16*, p. 24.

Crump, J. (1996, Spring). Environmental politics in Japan. *Environmental Politics*, *5*(1), 115–121.

Dauvergne, P. (1997). *Shadows in the forest: Japan and the politics of timber in Southeast Asia*. Cambridge: MIT Press.

Deguchi, M. (1998). *A comparative view of civil society*. By Japan-America Society of Washington, DC. Projection "Civil Society in Japan and America: Coping with Change." Available at http://www.us-japan.org/dc/civil/cspaper.deguchi

Dimitrov, R. (1997). The evolution of environmentalism in Japan. Unpublished term paper, University of Minnesota.

Dunlap, R., Gallup, G., Jr., and Gallup, A. (1992). *The health of the planet survey*. Princeton: George H. Gallup International Institute.

11 endocrine disrupters found in nation's water. (1998, December 8). *Japan Times*, p. 1.

Endocrine disrupters found at 90% of sites (Kankyo Horumon, Chosachiten No Kyuwari De Kenshutsu). (1998, December 7). *Asahi Newspaper*, evening edition, p. 1.

Environmental Agency of Japan (Kankyocho). (1987a). Environmental White Paper (Kankyo Hyakusho). Tokyo: Government of Japan.

Environmental Agency of Japan (Kankyocho). (1987b). Quality of the environment in Japan. Tokyo: Environmental Agency, Government of Japan.

Environmental Agency of Japan (Kankyocho). (1988). Environmental White Paper (Kankyo Hyakusho). Tokyo: Government of Japan.

Environmental Agency of Japan (Kankyocho). (1992). Environmental White Paper (Kankyo Hyakusho). Tokyo: Government of Japan.

Environmental hormones: National survey starts. (1998, July 22). *Yomiuri Newspaper*, p. 18.

Environmental assessment law passed (Kankyo Asesuho Seiritsu). (1997, June 9). *Asahi Newspaper*, evening edition, p. 1.

Fallows, J. (1989, May). Containing Japan. *Atlantic*.

Feinerman, J., and Fujikura, K. (1998). Japan: Consensus-based compliance. In E. B. Weiss and H. Jacobson, eds., *Engaging countries: Strengthening compliance with international environmental accords* (pp. 253–290). Cambridge: MIT Press.

Flavin, C., and Dunn, S. (1998). Responding to the threat of climate change. In L. Brown et al., eds., *State of the world* (pp. 113–130). New York: Norton.

Flavin, C., and Tunali, O. (1996). *Climate of hope: New strategies for stabilizing the world's atmosphere*. Washington, DC: Worldwatch Institute.

Flavin, C., and Young, J. (1993). Shaping the next industrial revolution. In L. Brown, ed., *State of the world* (pp. 180–199). New York: Norton.

Fukui, H. (1992). The Japanese state and economic development: A profile of a nationalist-paternalist capitalist state. In R. Appelbaum and J. Henderson, eds., *States and development in the Asian Pacific Rim* (pp. 199–225). Newbury Park, CA: Sage.

Funabashi, H. (1992, October). Environmental problems in Japanese society. *International Journal of Japanese Sociology, 1*, 3–18.

Funabashi, H., Hasegawa, K., Hatanaka, M., and Katsuta, H. (1985). *Shinkansen kogai: Kosoku bunmei no Shakai mondai* (Bullet train pollution: Social problems of a high-speed civilization). Tokyo: Yuhikaku.

George, T. S. (2001). *Minamata: Pollution and the struggle for democracy in postwar Japan.* Harvard East Asia Monographs, 194. Cambridge: Harvard University Press.

Gresser, J., Fujikura, K., and Morishima, A. (1981). *Environmental law in Japan.* Cambridge: MIT Press.

Hannigan, J. A. (1995). *Environmental sociology: A social constructivist perspective.* New York: Routledge.

Hasegawa, K. (1996). *Datsu genshiryoku shakai no sentaku* (The choice of a postnuclear Society: The age of a new energy revolution). Tokyo: Shinetsu Sha.

Hasegawa, K. (1998, July 26–August 1). Global climate change and Japanese nuclear policy. Paper presented at the World Congress of Sociology, Montreal.

Hashimoto, M. (1970). *Kogai wo kangaeru: Yori kagakuteki ni yori ningenteki ni* (Thinking about pollution; more scientifically and more humanely). Tokyo: Nihon Keizai Shimbunsha.

Hashimoto, M. (1987). Development of environmental policy and its institutional mechanisms of administration and finance. In United Nations Centre for Regional Development and United Nations Environment Programme, eds., *Environmental management for local and regional development: The Japanese experience* (pp. 57–105). Nagoya, Japan: United Nations Centre for Regional Development and United Nations Environment Programme.

Hashimoto, M. (1988). *Shishi kankyo gyosei* (Personal history: Environmental administration). Tokyo: Asahi Shimbunsha.

Herskowitz, A., and Salerni, E. (1987). *Garbage management in Japan: Leading the way.* New York: Inform, Inc.

Hiraishi, T. (1989). Control of chemicals. In S. Tsuru and H. Weidner, eds., *Environmental policy in Japan* (pp. 332–342). Berlin: edition sigma.

Hoshino, Y. (1992). Japan's post–Second World War environmental problems. In J. Ui, ed., *Industrial pollution in Japan* (pp. 64–76). Tokyo: United Nations University Press.

How can we bring the Environment Basic Law to life? (Kankyokihonho O Do Ikasuka). (1993, November 14). *Asahi Newspaper*, morning edition, p. 2.

Huddle, N., Reich, M., and Stiskin, N. (1975). *Island of dreams.* New York: Autumn Press.

Iijima, N. (1992). Social structures of pollution victims. In J. Ui, ed., *Industrial pollution in Japan* (pp. 154–172). Tokyo: United Nations University Press.

Iijima, N. (1993). *Kankyo shakaigaku* (Environmental sociology). Tokyo: Yuhikaku Books.

Iijima, N. (1998, July 26–August 1). Environmental deterioration and the interrelationship between global and local inequalities: Perspectives from Asia and Australia. Paper presented at the World Congress of Sociology, Montreal.

Imamura, T. (1989). Environmental responsibilities at the national level: The Environment Agency. In S. Tsuru and H. Weidner, eds., *Environmental Policy in Japan* (pp. 43–53). Berlin: edition sigma.

Imura, H. (Forthcoming). In World Bank Institute, ed., *Japanese environmental policy.* Washington, DC: World Bank.

Industrial Pollution Control Association of Japan. (1983). *Environmental protection in the industrial sector in Japan.* Tokyo: Industrial Pollution Control Association of Japan.

Inoguchi, T., and Iwai, T. (1987). Zokugiin no Kenkyu (Research on legislator tribes). Tokyo: Nihon Keizai Shimbunsha.

Intergovernmental Panel on Climate Change. (2001). Summary for policymakers: A report of Working Group 1 of the IPCC. Geneva: United Nations. Available at http://www.usgcrp.gov/ipcc/wg1spm.pdf

Ishimure, M. (1990). *Paradise in the sea of sorrow: Our Minamata Disease.* Kyoto: Yamaguchi Publishing House.

Ishino, K. (1989). Waste management. In S. Tsuru and H. Weidner, eds., *Environmental policy in Japan* (pp. 320–331). Berlin: edition sigma.

Jacobson, H., and Weiss, E. B. (1998). A framework for analysis. In E. B. Weiss and H. Jacobson, eds., *Engaging countries: Strengthening compliance with international environmental accords* (pp. 1–18). Cambridge: MIT Press.

Janicke, M. (1992). Conditions for environmental policy success: An international comparison. *The Environmentalist, 12*(1), 47–58.

"Japan joins ban on nory trade." (1989, October 31). *Washington Post*, p. A14.

Japan Center for a Sustainable Environment and Society. (1996). *Environment and sustainable development in official development assistance since the 1992 Earth Summit.* Tokyo: Japan Center for a Sustainable Environment and Society.

Japan Federation of Bar Associations (Nihon Bengoshi Rengokai, Kogai Taisaku-Kankyo Hozen Iinkai). (1995). *Nihon No Kogai Yushutsu to Kankyo Hakai: Tonan Ajiya Ni Okeru Kigyo Yushutsu to ODA* (Japan's export of pollution and environmental destruction: Industrial location and ODA in South East Asia). Tokyo: Nihon Hyoronsha.

Johnson, C. (1982). *MITI and the Japanese miracle.* Stanford: Stanford University Press.

Kajiyama, S., ed. (1995). *Gomi Mondai Funso Jiten* (A dictionary of garbage problem disputes). Tokyo: Resaikuru Bunkasha.

Kato, I., Kumamoto, N., and Matthews, W. H. (1981). *Environmental law and policy in the Pacific Basin area.* Tokyo: University of Tokyo Press.

Kato, S. (1989, November). *Kogai Taisaku Kihonho to Sono Sekai* (The world of the Basic Law against Pollution). *Gesuido Kyokaishi, 26*(306), 2–9.

Katz, R. (2001, February 26). Slow unraveling of convoy capitalism. *Asian Wall Street Journal*, pp. 14 op-ed.

Kawana, H. (1987). *Nihon No Kogai* (Japan's pollution). Tokyo: Ryokufu.

Kawana, H. (1988). *Dokyumento: Nihon No Kogai* (Documents: Japan's pollution). Tokyo: Ryokufu Shuppan.

Kawana, H. (1995a). *Dokyumento Nihon No Kogai, Dai / 11 / Maki: Kankyo Gyosei No Kiro* (Documents: The pollution of Japan, Vol. 11: Quandaries for environmental bureaucracy). Tokyo: Ryokufu Shuppan.

Kawana, H. (1995b). *Dokyumento: Nihon No Kogai, Dai / 12 / Maki: Chikyu Kankyo No Kiki* (Documents: The pollution of Japan, Vol. 12: Crisis in the global environment). Tokyo: Ryokufu Shuppan.

Kawana, H. (1998). *Kensho: Daokishin Osen* (Documents: Dioxin pollution). Tokyo: Rokufu Shuppan.

Kawashima, Y. (1997, Spring). A comparative analysis of the decision-making process of developed countries toward CO_2 emissions reduction targets. *International Environmental Affairs, 9*(2), 95–126.

Kawashima, Y. (1998). *Kisho Hendo Wakugumi Joyaku Dai 3 Kai Teiyaku Kaigi—Kosho Katei, Goi, Kongono Kadai* (Framework conditions for atmospheric change, third conditions legislative deliberations—process of negotiations, consensus, future issues) (Technical report). Tsukuba, Japan: Kokuritsu Kankyo Kenkyujo (National Environmental Research Institute).

Kelley, D. R., Stunkel, K. R., and Wescott, R. R. (1976). *The economic Superpowers and the environment: The US, the Soviet Union, and Japan.* San Francisco: Freeman.

Lewis, J. G. (1980). Civic protest in Mishima: Citizens' movements and the politics of the environment in contemporary Japan. In Kurt Steiner, ed., *Political opposition and local politics in Japan* (pp. 274–313). Princeton: Princeton University Press.

Masumi, J. (1995). *Contemporary politics in Japan* (L. Carlile, trans.). Berkeley: University of California Press.

Matsubara, H. (1971). *Kogai to chiiki shakai* (Pollution and regional society). Tokyo: Nihon Keizai Shimbunsha.

Matsuno, Yu (Forthcoming). In World Bank Institute, ed., *Japanese Environmental Policy.* Washington, DC: World Bank.

Matsuoka, N. (1989). Energy policy and the environment. In S. Tsuru and H. Weidner, eds., *Environmental policy in Japan* (pp. 437–450). Berlin: edition sigma.

McKean, M. (1977). Pollution and policymaking. In T. J. Pempel, ed., *Policy making in contemporary Japan* (pp. 201–238). Ithaca: Cornell University Press.

McKean, M. (1981). *Environmental protest and citizen politics in Japan.* Berkeley: University of California Press.

Meeting reaches accord to reduce greenhouse gasses. (1997, December 11). *New York Times*, p. 1.

Meyerson, A. (1980, September 5). Japan: Environmentalism with growth. *Wall Street Journal.*

Miller, A. (1989, July/August). Report on reports. *Environment.*

Miller, A., and Moore, C. (1990). *Japan and the global environment*: Center for Global Change. (1991, January). 44 PGS (monograph). Available at http://www.globalchange.org

Ministry of Foreign Affairs, Government of Japan. (1977). *Environmental policy of Japan.* Tokyo: Ministry of Foreign Affairs, Government of Japan.

Ministry of International Trade and Industry. (1988). *Japan's energy conservation policy.* Tokyo: Ministry of International Trade and Industry.

Mishima, A. (1992). *Bitter sea: The human cost of Minamata Disease.* Tokyo: Kosei Publishing Company.

Mitake industrial waste site 80% against (Mitake Sanpai Hantai Hachiwari). (1997, June 23). *Asahi Newspaper*, morning edition, p. 1.

Miyaoka, I. (1998). More than one way to save an elephant: Foreign pressure and the Japanese policy process. *Japan Forum, 10*(2), 167–179.

Mofson, P. (1997). Zimbabwe and CITES: Illustrating the reciprocal relationship between the state and the international regime. In M. Schreurs and E. Economy, eds., *The internationalization of environmental protection* (pp. 162–187). Cambridge: Cambridge University Press.

Mol, A., and Sonnenfeld, D., eds. (2000). *Ecological modernization around the world: Perspectives and critical debates.* London: Frank Cass.

Muramatsu, M. (1997). *Local power in the Japanese state.* Berkeley: University of California Press.

Nagayama, J. (1998). *Daiokishin osen retto: Nihon e no keikoku* (Dioxin pollution archipelago: A warning to Japan). Tokyo: Kanki Shuppan.

Nakemata, T., and du Florey, C. (1987). *Health effects of air pollution and the Japanese Compensation Law.* Columbus, OH: Batelle Press.

Nihon Kankyo Kaigi (Japan Environmental Council), ed. (1994). *Kankyo Kihonho Wo Kangaeru* (Thinking about the Basic Law on the Environment) (editorial). Tokyo: Jikkyo Shuppan Co.

Nishimura, H. (1989). *How to conquer air pollution: A Japanese experience.* Amsterdam: Elsevier.

No! Dokyumento Shimin Tohyo (No! Documents on Citizen Referendum). (1997). *Asahi Newspaper Nagoya* (Asahi Shinbun Nagoya Shakaibu). Nagoya: Fubosha Press.

Notehelfer, F. G. (1975, Spring). Japan's first pollution incident. *Journal of Japanese Studies, 1*(2), 351–383.

Okimoto, D. (1989). *Between MITI and the market: Japanese industrial policy for high technology.* Stanford: Stanford University Press.

Organization for Economic Cooperation and Development. (1977). *Environmental policies in Japan.* Paris: Organization for Economic Cooperation and Development.

Organization for Economic Cooperation and Development. (1993). *OECD environmental data compendium 1993.* Paris: Organization for Economic Cooperation and Development.

Organization for Economic Cooperation and Development. (1994). *Environmental performance reviews.* Paris: Organization for Economic Cooperation and Development.

Organization for Economic Cooperation and Development. (N.d.). *Environmental performance reviews.* Paris: Organization for Economic Cooperation and Development.

Pearce, D., and Warford, J. (1993). *World without end: Economics, environment and sustainable development.* Oxford: Oxford University Press.

Pekkanen, R. (2000, Winter). Japan's new politics: The case of the NPO Law. *Journal of Japanese Studies, 26*(1), 111–148.

Pempel, T. J. (1982). *Policy and politics in Japan: Creative conservatism.* Phildelphia: Temple University Press.

Pharr, S., and Badaracio, J. (1986). Coping with crisis: Environmental regulation. In T. McCraw, ed., *America versus Japan.* Boston: Harvard Business School Press.

Piven, F. F., and Cloward, R. (1971). *Regulating the poor.* New York: Pantheon.

Reed, S. (1986). *Japanese prefectures and policy-making.* Pittsburgh: University of Pittsburgh.

Reich, M. (1983). Environmental policy and Japanese society, Part II: Lessons about Japan and about policy. *International Journal of Environmental Studies, 20*, 199–207.

Reich, M. (1991). *Toxic politics.* Ithaca: Cornell University Press.

Ren, Y. (Forthcoming). In World Bank Institute, ed., *Japanese environmental policy.* Washington, DC: World Bank.

Risaikuru. (N.d). *Tokushu: Yugai kagaku busshitsu no risk shogen o kangaeru* (Special edition: Thinking about how to reduce the risk of toxic materials). *Risaikuru Bunka* (Recycle Culture), 57.

Sabatier, P. (1975). Social movements and regulatory agencies: Toward a more adequate—and less pessimistic—theory of "clientele capture." *Policy Sciences*, 6, 301–342.

Sakakibara, E. (1993). *Beyond capitalism: The Japanese model of market economics*. New York: University Press of America.

Sawai, M. (1998, September/October). Aomori allows spent fuel shipment to Rokkasho. *Nuke Info Tokyo*, *67*, pp. 1–2.

Schnaiberg, A. (1980). *The environment: From surplus to scarcity*. New York: Oxford University Press.

Schnaiberg, A., and Gould, K. A. (1994). *Environment and society: The enduring conflict*. New York: St. Martin's Press.

Schreurs, M. (1996). *Domestic institutions, international agendas, and global environmental protection in Japan and Germany*. Unpublished doctoral dissertation, University of Michigan, Ann Arbor.

Schreurs, M. (1997a). Domestic institutions and international environmental agendas in Japan and Germany. In M. Schruers and E. Economy, eds., *The internationalization of environmental protection* (pp. 134–161). Cambridge: Cambridge University Press.

Schreurs, M. (1997b). A political system's capacity for global environmental leadership: A case study of Japan. In L. Mez and H. Weidner, eds., *Umweltpolitik und Staatsversagen: Perspektiven und Grenzen der Umweltpolitikanalyse: Festschrift für Martin Janicke zum 60. Geburtstag* (pp. 323–331). Berlin: edition sigma.

Schreurs, M. (2000). Japan: Law, technology and aid. In W. Lafferty and J. Meadowcraft, eds., *Implementing sustainable development: Strategies and initiatives in high consumption societies: A comparative assessment of national strategies and initiatives*. Oxford: Oxford University Press.

Schreurs, M. (Forthcoming). In World Bank Institute, ed., *Japanese environmental policy*. Washington, DC: World Bank.

Schwartz, M. (1976). *Radical protest and social structure: The Southern farmers; Alliance and cotton tenancy, 1880–1890*. New York: Academic Press.

Scott, J. (1998). *Seeing like a state: How certain schemes to improve the human condition have failed*. New Haven: Yale University Press.

Shibata, T. (1989a). The influence of big industries on environmental policies: The case of car exhaust standards. In S. Tsuru and H. Weidner, eds., *Environmental policy in Japan* (pp. 99–108). Berlin: edition sigma.

Shibata, T. (1989b). Pollution control agreements: The case of Tokyo and other local authorities. In S. Tsuru and H. Weidner, eds., *Environmental policy in Japan* (pp. 246–251). Berlin: edition sigma.

Shoji, K., and Sugai, M. (1992). The Arsenic Milk Poisoning Incident. In J. Ui, ed., *Industrial pollution in Japan* (pp. 77–102). Tokyo: United Nations University Press.

Smith, P. (1997). *Japan: A reinterpretation*. New York: Pantheon Books.

Steiner, K., Krauss, E., and Flanagan, S., eds. (1980). *Political opposition and local politics in Japan*. Princeton: Princeton University Press.

Stone, A. (1975, Spring). The Japanese muckrakers. *Journal of Japanese Studies, 1*(2), 385–407.

Stripmining the seas. (1989, September 23). *Washington Post*, p. A22.

Strong, K. (1977). *Ox against the storm*. Vancouver: University of British Columbia Press.

Szasz, T. (1994). *Ecopopulism*. Minneapolis: University of Minnesota Press.

Taguchi, M. (1998). *Gomi Mondai Hyakka II: Soten to Tenbo* (Garbage problem Encyclopedia 2: Disputes and outlook). Tokyo: Shin Nihon Shuppansha.

Takeuchi, K. (1998). *Chikyu Ondanka No Seijigaku* (The politics of global warming). Tokyo: Asahi Sensho.

Takubo, Y. (1997). Emergence, development and the success of "Association for Doing the Referendum" in Maki. *Kankyo Shakaigaku Kenkyu, 3*, 31–48.

Tsuru, S. (1989). History of pollution control policy. In S. Tsuru and H. Weidner, eds., *Environmental policy in Japan* (pp. 15–42). Berlin: edition sigma.

Uchino, T. (1983). *Japan's postwar economy*. Tokyo: Kodansha International.

Ueda, H. (1998). *Oyakusyo kara daiokishin: Machigaidarake no kankyoseisaku e no shohosen* (Dioxin from the bureaucracy: A prescription for an environmental policy full of mistakes). Tokyo: Sairyusha.

Ueta, K. (Forthcoming). In World Bank Institute, ed., *Japanese environmental policy*. Washington, DC: World Bank.

Ui, J. (1968). *Kogai no seijigaku: Minamata Byo Wo Megutte* (The politics of pollution: On Minamata Disease). Tokyo: Sanseido.

Ui, J. (1989a). Anti-pollution movements and other grass-roots organizations. In S. Tsuru and H. Weidner, eds., *Environmental policy in Japan* (pp. 109–220). Berlin: edition sigma.

Ui, J. (1989b). Lessons for developing countries. In S. Tsuru and H. Weidner, eds., *Environmental policy in Japan* (pp. 553–570). Berlin: edition sigma.

Ui, J. (1989c). Pollution export. In S. Tsuru and H. Weidner, eds., *Environmental policy in Japan* (pp. 395–412). Berlin: edition sigma.

Ui, J. (1992a). Conclusions. In J. Ui, ed., *Industrial pollution in Japan* (pp. 173–183). Tokyo: United Nations University Press.

Ui, J. (1992b). Minamata Disease. In J. Ui, ed., *Industrial pollution in Japan* (pp. 173–183). Tokyo: United Nations University Press.

Ui, J., ed. (1992c). *Industrial pollution in Japan*. Tokyo: United Nations University Press.

Ui, J., ed. (1972). *Polluted Japan*. Tokyo: Jishu Koza.

Unease about nuclear power 90% (Genpatsu Ni Fuan Kyuwari). (1990, December 24). *Asahi Newspaper*, p. 3.

Upham, F. K. (1987). *Law and social change in postwar Japan.* Cambridge: Harvard University Press.

Use market methods on the environment (Kankyo Ni Keizai Shuho o . . .). (1994, May 31). *Asahi Newspaper* (Japanese), p. 1.

Vig, N., and Kraft, M. (1990). *Environmental policies in the 1990s USA.* Congressional Quarterly Press.

Wallace, D. (1995). *Environmental policy and industrial innovation: Strategies in Europe, the US and Japan.* London: Earthscan.

Watanuki, R. (1979). *Seimeikei no kiki* (Crisis in the life system). Tokyo: Anvil.

Weber, P. (1994). Safeguarding oceans. In L. Brown et al., eds., *State of the world* (pp. 41–60). New York: Norton.

Weidner, H. (1989a). An administrative compensation system for pollution-related health damage. In S. Tsuru and H. Weidner, eds., *Environmental policy in Japan* (pp. 139–165). Berlin: edition sigma.

Weidner, H. (1989b). Environmental monitoring and reporting by local government. In S. Tsuru and H. Weidner, eds., *Environmental policy in Japan* (pp. 461–476). Berlin: edition sigma.

Weidner, H. (1989c). Japanese environmental policy in an international perspective: Lessons for a preventive approach. In S. Tsuru and H. Weidner, eds., *Environmental policy in Japan* (pp. 479–552). Berlin: edition sigma.

Weiner, T., Engleberg, S., and Sterngold, J. (1994, October 9). C.I.A. spent millions to support Japanese Right in 50's and 60's. *New York Times*, p. 1.

Weiss, E. B. (1998). The five international treaties: A living history. In E. B. Weiss and H. Jacobson, eds., *Engaging countries: Strengthening compliance with international environmental accords* (pp. 89–172). Cambridge: MIT Press.

Wellock, T. R. (1998). *Critical masses: Opposition to nuclear power in California, 1958–1978.* Madison: University of Wisconsin Press.

Woodall, B. (1996). *Japan under construction.* Berkeley: University of California Press.

Yamaoka, Y. (1998). Shimin katsudo no zentaizo to shokatsudo (An overview of citizen activism). In Y. Yamaoka, ed., *NPO Kisokoza* 2 (pp. 1–28). Tokyo: Gyosei.

Yamauchi, N. (1998). *The non-profit sector in the Japanese economy: An overview.* Available at http://www.us-japan.org/dc/cs.yamauchi.paper.htm

Yoshida, F. (Forthcoming). In World Bank Institute, ed., *Japanese environmental policy.* Washington, DC: World Bank.

Yoshida, M., and Iguchi, T. (1998). *Kankyo horumon o tadashiku shiru hon* (The book of correct information about environmental hormones). Tokyo: Chukei Shuppan.

9

Institutional Profiles and Policy Performance: Summary and Conclusion

Uday Desai

In the introduction to this book, it was suggested that the course of public policy is influenced by policy history, policy process, and policy effectiveness. It is also influenced by many other factors—for example, by the physical character of the problem and by the limits of our existing knowledge. Chapters in this book have focused on the policy process, especially on how three specific institutions have influenced environmental policy in different industrialized countries. This concluding chapter first summarizes, based on the individual chapters, the institutional differences and similarities among the countries covered in the book. After discussing and contrasting the institutional frameworks of different nations, it discusses how these differences have shaped environmental policy and politics in these nations.

National Institutional Profiles

Intergovernmental and Interdepartmental Relations

Of the seven industrialized countries profiled in the book, four—Australia, Canada, Germany, and the United States—are federal states. In all four, there is a formal division of powers and responsibilities between the national government and state or provincial governments. While all four countries have a federal system, each has its own variety of federalism.

Kraft, in his chapter on the United States, argues that a distinctive characteristic of the U.S. system is the "constitutional specification for a wide distribution of authority." He characterizes the U.S. policy system as one of "institutional pluralism." This institutional pluralism includes the constitutional division of power among the three branches of the federal government, as well as between the federal government and the 50 states. According to Kraft, Congress has played the dominant role in federal policymaking on environmental matters for several decades, and the federal courts have also been extremely influential in this sphere. Besides citing the active role of Congress and the courts, Kraft identifies the important role of the states, especially in environmental policy implementation, as another distinctive feature of U.S. environmental policy and politics. He points out that many areas of environmental protection have always been and continue to be left to state and local governments. In addition, state governments have generally been given the responsibility for implementing almost all federal environmental policies and laws enacted in the last 30 years. The overall picture that emerges is one of a highly decentralized federal system, with perhaps extreme fragmentation of policy powers and processes. However, Kraft's chapter also implies a significant federal role, especially a regulatory role. Rabe and Lowry (1999, 264), comparing U.S. and Canadian varieties of federalism, make a similar point in observing that the U.S. federal government retains considerable control over many environmental policies.

In his chapter on Canada, Toner observes that both federal and provincial governments have considerable constitutional authority in the environmental policy area. The Canadian Constitution gives the federal government powers over coastal and inland fisheries, navigation, and federal lands and waters. This constitutional provision provided the basis for the Fisheries Act, the Navigable Waters Protection Act, the Arctic Waters Pollution Prevention Act, and the Canadian Environmental Protection Act. The Canadian Constitution also grants to the provinces numerous powers in the environmental policy area, including jurisdiction over lands and other resources owned by the provinces. This is a very important power since, except for the Northern territories, most lands in Canada are provincially owned. Rabe and Lowry (1999, 264), in their comparison of the institutional framework for environmental policy in the United States and Canada, assert that in Canada, "Constitutional mandates and political realities preclude a significant federal role in environmental issues." They characterize Canadian federalism as "a model of extreme decentralization among Western democracies" (Rabe and Lowry 1999, 264). While federal-provincial conflicts have played an important role in environmental politics and policy in Canada, Toner emphasizes the importance of conflicts between the federal environmental protection agency (Environment Canada) and economic and resource development agencies (e.g., Transport, Public Works, Natural Resources, Agriculture, Fisheries, and so on). Provincial governments, industry, and economic development agencies of the federal government constitute a strong alliance against environment ministries and environmental groups in this policy area. Relationships between government departments, Toner suggests, are just as important as relations between levels of governments in analyzing Canada's environmental policy performance. Rabe and Lowry have suggested that, because of its Westminster-style parliamentary system with its greater political party discipline, there is greater harmony among government departments in Canada than is the case in the United States, with its presidential system and weak political parties. On the other hand, Toner's detailed discussion of five major environmental policies paints a convincing picture of serious conflicts among government departments in Canada.

The German Constitution, known as the Basic Law, divides jurisdiction for environmental policy into a complex three-tier federal system. According to Weidner in his chapter on Germany, it is a system of "separated and overlapping powers." While powers to pass environmental laws are divided in complex and overlapping jurisdictions among the federal and state (Länder) levels, responsibility for environmental laws, like other laws, is usually conferred on the Länder. Länder, in turn, "rely on local government for this purpose" (Miebach 2001, 16). Thus "the preponderance of administrative responsibilities clearly lies with the Länder. They have jurisdiction for all areas of domestic administration and their system of public authorities is at the same time responsible for the implementation of most federal laws and federal statutory orders" (Miebach 2001, 16). In many ways, the German constitutional design of separated powers bears the distinct mark of the constitutional system of its major occupation power after the Second World War, the United States. However, the German Constitution also includes the principle of "cooperative federalism" that formally requires close cooperation between the federal and state (Länder) levels. This unique German type of federalism, according to Weidner, is quite different from the federalism of the United States or Canada. Environmental policy and its implementation in Germany are generally characterized by a high degree of cooperation between the federal and Länder governments, unlike the high level of conflict between the federal and state or provincial governments characteristic of U.S. and Canadian federalism. In practice, Weidner notes, this "cooperative federalism" meant in the past that policy elites from federal and state executive bodies had disproportionate power, with parliaments and local bodies playing a secondary role. This relatively closed system has changed in recent years, mainly as a result of the success of the green parties. Cooperative federalism has evolved into moderately competitive federalism, introducing more flexibility and plurality into the German environmental policy sphere. Many factors—"situational, structural, institutional and political"—and their evolving relationships condition the influence of federalism on environmental policy in Germany.

Walker, in his chapter on Australia, contends that the European notion of development has been a central force in Australian environmental policy and politics. The Australian economy and governments, both federal and state, are heavily dependent on the exploitation of natural resources and exports of primary products (i.e., raw materials). This "settler" or "dominion" capitalism has been especially detrimental to Australia's many unique ecosystems, resulting in serious ecological damage and extinction of record numbers of mammalian species. Australia has a Westminster-style federal system with judicial review added to the parliamentary system. The federal government has powers primarily in defense and external relations, with residual powers, including powers over the exploitation of natural resources, given to the states. And, like other federal states, the fiscal powers of the federal government and the court's interpretations of the constitutional division of powers have strengthened and extended the powers of the federal government, often at the expense of the states.

Australia's federalism is conditioned by the interplay of its historic "statist developmentalism" and its dominion or settler capitalist economy based on the exploitation and export of natural resources. Since the economies as well as the tax revenue of many states in Australia are heavily dependent on natural resources, its states are, as Walker points out, key agents of development. This dependence explains why states allow or even encourage exploitation of natural resources. As Walker notes, "powerful developmental premiers have been common" in the states. He refers to Queensland, especially under Bjelke-Petersen, its premier from 1968 to 1989, whose Country Party regime aggressively pursued a developmentalist agenda and "picked fights with conservationists and the federal government." Pressures for environmental protection and nature conservation are much stronger at the federal level. This results in conflicts and heightens differences between states and federal governments. Federalism here forces "environmental issues out into the open." Walker notes that federal-state conflict has been a constant feature of environmental policy. Devolution of responsibility for environmental protection to the states and local government in the 1990s further complicated the situation, since few states have the power or will to oppose major economic actors in their activities that harm the environment. In addition, the general lack of administrative capability at the state and local levels further constrains Australia's environmental policy and its implementation.

Three of the industrialized countries included in this book, Great Britain, Italy, and Japan, are not federal states. In all three, the national government is central. Local or regional governments in these three countries play varying roles and have varying degrees of influence on environmental politics and policy. In Italy, local governments have often been the first to oppose environmental pollution and degradation. As Lewanski and Liberatore point out in their chapter, Italian environmental policy began essentially as local policy in the mid-1960s, with local authorities playing a key role in initiating policies as well as in implementing laws. Lewanski and Liberatore further note that this continues, in some ways, to be the case even now. Indeed, the central government and its laws have often been more of a hindrance than a help to the local and regional authorities in their environmental protection efforts. Lewanski and Liberatore suggest that, overall, local and regional authorities have played a major role in promoting the development of environmental policy in Italy.

In both Japan and Britain, local governments have played a marginal, if any, role in environmental policy. In both countries, interministerial conflict and competition, primarily between economic growth–oriented ministries and the environment ministry, have characterized the governmental environmental policy dynamics. Broadbent, in his chapter on Japan, argues that the Ministry of International Trade and Industry (MITI) was the central player in the "growth coalition" that has dominated Japan's post–World War II public policy, including its environmental policy. Local governments have played a minimal role in Japan's environmental policy. Though local protest movements were instrumental in pushing the "ruling triad" of government bureaucrats, business leaders, and the Liberal Democratic Party (LDP) into enacting and implementing environmental pollution control policies, these local movements were self-limiting in their concerns for only local environmental problems, emphasizing only Not in My Backyard (NIMBY) issues. Japan's "ruling triad," or what Buruma (2001) has called the LDP state, was devoted to economic growth. It was not a true democratic state. Bureaucrats, industrialists, and politicians ran it in secret. In return, the LDP state promised and, for four decades, delivered peace, economic prosperity, and steady jobs to the Japanese people. Environmental protection was, by and large, a nonconcern. It acted to protect the environment only after serious local protest or international pressure, especially from the United States.

Britain has, according to McCormick, some of the oldest environmental laws and public agencies to implement them. To increase central government control over policies affecting the environment, in 1970, a new Department of the Environment (DOE) was created by combining the ministries of transport, housing and local government, and public buildings and works. However, McCormick argues that most of the staff and resources of this new department were devoted to non–environmental protection activities, such as housing and public works. Many other ministries, such as energy supply and trade, also play an important role in environmental policy. However, the Ministry of Agriculture, Fisheries and Food has an especially important role in environmental policy, since important issues with regard to rural environment, forestry, nature conservation, and protection of woodlands and hedgerows come under its jurisdiction. Britain has a vast array of government laws and agencies to protect the environment. However, these environmental agencies are highly decentralized, with much of the responsibility for environmental policy left to quasi-governmental agencies. Typically, administrative and policy responsibilities are divided among many specialized quasi-governmental agencies—for example, the Nature Conservancy Council, the Forestry Commission, and the Countryside Commissions. Policy implementation is carried out by "local authorities, quasi-governmental organization, and interest groups." Overall, McCormick finds "the institutional structure for environmental policy in Britain . . . fragmented and confusing" and its policymaking system deficient in both "coordination and direction."

Economic Organizations and Interests

Business and industry are central actors in environmental politics and policies in all industrialized countries. They and their trade associations and lobbying arms matter very much in determining environmental policy all over the world. In the United States, business and industry have generally opposed environmental policies on a broad front. Since the mid-1990s, they have had considerable influence in the Republican Party–controlled Congress, blocking reauthorization of many environmental laws enacted in earlier decades. With the election of George W. Bush to the presidency, the influence of business and industry on U.S. environmental policy has increased considerably. President Bush's decision to oppose the Kyoto treaty on global warming and his administration's energy policy task force, which recommended increasing oil and gas drilling on public lands, reflect these influences. In addition to business and industry, ranching, grazing, and farming organizations, especially in the Western United States, have had and continue to have considerable influence on environmental policies, especially concerning Western public lands and waters (Davis 1997; Wilkinson 1992). Kraft emphasizes that these groups have brought antienvironmental pressure to bear on policymakers for a decade or more. Their impact on environmental policy has been especially pronounced when the Republican Party has been in power, as noted earlier.

Canada, as Toner points out, is still highly dependent on the extraction and export of natural resources. Therefore, business and industry have been active participants in Canadian environmental politics and policy. They have forged a strong institutional coalition with the governments of Western provinces, especially Alberta, and with the economic departments of the federal government to resist strong environmental protection policies and regulations. Toner concludes that "the fossil-fuel sector, the government of Alberta, and NRCan (Department of Natural Resources Canada) have formed a formidable juggernaut against regulating economic activities to achieve reductions in greenhouse gases." Industry and federal resource departments have opposed stringent habitat protection rules; industry and their business-press allies and the right-wing Reform Party were active players in climate-change politics, attempting to cast doubt on its scientific basis. This convergence of industry, provincial governments, and economic or resource development departments of the federal government working to oppose environmental regulation has been the pattern across the spectrum of environmental policies. Yet Toner notes a significant change in business and industry's articulation of environmental issues since the late 1980s. They have "recognized that environmental protection can be both profitable and a job creator." Environmental industries, according to Toner, have been one of the fastest-growing segments of the Canadian economy.

The Australian economy, like that of Canada, is heavily dependent on extraction and export of natural resources. As noted earlier, it is characterized by Walker as a "dominion capitalist economy." He calls the Australian political economy "statist developmentalism." In such a system, governments and business are intimately intertwined, with business engaging in "state-subsidized profit taking." Big business in Australia has always had considerable impact on government even though it is poorly organized and often disunited. There is widespread, unquestioned belief among both major political parties (Labor and Liberals), the media, and policy elites that economic development and growth are good, which pushes environmental issues to the periphery. In the Australian developmental state, business and industry occupy a central place in environmental politics and policy.

Weidner characterizes the German policy process as a "neocorporatist system in which consensus on basic economic and social issues is pursued by the large business federations, trade unions, and the government." This arrangement fundamentally conditions the role of business and industry in environmental policy there. When the modern environmental movement got underway in the early 1970s, this neocorporatist system excluded environmental concerns, resulting in increasingly militant and sometimes violent mass protests. Along with the popular grassroots environmental protest movement, the proportional-representation electoral system eventually allowed environmentalists to enter Länder and federal parliaments, as well as government agencies. Weidner claims that the old neocorporatist mode of cooperation forged in the postwar period between state, business firms, and unions (the "iron triangle") is evolving into a network (the "green triangle") in which organizations representing environmental concerns are gaining a say in political decisions.

In addition to the inclusion of environmental concerns in the councils of national policymaking, there has been a gradual, but fundamental, shift in the paradigms and structures of established institutions to incorporate environmental concerns. Weidner calls it "ecologicalization" of established institutions, including business, trade unions, political parties, and churches. In Germany, there is now a "flourishing eco-industrial" and "green business" complex. Weidner even suggests that elements of the business sector have developed a growing interest in stringent environmental regulations, introducing new markets as well as expanding already-existing ones.

In postwar Japan, the LDP state single-mindedly pursued economic growth. The Ministry of International Trade and Industry (MITI) has been the coordinating focal point for the ruling triad of business, LDP leadership, and economic ministries. The close, often-incestuous relationships among the bureaucrats, the LDP politicians, and the business leaders allowed MITI to create consensus in the business sector to deal with environmental problems when it became politically necessary because of widespread citizen protest and media criticism. Unlike the U.S. system, there were no arm's-length or conflictual relations between the LDP state and business and industry. Therefore, once the LDP state, especially its bureaucrats in the economic ministries, decided that it was politically necessary to reduce air pollution or deal with a specific local environmental problem, the policies were adopted quickly and implemented effectively. As Broadbent points out, "when the Japanese government decided to 'compromise' and act, it did not need to rely on U.S.-style 'command-and-control,' arm's-length imposition of regulations." In Japan, government treats "businesses as national treasures, producing the lifeblood of the nation," in sharp contrast to the United States, where they are considered largely self-interested.

McCormick, in his chapter on Britain, points out that "public policy . . . traditionally has been worked out through a process that emphasizes consensus and consultation with affected interests and verges, in some places, on neocorporatism." Environmental policy is no exception. It is made by a small, closed community of affected private groups, and its enforcement relies heavily on voluntary compliance by industry with "decent" pollution control standards. In addition to business and industry, farmers, through the National Farmers Union, have been particularly influential. Their principal concern has been to ensure that environmental policy, especially air and water pollution and land management policies, "infringed as little as possible on their overriding concerns with efficiency and profits." However, according to McCormick, there have been fundamental changes in the last two decades. These changes have significantly undermined the influence of industrial and agricultural interests.

In Italy, Lewanski and Liberatore assert, "Particularistic interests of specific groups and clans, if not individuals, . . . dominate public policies." Consequently, a great many public policies are created to provide "immediate and direct gains to specific groups and corporations." This particularistic nature of public policy, including environmental policy, has sometimes become even "pathological." Recent scandals involving kickbacks to political parties and politicians in government public works contracts, including those for water pollution control and other environmental protection activities, are the latest examples of this extreme particularistic pathology. The influence of particular clienteles on environmental policy reaches beyond policymaking and into policy implementation, allowing them to "hinder implementation of unwelcome measures."

International Organizations

International organizations, multilateral and bilateral agreements, and treaty obligations are increasingly important influences on the environmental politics and policy in all seven industrialized countries discussed in this book. (For an excellent and thorough discussion of the politics of international environmental agreements, see Porter and Brown 1996.) Among the seven major industrial nations discussed in this book, U.S. environmental policymaking shows the most resistance to international organizations and agreements. However, international aspect has become an important element in U.S. environmental policy debates. Indeed, a good many of these debates center on U.S. participation in international environmental agreements or treaties. Vig and Kraft (1997, 375) have written that "international cooperation to prevent global climate change, protect the stratospheric ozone layer, and preserve biodiversity is becoming an increasingly important part of the national environmental agenda." The U.S. government has often been unwilling to ratify international environmental agreements. For instance, the Senate has yet to ratify the 1982 UN Convention on the Law of the Sea or the 1992 Biodiversity Treaty (though President Clinton signed both). There is no likelihood of either being ratified in the foreseeable future. The United States is alone among the industrial countries in its general opposition to global approaches to protecting the environment. The announcement by President George W. Bush that he opposes the Kyoto Protocol on climate change continues the isolation of the United States from the industrial West in the international environmental arena. In the United States, there is much less institutionalized cooperation between government and business and industry in national policymaking, including environmental policymaking, than in other nations, especially in Germany and in Japan. This is partly a reflection of the more competitive capitalism in the United States and partly a result of the historical relationship between government and the economy there. This lack of cooperation not only makes environmental policy in the United States more contentious domestically, but also severely affects the U.S. government's environmental agreements and treaties. The domestic politics of the environment critically influences its decisions on international environmental institutions. But, while the U.S. government has been often reluctant to participate in international environmental agreements, American environmental nongovernmental organizations (ENGOs) have been actively engaged in global environmental policies and projects. They often provide funding, technical expertise, and political support for ENGOs in poor, developing countries and former Communist countries of Eastern Europe.

While the U.S. government has been reluctant to participate in major international environmental agreements, it does not mean that domestic environmental policy has, as a consequence, been weak. It simply means that U.S. policy has been much less impacted by international considerations than has been the case with other industrialized nations. In the rest of the industrial West, the influence of international organizations on domestic environmental policy is much stronger. Canada, Toner observes, has been a strong supporter of United Nations environmental agencies, commissions, and agreements. It has signed a number of international bilateral and multilateral environmental agreements, including the UN Conventions on Biodiversity and Climate Change and the Treaty on Persistent Organic Pollution. Canada's National Task Force on Environment and Economy was created, Toner suggests, as a result of the 1986 visit by the World Commission on Environment and Development. This task force's work led to the creation of a statutory body, the National Round Table on the Economy and Environment. Toner describes in detail the numerous other ways in which international organizations and commissions have influenced both Canadian environmental institutions and policies.

Australia played a lead role in international agreements to prevent nuclear waste dumping in the Pacific and to prevent mineral exploration in Antarctica (Porter and Brown 1996). Australia also strongly supported international agreements on climate change and ozone-layer protection. Porter and Brown (1996, 36) attribute Australia's leadership role in these agreements, in part, to "the rise of Australia's environmental movement as a crucial factor in Australian elections." The green vote was credited with electing the Labor Party parliamentary majority in the 1987 election. However, Walker finds that "environmental groups are at best marginal to the dominant policy communities." He also finds that, while science is revered, it is not always heeded. Its findings are often ignored.

Germany, Weidner contends, has been a pioneer in environmental policy. Porter and Brown (1996) point out that Germany, along with France, has been a leader in negotiations for the Framework Convention on Climate Change. They attribute Germany's leadership in international environmental agreements, conventions, and organizations partly to strong public support for environmental protection by German voters, reflected in the electoral performance of the German Green Party (Porter and Brown 1996, 36). Germany's environmental policy has been and continues to be significantly affected by its founding membership in the European Union (EU). Weidner points out that since the Treaty of Maastricht, EU policy has increasingly required member states to integrate environmental considerations into all policy sectors. He believes that the new EU policy style favoring "flexible, incentive-based instrument . . . and multistakeholder involvement and cooperation" poses a big challenge to Germany's neocorporatist, fragmented, sectoral, multilevel federal policy system. The EU policy has already led to changes in administrative structures and processes, according to Weidner. Clearly, EU policies have been a major, positive influence, he believes, on German environmental policy.

The EU has largely driven Britain's environmental policy in recent decades. McCormick lists several examples of recent environmental policy initiatives in Britain that resulted not from domestic, but from EU pressure. He quotes another scholar who argues that "the most striking feature of the government's policy on pollution is the extent to which it is dictated by [EU] directives." McCormick suggests that EU member states are increasingly losing power over policy development to the EU. More and more often, the member-state governments are left with the responsibility of implementing and enforcing EU policies. McCormick predicts that this EU policy dominance will continue and that, in the future, the British government will become "increasingly involved in negotiating policies with its EU partners" and will be less responsible "for policy initiation than for implementation and enforcement."

Lewanski and Liberatore's account of the influence of the EU on Italian environmental policy development indicates that it has been even more profound than in the case of Britain. Italy was a "laggard" in this area among EU member states. It was only after a 1987 law gave the Italian government power to enforce EC (later EU) legislation that Italy began to improve on its "laggard" status in environmental protection. A large corpus of Italian environmental legislations and regulations has been the result of pressure to comply with EU directives. In recent years, Lewanski and Liberatore argue, Italy has even started to play an active role in initiating environmental policy in the EU, as well as in the international environmental conventions, agreements, and organizations. Eagerness to create a positive international image by active participation in environmental agreements and organizations at the international and EU levels, Lewanski and Liberatore believe, is likely to provide "a powerful stimulus to improvement at the domestic level as well."

From Broadbent's detailed discussion of the history of Japan's environmental policy and politics, it is clear that international criticisms and pressures, along with grassroots protests, have been driving forces. It was only in the mid-1960s, after more than a decade of single-minded pursuit of economic growth after the Second World War, that Japan's ruling triad, in response to domestic and international criticism, began to take some action to deal with environmental pollution. The response, led by MITI, allocated substantial financial resources to environmental pollution control and achieved considerable success. However, it was a largely technocratic response that allowed little citizen participation. By the mid-1970s, both domestic protests and foreign pressures declined significantly. In the 1970s, Japan also began "exporting" its highly polluting oil, chemical, aluminum, and other industrial plants to developing countries. All through the 1980s, the Japanese government generally refused to participate in or abide by international agreements and conventions. While Japan signed the London Convention on Marine Pollution, for example, it refused to abide by it. The Japanese government also refused to participate in the 1985 Vienna Convention on reducing ozone-depleting CFC production and use. Only intense international criticism and pressure, especially from the United States and Western European nations, finally prompted the Japanese government, in the late 1980s, to sign and enforce a number of international agreements, including the Vienna and Montreal protocols on the protection of the ozone layer, a ban on the import of ivory, and reduction in drift-net fishing boats. Since the early 1990s, Japan has been more actively engaged in international environmental conventions and organizations. Unlike the United States, Japan has signed Agenda 21, the Framework Convention on Climate Change and Biodiversity Treaty. Clearly, there have been significant changes in Japan's engagement in international organizations and agreements over the last four decades. The international community's influence on Japan's environmental policy has increased significantly.

Institutions and Environmental Policy

This book is animated by three basic concerns. The first of these concerns is when and how the state in the industrialized countries has intervened in the last four decades to protect the environment. The second concern involves the role of three specific institutions in influencing these interventions and their implementation. The third concern is how effective these interventions have been in these countries. The preceding chapters reveal interesting and important national differences. However, they do not advance or test causal propositions about variables that determine when and how the state intervenes in environmental matters or how effectively. Instead, they have assumed that institutional design does matter in when and how the state intervenes and in how effective such interventions turn out to be. While it is recognized that interests play a role in influencing environmental policies, it is assumed that institutions mediate and channel these interests. It is also recognized that social and cultural factors particular to each country affect environmental policy responses and effectiveness. Again, while assuming that the three types of institutions discussed in this book mediate and channel the social and cultural factors, it is also recognized that these three institutions change and adapt as a result of their interactions with other formal and informal institutions and as a result of changing social and political values and preferences.

The foregoing discussion of the roles of government, business, and international institutions in environmental protection in the seven countries advances our understanding of factors that influence these institutions in environmental matters. Some theoretical insights and extensions that may be derived from the preceding chapters are discussed below.

That the institutional structure of relationships between national and subnational governments influences environmental policies and their implementation is generally conceded. However, the chapters in this book show that there are significant differences in these influences among the seven coutries. In three of the four countries with federal systems (the United States, Canada, and Australia), the relationships between federal and state or provincial governments are often characterized more by conflict than by cooperation. The state or provincial governments generally act as a constraint on the federal government's environmental policies. They play an important role in articulating and representing economic values and interests. In all three countries, business and industry and the state or provincial governments closely collaborate to resist federal environmental policies and regulations. The dependence of the state or province's economy on the extraction and use of natural resources significantly strengthens the nexus between business and industry and state or provincial government. The pervasiveness of "developmentalist" ideology further strengthens this nexus. The long tradition of citizen preference for weak central government and the settlement colony history of the three countries underpin and sustain the intergovernmental tensions in their environmental policies. In contrast to the situation in the United States, Canada, and Australia, environmental policy in the German federal system is characterized more by cooperation between federal and Länder governments. Germany does not have a "settlement colony" history. The dependence of its states' economies and governments on extraction of natural resources is low. Its history and political culture are not at all characterized by a preference for weak central government. Indeed, Germany's federal system, in its current form, is the result of a constitutional scheme imposed on it by the occupation powers, especially the United States, after the Second World War.

In addition to economic and historical factors, electoral rules also play an important role. In the three English-speaking federal systems above, the candidate with the most votes wins the election. This structure favors the two-party system and makes it very difficult for third parties to take hold. In contrast, the German electoral system of proportional representation allows and even facilitates the emergence of third parties. This has allowed German "greens" first to establish a political presence at the Länder level, and later to emerge as a political force at the national level.

Federal systems are generally characterized more by conflict than collaboration in environmental matters. However, the four country cases show that federalism is conditioned by the extent of state economic dependence on extraction of natural resources, developmentalist ideology, settlement-colony history, historical preferences for a weak central government, and the electoral system, among other factors. Federalism—that is, the structure of federal and state or provincial relations—conditions when and how the state acts to protect the environment. However, the dynamics of this structure, its actual playing out in each country, is conditioned and mediated by the various formal and informal institutions and social and political forces mentioned here.

Three countries discussed in this book, Britain, Italy, and Japan, are not federal states. In two, Britain and Japan, local governments have had a marginal role in environmental policy. The chapter on Britain makes very little reference to local governments. In Japan, according to Broadbent, the local governments, especially those controlled by opposition (non-LDP) parties, periodically agitated to get the national government to mitigate their local environmental pollution problems, but were not active or influential in shaping the national environmental policy. However, unlike states and provinces in the three countries with a federal system discussed above, they have generally been a force in favor of, rather than against, national government actions to protect the environment. Local and regional governments have been more active and influential in Italian environmental policy. They have actually been one of the two major driving forces in shaping environmental policy in Italy, the other being the European Union. To a considerable extent, they have played a role similar to that played by the states and provinces in a federal system, except that rather than resisting environmental action, they have pushed the Italian national government to take stronger actions to protect the environment.

It may be hypothesized, based on the chapters in this book, that federal systems with a long tradition of a weak central government, winner-take-all electoral systems, developmentalist ideology, and states or provinces economically heavily dependent on exploitation of natural resources are likely to be characterized more by conflict than by collaboration among national and state or provincial governments. On the other hand, federal systems with a neocorporatist policymaking system and a proportional representation electoral system are likely to have more collaborative federal-state relations. In a nonfederal unitary system, the local government is likely to represent its citizens' voice for local environmental protection. Local governments are likely to be an important influence on the central government for stronger policy actions to protect the environment.

The relative power of different agencies in the government significantly influences environmental politics in all industrialized countries. The importance of this fact is seldom fully recognized. Government agencies and ministries concerned with economic growth—for example, ministries of industry, trade, power, public works, agriculture, and fisheries—usually have more power and influence on government policy than the ministry concerned with the environment. And these agencies generally argue against strict environmental laws and regulations on the grounds that such policies reduce economic growth. Business, industry, farming, and other economic organizations opposed to environmental regulations support these agencies in their fight against environmental regulation. Environmental agencies are relatively new additions to the government and often have low status and power among the other government agencies or ministries. Strong public opinion, grassroots environmental organizations, and the scientific community in support of strong environmental laws and regulations form the principal base of support for environmental agencies in the government. While the influence of the economic ministries varies considerably among the seven countries studied in this book, with Japan perhaps an extreme case of strong influence, in all seven countries the economic ministries are more powerful than the environmental agencies. Consequently, they have much greater influence on environmental policy than do the agencies with specific environmental protection responsibilities. A more systematic study of the factors that determine levels of influence of various agencies on environmental policies and the consequences of these differences for environmental policies is needed.

In all seven countries, economic organizations (business, industry, farming, fisheries, and so on) are central actors in environmental politics. In the highly fragmented, pluralist U.S. policy system, these economic organizations have often allied themselves with the Republican Party. Whenever the balance of power tips in favor of the Republican Party in national and state governments, the influence of business and industry increases significantly. However, the strong environmental groups in the United States combined with the relatively open policy process and an interventionist judiciary have meant a great deal of open conflict between environmental protection forces and business and industry. In Japan, on the other hand, with its closed "ruling triad" of business leaders, government bureaucrats, and LDP party politicians, there has been relatively little challenge to the influence of business and industry. In Germany and Britain, the inclusion of environmental protection forces in government environmental policymaking councils in recent decades has provided an opportunity for them to challenge business and industry's influence and increase their own influence on policy.

While the presence of developmentalist ideology in all seven countries clearly plays a major role in ensuring the significant influence of economic organizations on environmental policy, it is not clear from the chapters in this book what other factors determine the influence of business and industry on environmental policy. The particular mix of industry (whether resource extraction, service, agriculture, and so on), the extent of the government ownership or regulation of industry, the existence and dominance of energy, manufacturing, and other highly polluting industries, levels of economic wealth, wide acceptance of environmental values, and many other factors are all likely to determine, to some degree, the influence of business and industry on environmental policy.

International organizations and conventions have a significant influence on the environmental politics and policies in all seven countries. It has even been argued that national environmental policies are driven less by national-level forces than by the accepted international norm that nation-states are responsible for protecting the environment (Frank, Hironaka, and Schofer 2000). The increasing formative role of the EU in establishing environmental policies for its member countries is evident in all three European countries discussed in this book. The importance of international pressures in shaping Japanese environmental policy is also unmistakable. Canada and Australia seem to both lead and respond to the internationalization of environmental policy as well. While the United States, at present, appears to be the least affected by international pressures, this is perhaps more a reflection of recent official positions concerning global warming and biodiversity agreements than a true picture. American environmental groups are actively engaged in international environmental organizations and conventions, as well as in supporting environmental programs and NGOs across the world. A great deal of environmental policy debate in the United States now concerns the global environment and international agreements and conventions. It is apparent from the chapters in this book that it is impossible to understand or explain environmental politics in the industrialized countries without fully integrating the internationalization of environmental politics and policies in such explanations.

Policy Performance

Judging the overall effectiveness of any country's environmental policy is a difficult and highly contested matter. Kraft, in his chapter, points out that accurate assessment both of environmental policy implementation and of the quality of the environment itself is still rare. Nevertheless, the authors of the preceding chapters, with perhaps the exception of the Australia chapter, find that there has been considerable improvement in environmental quality in these countries since 1970, as a result of environmental policies and regulations.

Kraft concludes that, over the last three decades, there has been a significant reduction in major air pollutants and toxic waste and improvement in water quality. These improvements suggest to him that U.S. environmental policies have been more successful than might have been assumed. Toner reviews two international studies that assess Canada's environmental performance. One study ranked Canada third among 122 countries in its overall environmental performance. The second study found Canada in the middle category among a number of OECD countries. A 1997 report, by Canada's own Commissioner of Environment and Sustainable Development, judged that "although progress has been made in a number of areas, it has not been uniform." Toner concludes that some progress has been made despite setbacks.

In Britain, overall environmental trends have been positive, according to McCormick. Positive developments can be discerned with respect to land use, and emission of major air pollutants has declined over the last three decades. The record on water pollution, however, has not been as good. Thus, while surface-water quality has deteriorated slightly in the last 20 years, McCormick sees more positive than negative trends. Weidner finds that Germany has made remarkable progress in several environmental areas. There has been a substantial reduction in emissions of major air pollutants, with improvement in ambient-air quality. Improvement in water quality has been mixed. It has improved, in some respects, as a result of massive investments in municipal and industrial wastewater treatment plants. However, it has deteriorated, in other respects, as a result of agricultural runoffs. The generation of solid waste has been stabilized and the rate of recycling increased.

Australian environmental policy performance has been rather poor, according to Walker. He finds a general ignorance of environmental issues on the part of politicians and senior bureaucrats. Regulation has received less attention than it should have and, not surprisingly, has declined to some extent in effectiveness. Walker contends that an effective takeover of Australian public policy by economic "rationalists" since the 1970s has led to a disproportionate emphasis on economic growth as the criterion of policy performance, at the expense of environmental issues. Lewanski notes that in Italy, emissions of several air pollutants (e.g., sulfor oxides and particulate matters) have declined or stabilized over the last 20 years or so, while emission of greenhouse gases has continued to increase. In the case of Japan, it seems that environmental policies have dealt with the obvious sources of air and water pollution, especially pollution that poses a clear and present threat to human health. There has been much less focus on long-term threats to the environment or to the quality of life.

The environmental policy performance of all seven countries may well be best summarized as mixed with advances in some areas, no or marginal improvements in others, and even backsliding in some areas. Judging from the seven cases, it seems that in countries with highly fragmented and decentralized policy and administrative systems, there is greater likelihood that national environmental policy will lurch along, advancing in some areas and retreating in others. Where the center is strong, environmental policy is likely to have a clearer direction, though not necessarily a pro-environmentalist direction. The German case may indicate that cooperative federalism (i.e., close collaborative relations between the national and state or provincial governments), paired with a proportional representational electoral system (facilitating the emergence of green parties) and significant pro-environmental popular sentiment, is likely to lead to both effective environmental policy and implementation. The result is high levels of environmental protection and pollution control.

Overall, the chapters in the book do not show that any particular characteristics or relationships of the three major institutions result in demonstrably superior overall environmental policy performance. Without much empirical study involving many more countries, we cannot generalize about causal relationships between specific institutional characteristics and environmental policy performance. Much work remains to be done to fully understand the causal connections between institutions, interests, values, and so on, on the one hand, and the differences in environmental policies and their outcomes in different nations, on the other.

Conclusion

Over the last three decades, progress has been made in softening the impact on the environment of the high-production and high-consumption lifestyle in the developed countries. As the foregoing chapters point out, most of the developed countries have made some modest progress in reducing air and water pollution, in cleaning up rivers and waterways, and in protecting the forests, wetlands, and wildlife and their habitats. However, as the chapters in this book also show, the desire for economic growth and material consumption remains strong in industrialized countries. Business and industry, trade and economic ministries, political organizations, and other institutions representing pro-growth and pro-consumption forces and values remain powerful in all of these countries. But there is considerable variation in their power and influence on environmental policy. Among the seven countries discussed in the book, Japan is a case of more influence by pro-growth forces, while Germany appears to be at the other end, with less influence by these forces.

While pro-growth desires remain strong, there has also been growing recognition and acceptance of the idea of sustainable development. There is increasing awareness that growth in material prosperity of the present must be balanced with the needs of future generations. The 1992 United Nations Conference on Environment and Development (Rio Earth Summit) and its Agenda 21 made the idea of sustainable development a centerpiece of international cooperation and action for global environmental protection. The actual impact of the concept of sustainable development on specific policies in industrialized countries is difficult to ascertain at present. However, the concept of sustainable development has become part of the environmental policy and politics rhetoric in the developed world. It is employed by business and industry, as well as by scientists, politicians, policy analysts, and, with some trepidation, by the environmentalists. Even in the United States, as Kraft points out in his chapter, corporate leaders, environmental-group executives, and politicians have embraced the sustainable development idea. The concept of sustainability may provide a bridge between the pro-growth and pro-environment forces. With new technologies (e.g., alternative-energy and energy-efficiency technologies) making an ecofriendly, high-consumption lifestyle more feasible, sustainable development may become an important concept in environmental politics and policy. The institutional framework within which environmental policies are framed and carried out is likely to play a crucial role in determining the importance of the sustainable development concept in industrialized countries. The chapters above provide a good understanding of the institutional context in the industrialized countries within which sustainable development policies will be debated, developed, and implemented.

The global environmental footprint of the industrialized nations will continue to be immense. Pressure to maintain, even increase, high consumption levels is likely to continue unabated. And the debates and conflicts over environmental policies in both the developed and developing worlds will also continue and even intensify in the twenty-first century. Global warming, destruction of forests and biodiversity, the "hole" in the ozone layer, rapid increases in toxic and hazardous wastes, and numerous other developments pose global environmental threats. They affect many, if not all, nations in the world. Environmental policies and practices of individual nations have serious consequences for the well-being of the entire earth. Research to increase our understanding of the influence of international institutions on national environmental policy and politics is important. What factors play a critical role in determining the likelihood of international cooperation? Under what conditions is such cooperation likely to succeed or fail? What types of international institutions are likely to emerge, to succeed? Under what national conditions (cultural, institutional, historical) are international institutions likely to influence domestic environmental policy as well as politics? Careful comparative studies of a number of countries will contribute much toward answers to these questions.

The environment and the economy are inseparable. Economic organizations and interests have a strong influence on environmental policy in all nations. Comparative studies of how economic forces shape environmental policies in industrialized nations contribute much to our understanding of the future of environmental protection. Such studies show how differences in economic organizations and changes in their structures and values influence environmental politics and their evolution.

Governments play an important role in protecting the environment. Many factors condition the role of government in environmental policy. Comparative studies of these factors, and of relationships between governmental institutions and environmental policy performance in industrialized countries, would provide useful guidance for institutional design in other countries to enhance their environmental policies and performance.

These studies, building on the insights on the environmental policy process provided by the chapters in this book, would increase our understanding, and thus our ability to develop institutions and policies to better protect the environment of our global village.

References

Buruma, I. (2001). The Japanese malaise. *New York Review of Books*, *48*(11), 39–41.

Davis, C., ed. (1997). *Western public lands and environmental politics*. Boulder: Westview Press.

Frank, D. J., Hironaka, A., and Schofer, E. (2000). The nation-state and the natural environment over the twentieth century. *American Sociological Review*, *65*(1), 96–116.

Miebach, K. M. (2001). *Federalism in Germany*. Basic-Info. 11-2201/Home Affairs. Code No. 710 Q1116. Bonn, Germany: Inter Nationes.

Porter, G., and Brown, J. W. (1996). *Global environmental politics*. Boulder: Westview Press.

Rabe, B. G., and Lowry, W. R. (1999). Comparative analysis of Canadian and American environmental policy: An introduction to the symposium. *Policy Studies Journal*, *27*(2): 263–266.

Vig, N. G., and Kraft, M. E. (1997). The new environmental agenda. In N. G. Vig and M. E. Kraft, eds., *Environmental policy in the 1990s* (pp. 365–389). Washington, DC: Congressional Quarterly Press.

Wilkinson, C. (1992). *Crossing the next meridian: Land, water, and the American West*. Washington, DC: Island Press.

Contributors

Jeffrey Broadbent received his Ph.D. from Harvard University in 1982 and is associate professor of sociology at the University of Minnesota. Recent publications include *Environmental Politics in Japan: Networks of Power and Protest* (Cambridge 1998), winner of the Outstanding Publication Award 2000 from the Section on Environment and Technology of the American Sociological Association; and the Masayoshi Ohira Memorial Prize 2001 for contributions to public policy; *Comparing Policy Networks: Labor Politics in the US, Germany and Japan* (with three coauthors) (1996); and "Social Capital and Labor Politics in Japan: Cooperation or Cooptation?" in *Policy Sciences* 33 (3–4), December 2000.

Uday Desai is professor in and chair of the political science department at Southern Illinois University Carbondale. His last book, *Ecological Policy and Politics in Developing Countries*, was published by State University of New York Press in 1998. He has published over two dozen articles in refereed journals and six book chapters and is the senior coeditor of *Policy Studies Journal.*

Michael E. Kraft is professor and chair of Public and Environmental Affairs and Herbert Fisk Johnson Professor of Environmental Studies at the University of Wisconsin, Green Bay. Among other works, he is author of *Environmental Policy and Politics: Toward the 21st Century* (1996) and coeditor and contributing author of *Environmental Policy in the 1990s: Reform or Reaction?* (1997) and *Public Reactions to Nuclear Waste* (1993). He is completing work on *Toward Sustainable Communities: Transition and Transformations in Environmental Policy*, with Daniel Mazmanian.

Rudolfo Lewanski is an associate professor at the Faculty of Political Science, University of Bologna, Italy. He has also taught at the European University Institute and Dickinson College (Bologna) and was a visiting scholar in the Department of Political Science at the University of Calgary in 1999. His main research interests are in the fields of public, environmental policy, and public administration, with specific emphasis on Italian issues. He has written extensively on Italian environmental policy.

Angela Liberatore works in the Directorate General for Research of the European Commission on issue related to governance and citizenship. Earlier she worked on social and institutional aspects of sustainable development and global change. She holds a Ph.D. in political and social sciences and a degree in philosophy. Her publications are in the fields of environmental policy, European integration, science/policy interface, and risk management, and include the book *The Management of Uncertainty: Learning from Chernobyl.*

John McCormick is an associate professor of political science at the Indianapolis campus of Indiana University. Before completing his Ph.D. at Indiana University in Bloomington, he worked for several years for environmental interest groups in Britain, including the World Wide Fund for Nature. His teaching and research specialties include comparative politics, environmental policy, and the politics of the European Union. Recent publications include *The Global Environmental movement* (1995), *The European Union: Politics and policies* (1996), and *Acid Earth: The Politics of Acid Pollution* (1997).

Glen Toner was educated at the Universities of Saskatchewan and Alberta and at Carleton University. He is a professor in the School of Public Administration at Carleton University in Ottawa and a senior research officer in CRUISE, the Carleton Research Unit in Innovation, Science and Environment Policy. He is the author of numerous publications on energy and environment policy. Cofounder of the New Directions Group of senior industrialists and environmentalists, he has also been an advisor to Environment Canada and other departments of the Government of Canada.

K. J. Walker is an independent researcher. According to a recent reviewer, he "pioneered the study of environmental politics in Australia." He edited the collection *Australian Environmental Policy* and is author of the textbook *The Political Ecology of Environmental Policy: An Australian Introduction* (1992, 1994). He has taught at Melbourne, Monash, Flinders, Adelaide, Case Western Reserve and Cleveland State Universities. From 1976 to 1995 he played a major role in developing interdisciplinary courses in environmental policy and politics at the Environmental Studies School at Griffith University in Brisbane. He is currently at work on a book, *The State in Environmental Management*, and a number of papers.

Helmut Weidner, Ph.D. and Diplom-Politologe (Dipl.-Pol.), was a lecturer at the department of political science, Free University Berlin from 1975 to 1978. He is a senior research fellow and a codirector of an international comparative research project on mediation approaches in waste management conflicts at the Environment Unit of the Science Center for Social Research Berlin (Wissenschaftszentrum Berlinfiir Sozialforschung). Earlier he codirected a cross-national comparative study on environmental policies in E.C. member countries (1979–1984) and directed a research project on environmental policy in Japan (1985–1992). He has authored several books and articles on environmental policy, including *Environmental Policies and Politics in Japan* (1989), coedited with Shigeto Tsuru; *Clean Air Policy in Great Britain* (1987); *Umwelt-lnformation. Berichterstattung und Informationssvsteme in 12 Landern* (1992), coedited with Roland Zieschank and Peter Knoepfel.

Index

Coal, 102